The Science of Deception

The Science of Deception

Psychology and Commerce in America

MICHAEL PETTIT

THE UNIVERSITY OF CHICAGO PRESS CHICAGO AND LONDON

MICHAEL PETTIT is assistant professor of the history and theory of psychology at York University in Toronto.

The University of Chicago Press, Chicago 60637
The University of Chicago Press, Ltd., London
© 2013 by The University of Chicago
All rights reserved. Published 2013.
Printed in the United States of America

22 21 20 19 18 17 16 15 14 13   1 2 3 4 5

ISBN-13: 978-0-226-92374-1 (cloth)
ISBN-13: 978-0-226-92375-8 (e-book)
ISBN-10: 0-226-92374-6 (cloth)
ISBN-10: 0-226-92375-4 (e-book)

Library of Congress Cataloging-in-Publication Data

Pettit, Michael (Michael John), author.
  The science of deception : psychology and commerce in America / Michael Pettit.
      pages cm
  Includes bibliographical references and index.
    ISBN 978-0-226-92374-1 (cloth : alkaline paper)
    ISBN 0-226-92374-6 (cloth : alkaline paper)
    ISBN 978-0-226-92375-8 (e-book)
    ISBN 0-226-92375-4 (e-book)   1. Deception. 2. Psychology—United States—History.   I. Title.
  BF637.D42P48 2013
  155.9'2—dc23

                                                    2012013095

♾ This paper meets the requirements of ANSI/NISO Z39.48-1992 (Permanence of Paper).

Contents

ACKNOWLEDGMENTS   vii

INTRODUCTION   1

CHAPTER 1.   "Graft Is the Worst Form of Despotism":
             Swindlers, Commercial Culture, and the Deceivable
             Self   21

CHAPTER 2.   Hunting Duck-Rabbits:
             Illusions, Mass Culture, and the Law of Economy   49

CHAPTER 3.   "Not Our Houses but Our Brains Are Haunted":
             The Arts of Exposure at the Boundaries of Credulity   84

CHAPTER 4.   The Unwary Purchaser:
             Trademark Infringement, the Deceivable Self,
             and the Subject of Consumption   121

CHAPTER 5.   Diagnosing Deception:
             Pathological Lying, Lie Detectors, and the Normality
             of the Deceitful Self   156

CHAPTER 6.   *Studies in Deceit*: Personality Testing and the Character
             of Experiments   194

CONCLUSION. Barnum's Ghost Gives an Encore Performance   228

Notes   237

Bibliography   287

Index   305

Acknowledgments

As this is a book about scams, fraud, and trickery, it is only proper that I take delight in exposing my own confederates and sincerely acknowledge a host of debts owed.

I have many institutions to thank for their support of this work over the years. Funding for this project came from the Social Sciences and Humanities Research Council of Canada, an Ontario Graduate Scholarship, a Lupina Foundation Doctoral Fellowship from the Comparative Program on Health and Society at the University of Toronto, the Associates of the University of Toronto Award for the Study of the United States, and the Faculty of Health at York University. For greatly facilitating my research, I would like to thank the librarians and archivists at the Library of Congress; the Boston Public Library; the research center at the Chicago History Museum; the Houghton Library of Harvard University; the Divinity School Library and the Sterling Memorial Library at Yale University; the Bancroft Library at the University of California, Berkeley; the Stanford University Archives; the special collections at Duke University; the Northwestern University Archives; the Rockefeller Archive Center; and the National Archives in College Park, Maryland.

Parts of chapters 2 and 4 previously appeared as "Joseph Jastrow, the Psychology of Deception, and the Racial Economy of Observation," *Journal of the History of the Behavioral Sciences* 43, no. 2 (2007): 159–75, and "The Unwary Purchaser: Consumer Psychology and the Regulation of Commerce in America," *Journal of the History of the Behavioral Sciences* 43, no. 4 (2007): 379–99.

I remain greatly indebted to the scholarly community at the University of Toronto. I am grateful to Michelle Murphy for her guiding influence in making my claims simultaneously bolder and more precise. Elspeth Brown pushed me constantly to think of the big picture while simultaneously focusing on the micro level of honing my prose. Mariana Valverde introduced me to a different way of thinking about the law, but more importantly, offered valued advice and support at crucial turning points during my academic career. Dan Bender and Theodore Porter offered valuable feedback on an earlier version of the book. I have enjoyed the community of insightful and generous people I first met at the University of Toronto, namely, Sarah Amato, Ariel Beaujot, Erin Black, Jeff Bowersox, Robyn Brown, Liviya Mendelsohn, Amy Milne-Smith, Ruth Percy, Tim Sedo, Nathan Smith, Heather Welland, and the irreplaceable Carol Ali.

I have been fortunate to have York University's rather unique History and Theory of Psychology program as my institutional home as I revised and completed this book. The Toronto area likely has more historians of psychology than anywhere else in the world, and I have benefited tremendously from spirited conversations with Chris Green, Kenton Kroker, Wade Pickren, Alex Rutherford, Mark Solovey, Thomas Teo, Marga Vicedo, and Andrew Winston. My colleagues and students in both the Department of Psychology and the program in Science and Technology Studies at York University created a friendly and stimulating setting in which to work.

As I have traveled with this project for close to a decade, numerous friends and colleagues have listened to portions as talks and read chapter drafts. I want to thank Ken Alder, Betty Bayer, John Carson, Fran Cherry, Joseph Gabriel, Thomas Gieryn, Peter Hegarty, Sarah Igo, Bernard Lightman, Jill Morawski, Trevor Pearce, David Serlin, Michael Sokal, and Laura Stark for sharing their ideas and clarifying my understanding of the issues involved in this book. Their advice and suggestions are always greatly appreciated, even though I did not always succeed in meeting their expectations.

As the book neared completion a number people helped get me over the finishing line. Karen Darling shepherded the manuscript through the review process at the Press. Therese Boyd and Mary Gehl gave the manuscript careful attention. Pedram Mossallanejad provided valued assistance in the preparing the manuscript for final submission.

Works of this length require a commitment that could not be sustained without moral as well as institutional support, not to mention necessary opportunities for escape and diversion. My mother, Kathie Pettit, has been unwavering in her support over the years. The Yates family has provided much encouragement and fun. Whether in Toronto, Chicago, Paris, or Boston, my partner in crime has been Alexia Yates. She has provided much that is genuine in this book and joyful in my life.

Introduction

Writing at the height of the Great Depression, Chicago-based private investigator Frank Dalton O'Sullivan penned a blistering attack on what he perceived as one of the day's greatest threats to the social good: the Better Business Bureau (BBB). According to O'Sullivan, although the BBB claimed to back efforts to enforce "truth in advertising" and corporate accountability, the organization was nothing more than an illegal racket that encouraged the commercial behavior responsible for the financial collapse of 1929. The BBB, he claimed, protected its members rather than the public, usurping the right to investigate charges of fraud from the police and private detectives while also refusing to be licensed or bonded by the state.[1] In the wake of the stock market crash, he felt that there was an urgent need to ensure that the claims made by businesses were reliable and trustworthy, but instead of offering assurances of truth in advertising a corrupt BBB had foisted itself upon an unsuspecting purchasing public. O'Sullivan argued that the bureau had originally been created in the 1910s "to patrol the avenues of business, keeping them clear of fraud and deception, without the necessity for intervention by government agencies." He contended that, around 1919, there had been a significant turn in the way the bureaus functioned in favor of the reckless stock market promotion.[2] In O'Sullivan's eyes, the national organization had become a mere front for the New York stock exchange, harassing small, nonmember businesses that refused to pay for protection. The business practices of the city's financial elites, he concluded, had become indistinguishable from the criminal operations of the gangsters and frauds they claimed to combat.[3]

During the course of his exposé, O'Sullivan drew a parallel between two relatively novel institutions. From his perspective, the BBB and the recently created Scientific Crime Detection Laboratory, based out of Northwestern University, were analogous commercial rackets that nurtured criminality in the name of preventing it. The most damning piece of evidence that revealed the laboratory's true nature was its latest piece of pseudoscientific equipment, the lie detector, a machine whose advocates claimed could replace the trained judgment of the veteran detective. O'Sullivan scoffed at the machine's prominent place at that year's Century of Progress exhibition, arguing that its ineffectiveness would merely invite increased criminal activity. O'Sullivan described the crime laboratory as simply the BBB in a slightly altered form. Both were misleadingly named institutions that would ultimately defraud the public with their false promises and lackluster methods.[4]

Undoubtedly O'Sullivan's position as a Chicago private investigator made him particularly sensitive to the encroachments of these institutions upon his occupation.[5] Nevertheless, because of his particular perspective, he noticed how the scientific study of deception was intimately bound to legal projects for governing commercial life. This connection was not an aberration unique to the rise of the lie detector.[6] From the late nineteenth to the early twentieth century, the scientific study of potentially deceitful persons and deceptive things became closely linked to ways of being in, and governing, the marketplace.

Toward a History of Deception

Deception has a history, a history that consists of shifts in the activities, objects, and bodily processes that people associate with deception, the concept's moral valences, and the sites where it is both performed and exposed. Duplicity is undoubtedly an ancient human behavior, but how people have explained and especially policed the experience of both erring and lying has changed over time. I will attempt here to elucidate what is historical about the experience of deception.[7] Undertaking such a history might seem like a foolhardy task. The sixteenth-century humanist Michel de Montaigne remarked, "If falsehood had, like truth, but one face only, we should be upon better terms; for we should then take for certain the contrary to what the liar says: but the reverse of truth has a hundred thousand forms, and a field indefinite, without bound or limit." In a similar vein, the nineteenth-century American legal theorist Melville Bigelow ventured the opinion that "fraud is manifested in such endless variety of form and phase, in such manifold and ever-changing disguises and colors, that a definition might at first indeed appear hopeless."[8] Three centuries apart, Montaigne and Bigelow issue a common warning: defining what counts as truth is a fairly straightforward affair but its opposite is an unwieldy web of behaviors and meanings. Even today, there is little agreement on the contours of what counts as deception as it elides the certainty, stability, and transparency promised by honesty.[9] I have tried to trace how shifting assemblages of peoples, things, and practices lead to differing perceptions of what exactly constitutes deception.[10] This is a history of where deception is understood to reside: whether in certain persons, words, things, actions, or their relations.

Dishonesty has long been the purview of moralists, but this book traces the emergence of a naturalistic investigation of deception in the human sciences, administration, and the law. These thinkers grappled with the human machinery of deception: why and under what circumstances people did lie and how a person was deceived. They pursued this naturalism in the name of detecting deception in particular persons, things, and situations. The history of psychology offers a particularly salient means of tracking the history of deception in the modern world. As a scientific discipline, psychology is typically depicted as an outcome of the growth of knowledge. It is said to have resulted from subjecting once-philosophical questions to the research methods of the physiologist. I take a different tack by recasting the emergence of psychology in relationship to an economy of uncertainty, characteristic of modern life. The rules of capitalist exchange demanded greater attention to the particulars of the world of goods and a transparency among persons as they conducted business with strangers over greater distances.[11] Despite these incentives to protect one's reputation, the same system also rewarded people for their dishonest acts and insincere behavior. The late nineteenth century was saturated with new financial arrangements and commercial objects that depended upon, yet undercut, individual confidence and cognitive independence. These commercial deceptions were captivating in the dual sense of eliciting fascination and ensnaring people in processes beyond their immediate control. The tension at the heart of the commercial life between demands for honesty and the prevalence of deceptive appearances cultivated an abundance of uncertainty that seeded the cultural ground for the psychological self of the twentieth century.

Commercial Culture and the Captivation of Deception

The very novelty of the organizations decried by O'Sullivan provides an important clue for illuminating the contours of deception's history. They appeared in the wake of significant changes in the nature of American capitalism following the Civil War. Due in no small part to the federal legislation passed during the war, the United States became a more tightly integrated nation whose economic productivity centered on the factory and the corporation rather than the plantation and farm. Summarizing an influential line of historical interpretation, Louis Galambos suggested that this period is best understood in terms of "a shift from small-scale, informal, locally or regionally oriented groups to large-scale, national formal organizations."[12] The emergence of a national "mass market" was undoubtedly the most prominent marker of this new organizational society. In 1907 economist Simon Patten declared that the United States was witnessing a revolution as significant as the political one that had created the country. If the story of human history thus far had consisted of conflicts over scarce resources, Patten predicted that the story of the twentieth century would be one of the effects of abundance. The widespread consumption of cheap food and consumer goods would provide a "new basis of civilization." Changes in agriculture and transportation had facilitated this transformation and Patten called upon his contemporaries to initiate a revolution in their "mental habits" to realize the possibilities proffered by new levels of material abundance.[13]

As Patten acutely observed, the creation of this new national mass market involved changes not only in the production of goods, but also in their distribution. With the establishment of nationally integrated telegraph (1861) and railroad (1869) networks, changes in communication and transportation quickened the flow of information and goods and this compressed notions of distance.[14] Through the process of vertical integration, individual businesses acquired formerly independent firms in order to control each stage in the production and distribution of goods, from their extraction as raw materials to their delivery to consumers. These consumer goods became associated with specific brand names, whose recognition was promoted by advertisements placed in print media. Consumers increasingly received information about the goods they purchased from the newly formed bodies of advertising experts through the features they placed in mass-circulation periodicals and daily newspapers.[15]

In talking about an economy of uncertainty, I am not simply referring to an abstract sense of anxiety that accompanied the shift from an interconnected community to an anonymous society. Rather, I mean those nodes of commercial activity where deception was a central feature. The problem of deception and fraud intersected with these new patterns of commercial life on multiple levels. Historian Richard White, for instance, has documented how the consolidation of a national railroad network depended upon the fraudulent manipulation of the financial information provided to small investors by captains of industry. The construction of the railroad confronted Gilded Age Americans with the fact that the nation's transportation network had come at the cost of deceiving and defrauding countless small investors. Corruption had been a political concern since the founding of the republic, but what made the Gilded Age situation different was that new communication technologies like the telegraph and inexpensive newspapers "magnified the results of relatively small actions and yielded disproportionate results."[16] This transformation in scale meant that people understood the deceptions of old as having new significance and effects.

It was not simply the organizational structures of the new economy that became associated with fraud but also the commodities circulating within it. The final quarter of the nineteenth century witnessed a host of legal activity designed to curtail counterfeiting and adulteration. In discussing the French case, Alessandro Stanziani has suggested that this legislative activity had much to do with an uneven definition of technological change in the production of consumer goods. What people branded as innovation in certain realms (such as the manufacturing of guns or bicycles), they labeled fraudulent adulteration when it came to the production of food, drink, or medication.[17] In other words, a host of genuinely novel yet seeming deceptive things confronted the late nineteenth-century purchaser. Moreover, the success of nationally recognized brands soon led to their counterfeit in the form of trademark infringement from rival companies.

Starting with the 1870 copyright law, which included provisions for the definition of what constituted a trademark, the federal government passed a series of acts to increase its police power to control deceitful persons and deceptive things. The next fifty years witnessed such legislation as the creation of the Interstate Commerce Commission (1887), supposedly to prevent exorbitant railroad duties; the Sherman Anti-Trust Act (1890), to monitor and prevent industrial monopolies; and the Pure Food and Drugs Act (1906), which aimed for more transparent labeling of consumables.[18] During the so-called butter wars starting in the 1880s, state and federal legislatures responded to the intense lobbying of the dairy industry by passing numerous acts severely taxing or prohibiting the then-novel substitute oleomargarine, owing to its deceptive appearance. Originally justifying this with claims about the ill effects of these products on consumer health, lawmakers ultimately insisted that they were protecting an unwary purchaser likely to be confused and deceived into believing that margarine was true butter. This legal activity culminated in the permanent bureaucratization of the state's police power to eliminate unfair competition with the creation of the Federal Trade Commission (FTC) in 1914. Other measures arose from the nation's advertisers clubs, such as the 1911 campaign for "Truth in Advertising" organized by the leading trade journal *Printer's Ink* and the subsequent creation of those Better Business Bureaus that would later inspire O'Sullivan's ire.

This legal activity was almost exclusively the product of the big business–oriented political culture of the Gilded Age.[19] The primary aim of this legislation was the protection of the financial interests of established companies and industries. But by grounding regulatory logic in the perceptual vulnerabilities of the purchaser, these decisions opened up a new legal space for a more interventionist state.[20] They laid the foundation for a very different form of statecraft in the twentieth century, a progressive version of liberalism predicated on the notion of consumer citizenship.[21]

Setbacks did occur. The government's policing of those forms of commerce intending to deceive the purchasing public met a severe obstacle with *FTC v. Raladam Co.* (1931). In this case, the FTC sought to prevent the deceptive advertising claims made about the weight-loss remedy Marmola. The Supreme Court determined that, unless the FTC could prove that the company's fallacious claims about the product's effectiveness actually harmed the trade of any rivals, the commission had overstepped its authority. As the majority in the Supreme Court argued, these police powers were originally designed for the protection of open commerce, not consumers. Despite the impediment of decisions like *Raladam*, consumer rights advocates continued to lobby for a more robust version of the FTC, which came into being in 1938 with the Wheeler-Lea Act.[22]

This book is not primarily concerned with this economic and legislative activity but, rather, excavates the contemporaneous understandings of the self articulated in relationship to these commercial and judicial practices. I argue that the rise of market culture was the most important element in explaining the heightened captivation with deception, but its importance was never exclusive. Concerns about deception also intersected with questions about religious belief, pedagogical strategy, and the design of scientific experiments. Indeed, one of the most durable outcomes of this captivation with deception was the institutionalization of the scientific discipline of psychology as a particularly prominent feature on the American cultural landscape. At the heart of this book, then, are two interrelated questions. How did psychology take root in a culture fascinated by robber barons and confidence men, national brands and their counterfeit, yellow journalism and muckraking exposés? How did the growing presence of psychology on the American cultural landscape transform these concerns about deception?

Deception and the Americanization of Psychology

A broader captivation with deceitful persons and deceptive things helped frame the integration of scientific psychology into American culture and shaped the particular contours of the science in that country. The concept of indigenization has been used to emphasize how specific aspects of an imported intellectual tradition are privileged or transformed because of their resonances with local conditions. Used to discuss the exportation of mid-twentieth-century American psychology, it is also a fruitful analytic for understanding how scientific psychology was first received in the United States.[23] Understanding the historical development of American psychology in this way helps explain how what became the globally dominant version of the science during the twentieth century was patched together in the United States from a host of heterogeneous European traditions. American psychology represented a hybrid of German experimental practice, a French tradition of clinical observation, and British mental measurement.[24] American scientists and their publics selectively adopted and recombined features of each research tradition based on their salience to local demands. The studies of illusions, testimony, and psychotherapy were far from uniquely American developments, but each topic acquired new cultural meanings in the United States.

Late nineteenth-century advocates of the "new psychology" crafted an intellectual and cultural space in large part through the study of how humans were deceivable and why they deceived. Such concerns flowed through and gave some unity to their disparate interests in psychophysics, child development, and evolutionary narratives of human difference. Furthermore, the pursuit of deception as an object of investigation led to the formation of laboratory spaces and experimental practices, policing the boundaries of the discipline's expertise, the writing of popular science, and the search for legal and industrial patronage. Talk about deception in terms of affect and cognition was not limited to the discipline of psychology; it also flourished in the form of a "common sense" knowledge deployed by jurists and journalists, among others.

While the threat posed by deceivable and deceitful persons provided one of the major tropes around which psychologists initially built the public image of their discipline, this changed by the early twentieth century. Rather than admonishing its existence, they came to argue that deception was a necessary and normal aspect of human nature, one that could be best managed through their own diagnostic instruments and therapeutic techniques. Psychologists offered an array of tools for navigating the deceptive landscape of modern life. These ranged from self-help manuals for overcoming perceptual limitations, published exposés of commercial charlatans, scales for detecting the likelihood of trademark infringement, machines to detect liars, and therapeutic techniques to reveal self-deceptions. These interventions did not eliminate deception from the public sphere or private life but were instead predicated on the acceptance of its existence. One of the central tenets of much twentieth-century psychology, from psychoanalysis to cognitive dissonance research to behavioral economics, was that humans deceived themselves as to their own true motivations. Part of the reason that psychologists normalized deception was that it had become an indispensable part of their own investigative enterprise. Fearing both resistant and overly complicit subjects whose behavior would threaten the investigations' validity, psychologists turned to deception to camouflage their intentions from participants.

This dual focus on the psychological sciences and the role played by consumer culture requires understanding the intertwined histories of the *deceitful self* and the *deceivable self*. In introducing these two terms, I do not mean to imply the existence of these "selves" as specific kinds of people. I use the notion of the "self" to mean widely ascribed and normative characteristics of personhood found in a given historical epoch and how individuals grow into and come to inhabit such discursive formations.[25] In my typology, the *deceitful self* refers to an opaque form of subjectivity whose activities include lying, cheating, stealing, and the manipulation of others, usually in the name of personal gain. I use the notion of the *deceivable self* to underscore a view of the individual that emphasizes the unavoidability of humanity's error-prone, suggestible, and fallible nature. By and large, the literature on the history of forensics (anthropometry, fingerprints, lie detectors) has given greater attention to the former, while historians of visual culture have grappled with the latter.[26] I build upon and reorient this extensive literature by tracing how scientific interpretations from both these modes of selfhood were co-produced in reference to each other.

Seen from this perspective, the production and policing of deception served as an important impetus for the emergence of what has been called the "psychological society." Historian Roger Smith describes the "significant sense in which everyone in the twentieth century learned to be a psychologist; everyone became her or his own psychologist, able and willing to describe life in psychological terms. The twentieth century was a psychological age, and in this it differed from earlier ages."[27] During this period, not only did psychology become a recognized academic discipline but its practices also saturated everyday life and the self-understanding of a wide swath of the population. When accounting for this transformation, historians have tended to emphasize the importance of wartime conditions. During the First World War, psychologists administered intelligence tests to a large number of the army recruits in a failed attempt to sort the military's personnel. This experience of mass testing exposed a large number of people to the methods of psychology and the postwar debates over the tests' insights into the mental stock of the American populace led to greater public attention being granted to the new science.[28] The Second World War was even more momentous, with a wide array of psychological specialties mobilized to assist in the war effort. There was considerable demand for psychologically informed experts who would assist with mental rehabilitation to veterans and this led to new forms of government patronage for training, rapidly expanding the discipline into a mass, helping profession.[29] Without doubt these historical events created new opportunities for psychologists, but a fascination with the confidence games of an earlier era inspired a strain of talk among psychologists and others outside the discipline about the self's affective and cognitive vulnerabilities.

The Staging of Uncertainty

In the years before academic psychologists began to speak about deception, no individual more artfully exploited the American captivation with fraud than the celebrated showman P. T. Barnum. In important ways, he crafted the cultural scripts with which psychologists had to contend. Often reluctantly, psychologists did acknowledge that in certain respects they existed as the showman's heirs.[30] Barnum began his career in 1834 by exhibiting a woman named Joice Heth, who regaled her audiences with fictitious recollections of having served as a young George Washington's enslaved nanny. An integral element of the success of the Heth exhibit was Barnum's artful manipulation of the uncertainty surrounding her authenticity. He pursued an anonymous promotional campaign that in turn raised doubts about her legitimacy and then proclaimed her genuineness. The campaign called upon ordinary Americans to pay the viewing fee so that each individual could assess Heth's true nature.

Barnum's marketing strategy hinged on societal norms that placed relatively little value on the final authority of experts. Although historians have challenged the reality of participatory democracy, it was in ascendancy as a political ideal during Barnum's early career.[31] Just as the "ordinary man" was expected to participate in the politics of the republic, the political culture demanded that the individual observer take responsibility for assessing the authenticity of these sorts of spectacles. For much of the nineteenth century, the more recent distinction between high and popular culture did not exist. Attending a performance of Shakespeare's plays was a spectacle for both the merchant and the laborer. Plebian attendees had distinct expectations about such performances and they passionately expressed their views. An actor who confounded them risked a barrage of vocal criticism or, in the case of New York City's Astor Palace Riots (1849), outright violence.[32]

Barnum used the profits from the Heth hoax to purchase the American Museum in New York City. Within its walls, he mingled curious natural objects with deceptive artifacts, known as humbugs, and asked the audience to distinguish one from the other. Barnum made a fortune displaying such dubious objects as the Feejee Mermaid, in fact a monkey's torso carefully stitched onto the body of a fish, and the What Is It?, an African American man cloaked in furs and presented as a potential missing link between humans and animals. As he did with Heth, Barnum would anonymously question the authenticity of these objects in his promotional materials and call upon the audience to decide their worth. Their involvement required that the audience serve as active participants rather than simply passive viewers. Historians have connected people's engagement with this kind of activity to the market revolution that was transforming their daily lives.[33]

Changes in transportation and the elimination of trade barriers between states meant that the former excess production of rural economies could be sold in more far-flung markets and the profits from these commodities could be stored in the abstract form of money.[34] This transformation began the shift in the image of the market from a bounded, local place to something more abstract and better seen as national in scope. In the face of these abstractions, the act of detecting one of Barnum's humbugs resonated with the process of evaluating the reliability and worth of commercial goods. His exhibitions also made tangible the power and limits of the legal imperative of *caveat emptor* or "buyer beware," a central tenet of contract law.[35]

Despite its one-time popularity, by the century's end Barnum's brand of spectacular deception was, in the words of historian Janet Davis, an "ostensibly dying cultural form."[36] The seeming disappearance of humbuggery as a prevalent cultural practice raises the question of what succeeded it. In terms of entertainment, the answer includes the vaudeville stage, cinema, radio, and television, but what about the cultural work that these spectacles performed in training people to look for deception in the commercial realm? In large part, psychologists took over the cultural conversation about deception at the end of the age of Barnum.

The Work of Deceit in the Age of Mechanical Objectivity

Those who crafted the sciences of deception did not simply contend with Barnum's legacy, but also participated in conversations about the nature of science and its social place. Indeed, with their concerted attention to questions of credulity, belief, and the very capacity to observe, attempts at studying deception scientifically after 1870 represented an important moment in the much longer history of truth-telling. Standards for what constitutes truthfulness have a history and tracking changing understandings of deception is crucial to this project of historicizing objectivity.[37] After all, as Miles Orvell has argued, self-conscious identification of authenticity is predicated on the existence of the counterfeit and fraudulent which violate the original.[38]

Objectivity, understood as the effacement of one's idiosyncrasies and the cultivation of disinterestedness through adherence to mechanical rules, has not always been the norm for scientific practice. During the mid-seventeenth century, at the height of the creation of new forms of scientific sociability and the foundation of new learned societies, being overly incredulous about the possibility that wonders could exist was seen as a marker of an insufficient worldliness. In an age of global voyaging, the cultivation of a trusting curiosity became an essential moral quality among

Europe's most cosmopolitan natural philosophers. Respected learned publications were filled with reports of wondrous phenomena; objects were collected to challenge Aristotelian orthodoxy in order to formulate more rigorous scientific laws. Starting in the 1730s, attempts to rein in this natural history of the exceptional began together with the juvenilization of wonder as a sense of inquiry. To be credulous had become unscientific, merely simple and childlike.[39]

Only in the second half of the nineteenth century did scientists operating in observatories and laboratories truly embrace the notion of mechanical objectivity. Modeled on the regularity of the industrial machine, objectivity required the restraint of one's judgments in the hope of making the body into yet another recording instrument.[40] The form that psychology took as an autonomous scientific discipline owed much to the advent of mechanical objectivity as researchers attempted to understand the subjective through the means of machine-modeled precision. This exacting regime was more than a means of acquiring truth; for some, it served as the very foundation for reorganizing society.

The roots of the sciences of deception lay in the restaging of this decoupling of wonder and objectivity. During the ascent of mechanical objectivity, scientists frequently placed wonder into an evolutionary narrative in which it became closely linked to the feminine and the "uncivilized" races. When expressed within the confines of modernity, this once-esteemed public sentiment became understood as a marker of racial recapitulation. For example, an anonymous editorial in the second volume of the journal *Science* (1883) set out to distinguish humbug, superstition, and science. It began by locating humbug in a surprising place: the British colonization of India. According to the author, because of Britain's superior technological power, the "ignorant" natives "exalt them [the British] as more than human beings" so that the Indian "turns from its own hazy gods to new and visible wonder-workers."[41] Thus, humbug was not identical to religion or superstition; it was an intermediate stage between superstition and the attainment of true scientific knowledge. Unlike the ancient gods, these new Western-inspired deities were at least material and visible in this world. In light of the discourse of civilization that was so prominent in the late nineteenth century, such a rhetorical strategy was significant. For the audience reading *Science*, civilization was coded as refined, respectable, and white, though its gains remained insecure because of individual and racial degeneration.[42] For those who may have felt that Barnum's humbugs were harmless fun, the association with primitives would have been disconcerting. That they could be so easily duped challenged the notion of racial superiority that was central to their identity. While the author reassured the reader that the scientific charlatan is "not dangerous" and "nearly always amusing," the connection with a colonized, unfree people sent a different message. In this sense, the article was a call to expel humbug from western society in the name of the highly prized goals of rationality and civilization.

In an early issue of *Popular Science Monthly* (1874), editor Edward L. Youmans connected proper scientific training to the preservation of democracy in the face of rampant commercial and political fraud. While avoiding claims that his own era was entirely unique in this regard, he insisted that "at home and on the street, in the cars, at public assemblies, and in all the relations of life, we are beset and imposed upon by designing knaves of every shade. We eat the falsified foods of the grocer, and wear the swindling textures of the dry-goods man." He equated the deceptions perpetrated by corrupt political machines and deceitful entrepreneurs with medieval despotism. Both hinged upon "the submission of the mind to beliefs imposed on it by authority, and interpreted by authority, the effect of which was to make blind credence the universal mental habit." Less than ten years after the Civil War, he invoked the specter of a form of resurgent "slavery" that would ensnare all Americans regardless of race. For Youmans, the glory of modern science resided in its historical role in emancipating the mind from such habits: "There is but one thing that can protect people against the thousand-fold insidious and plausible impostures to which they are continually and everywhere exposed, and that is a resolute mood of scepticism, and an intelligent habit of sifting evidence that shall become a daily and constant practice." With deception coming to define the national character, Youmans argued that freedom itself required the cultivation of scientific skepticism as a means of combating the slavery of despotic psychological habits.[43]

While these ideals of skepticism and mechanical objectivity were designed to ensure more accurate observations, truthfulness was connected to the sincerity with which people expressed their views. The philosopher Bernard Williams has traced the emergence of this imperative to the eighteenth-century thinker Jean-Jacques Rousseau, who built an influential philosophy around the hope of reclaiming a state of reciprocal transparency among humans, a condition he felt was lost with the acquisition of the decorum demanded by civilization. Rousseau's ideal was the unmediated and full expression of his innermost feelings, opinions, and motivations. Indeed, the charges of hypocrisy frequently leveled against Rousseau derived from his own exacting standard for truth.[44]

In the United States, this demand for sincerity found its most eloquent spokesperson at the end of the nineteenth century in the figure of the humorist Mark Twain. In his novels, he skewered the artifice of courtly manners, the arbitrariness of nobility by birthright, the hypocrisy of racial prejudice, and the financial corruption prevalent in the nation's capital. Twain also believed in the necessity of deception; he argued that lying was virtuous insofar as it provided the glue for all interpersonal intercourse. Twain held to an exacting standard for truth: nothing but a full and sincere expression of one's innermost thoughts counted as honesty and any deviation from one's true feelings constituted a lie. For Twain, dishonesty included remaining silent and not actively expressing one's feelings or opinions. He dismissed moralistic condemnations of minor spoken lies to accentuate the horror of large-scale shared lies of silent omission on the day's most pressing social issues. When he wrote disparagingly about the "decay" in the art of lying, he referred to what he perceived as the increased use of deception for personal gain rather than the common good.[45]

Starting in Twain's lifetime, a small cohort of scientists transformed this imperative toward truth-telling into a set of experimental facts about the mind and body. These scientists labored to craft techniques and tools for making people both more accurate and sincere in their actions. The scientist's performances, both in person and in print, in which they used their tools to assess a seemingly deceptive person or unmake a potentially fraudulent artifact, were critical in building the new discipline's public image. Such demonstrations allowed for the visible production of the boundaries between the credulous and the trustworthy. The study of deception also facilitated the forays of the psychologist as an expert in human fallibility into industrial organization and the legal system. Their involvement in these areas became a particularly acute concern for psychologists as they attempted to gain a monopoly over talk about the mind from psychical researchers, spiritualists, Christian Scientists, and others.[46]

Instead of trying to realize the ideal of mechanical objectivity, scientists grappling with deception came to embrace the image of an "unobserved but observing observer," a performance deploying their acumen for ingenuity, guile, and trickery to reveal the truth about human nature.[47] This standpoint should not be confused with the *absence* characteristic of the "view from nowhere," or what Donna Haraway calls the "God trick."[48]

Rather it was carefully managed *presence*, disguised from those under observation through the scientists' own deceits. It was an epistemic ideal that helped resolve the tension between the demands of the laboratory's exactitude and the theatrics of public science.[49] It replaced the minimalist self-privileged by mechanical objectivity with a hard-boiled insincerity that borrowed freely from the masculine personae of the muckraking journalist, the private detective, and the stage magician.[50]

At the same moment that the influential sociologist Robert K. Merton (himself trained as a stage conjuror in his youth) heralded "intellectual honesty" as a central platform of the scientific ethos, psychologists increasingly saw their own brand of deception as operating in a symbiotic rather than antagonist relationship with scientific objectivity.[51] The scientists' deployment of deception became an important part of their truth-telling strategies, an ironic marker of their objectivity. While psychologists condemned the fraudulent fabrication of data, other forms of deceit formed a core of how they constituted their discipline as an objective science.[52] When it came to their own scientific practice, they championed the importance of accuracy while denying the desirability of sincerity. Nevertheless, this untruthfully acquired truth led to acrimonious disputes within the discipline and compromised how the science was presented in the public sphere.

Overview of the Book

The first chapter explores the narratives written about confidence men during "the age of emancipation" that followed the Civil War. With their proliferation in numerous national contexts at various times, confidence games seem to resist classification in terms of a single historical epoch.[53] This chapter addresses this conundrum by investigating how admirers and critics depicted the exploits of the *individual* swindler in an era synonymous with large-scale financial corruption and corporate scandals. This era's exposé and confessional literature centered on the relationship between the deceitful and deceivable self. The swindler's schemes posed a threat to individual liberty and independence in a society where such ideals were highly prized as markers of manliness. Whereas antebellum portrayals of the confidence man tended to hold that the mark was an unwary innocent, later exposés warned of the victim's complicity. In other words, deceivable persons became victims because, in actuality, they were deceitful selves. Such an understanding was central to the novel and controversial ruling in *People v. McCord* (1871), in which the New York Supreme Court decided that it would not prosecute swindlers whose only victims were likewise trying to defraud another person. Furthermore, turn-of-the-century ethnographic studies of what they called the criminal "underworld" provided American journalists and sociologists with a new vocabulary for understanding the relationship among the individual, politics, and the marketplace. With the circulation of the concept of "graft," the figure of the confidence man came to serve as a symbol for the period as a whole.

The second chapter charts the materialization of the deceivable self within the psychology laboratories that proliferated in American universities at the end of the nineteenth century. In the 1880s a shared interest in experimentation, statistical analysis, and evolutionary narratives coalesced into a "psychology of deception." This topic became so central to the new discipline that psychologist Edwin Boring called the 1890s "the decade of the illusion."[54] During this decade, experimental psychologists such as Joseph Jastrow, Norman Triplett, E. W. Scripture, Carl Seashore, and George Stratton articulated a vision of the self governed by what William James called the "law of economy." According to this mental law, habits and beliefs alleviated the burden of having to constantly form conscious choices, but consigned the individual to the comfortable pathway of the most probable. The individual risked self-deception caused by an unjustified classification of sensations in a novel situation, a predicament made increasingly common in the new economy of consumer abundance. Indeed, this chapter argues that those psychologists captivated by deception are best understood as both producers and consumers of the Gilded Age's visual and material culture. The numerous channels of exchange that flowed between the university and the burgeoning consumer culture helped in the formulation and sustenance of their scientific theories.

The resulting psychology of deception, especially as espoused in mass-circulation magazines, newspapers, and popular books, opened two pathways for transforming the observational habits of the consuming public. The psychologists mentioned above preached a gospel of self-reliance and self-control to overcome both the habituated perceptual errors bestowed upon the "race" through natural selection and the incorporating, "socialistic" tendencies believed to be at work in social organizations. Another set of psychologists, among them Hugo Münsterberg and Walter Dill Scott, argued that the proper solutions to the problem of deception were psychotechnic interventions into industrial organization and the law. The comparative success of these two competing visions is a recurring theme throughout the remaining chapters.

Spiritualists, psychics, and other mental wonder-workers were the commercial actors who particularly troubled psychologists, and these scientists publicly insisted that the revelation of the true deceitful nature of such hucksters was a fairly straightforward affair. Yet, chapter 3 investigates the complex negotiations that went into producing successful exposés and the tensions and conflicts that pervaded among supposedly likeminded debunkers. It considers the lessons that psychologists learned about the design of experiments and the arts of observation from their collaboration with stage magicians. Psychologists claimed that exposure of deceitful persons required deceit on their own part. These demonstrations led to a gendered understanding of deception as male psychologists demanded full transparency from their female subjects while priding themselves on their masculine guile in besting their opponents. This embrace of deceitful insincerity as a means of achieving objective results led to acrimonious disputes among scientists and had a profound influence on the subsequent shape of the discipline.

The intersection of psychology, consumption, and governance came to fruition in the legal fiction of the "unwary purchaser" prone to deception, the topic of chapter 4. In most nineteenth-century jurisprudence, the ideal of individual self-reliance and responsibility was paramount. Cases of trademark infringement and deceptive advertising, however, consisted of a legal tradition that tended to limit personal responsibility and suspend the notion of caveat emptor in order to protect the interests of established businesses. At the heart of these legal decisions was an understanding of the consumer closely resembling the psychologist's deceivable self. Although unevenly applied, the vision of the ordinary purchaser as an unwary and easily deceived consumer played a critical role in how federal judges regulated the economy. Psychologists, especially those at Harvard and Columbia, sought to provide an experimental foundation for this legal assemblage of the consuming subject. Using an array of laboratory tests, psychologists attempted to construct a standardized, objective measure for the point at which the average consumer was likely to be deceived. In the end, judges dismissed these experts. The legal, rather than the experimental, definition of the unwary purchaser became institutionalized in the regulatory apparatus of the American state, particularly with the creation of the Federal Trade Commission.

The fifth chapter details the tentative institutionalization of two psychological tools for diagnosing deception within urban police work and the judicial system. It juxtaposes the story of the technological wonder known as the lie detector with the reception of psychotherapy among American criminologists. The chapter initially details how William Healy introduced dynamic psychology into the study of delinquency as part of his work for the juvenile court in Chicago. Healy argued for a view of the criminal as an individual who *happened* to commit crimes rather trying to identify fixed *types*. Despite this commitment, in 1915 he suggested that he had uncovered a particularly troublesome kind of delinquent, the pathological liar. The chapter traces the emergence of the lie detector out of the Harvard psychology laboratory, the Berkeley police department, and the Scientific Crime Detection Laboratory at Northwestern University. By combining these two narratives, this chapter illuminates the intertwined history of psychotherapy and lie detection as distinct yet complementary technologies. Both conceived of an honest body that betrayed the mind's deceits. From the outset, people leveled charges that both techniques were a brand of fraudulent patent medicine foisted upon the consuming public. While people rejected many of the particulars of each technique, the debates about them helped promote a view of the normal self as governed by deceitfulness and self-deception.

The final chapter analyzes the ironies at the heart of attempts to study deception scientifically. Taking the Character Education Inquiry as its focal point, it investigates the use of experimentation to arrive at an explanation of deceptive behavior among schoolchildren. Marked by a more general concern with the moral corruption unleashed by a consumer society, a number of religiously minded psychologists devised a series of situations simulating everyday life to isolate the causes of deceit. The resulting volume, *Studies in Deceit*, shifted the locus of behavior from the individual toward the situation. Eliciting authentic responses from their subjects by catching them unawares required that these psychologists themselves deceive their subjects as a central tenet of their methodology. While their fellow psychologists largely rejected the interpretation of behavior offered by the members of the Character Education Inquiry, they praised the ingenuity in designing experiments. The critical insight of *Studies in Deceit* was clear: deception was not aberration but, rather, a normal attribute that manifested itself when the situation required it. One situation that especially required deception was the psychological experiment itself.

The historical mutations in the meanings attributed to deception remained tied to transformations in the marketplace. Barnum's trickster games, carefully stage-managed to simultaneously stimulate and satiate concerns about commercial fraud, represented an initial response to the perceptual challenges of a society increasingly organized around anonymous commercial exchange. Although this artful deception remained an important resource until the century's end, individuals sensed its inadequacy already by the 1870s.

The psychology of deception championed between the 1880s and the 1910s held that the unaided judgment of individuals could no longer manage the problem of fraudulent appearances. Despite this consensus among psychologists, they differed over which approach to take. One set valorized the importance of individuality in the face of the collectivist and incorporating trends of commercial imperatives. They emphasized how the consumption of the popular science writings of leading researchers could provide individuals with the necessary tools to manage their sense perception and detect deception. Another group held that certified expertise and technocratic solutions, inspired by laboratory research, could successfully root out the unwanted, deceptive aspects of the social. Both groups of psychologists struggled to fashion themselves as outside of the market and its corrosive influences while simultaneously seeking to use its mechanisms to publicize their approach to the mind. Psychologists offered their science as a means of more rationally navigating the vagaries of commercial life, as a means of learning how to separate the fraudulent from the authentic. Such easy demarcations were always difficult to maintain and this form of public science largely failed in its stated aims.

In the end, there never emerged a science of deception in the singular. Rather, a recurring preoccupation with deception structured the content and methods of psychology in the United States and the vision of selfhood it offered the American public. From the 1920s onwards, psychologists increasingly understood deception as an unavoidable, perhaps even a necessary and beneficial element of everyday life. This conviction cut across psychotherapy, the physiological study of emotions, and the social psychological understanding of moral behavior. The lie detector was predicated not on the search for a dishonest personality type, but on measuring the changing physiology of individuals as they momentarily deceived. With its carefully crafted experimental situations, *Studies in Deceit* made such an interpretation explicit. Not even the human sciences were outside the bounds of a world where the use of deception in the proper situation was deemed a necessity. At the same time as they outsmarted charlatans and commercial frauds, psychologists were also engaged in making the advertising industry more efficient through their knowledge about perception and trickery. The history of the scientific study of deception was ultimately that of normalizing what once seemed immoral and dangerous. The outcome of rendering deception psychological was a vision of the world where the purportedly passionless and aloof scientist's responsibility was to explain expected deceit for its perpetual management rather than to eliminate it.

CHAPTER ONE

"Graft Is the Worst Form of Despotism": Swindlers, Commercial Culture, and the Deceivable Self

The "grafter" attracted J. P. Johnston's attention soon upon entering his jewelry clearance store in late nineteenth-century Chicago. The tall man dressed in overalls, a dingy blouse, boots, and a cap had the appearance of a casual laborer, but Johnston recognized him as a former business acquaintance. The young gentlemen came from a reputable family, had graduated from Yale, and had previously held a managerial position of some responsibility with a dry goods firm. For these reasons, Johnston, himself an itinerant merchant and self-described hustler, was taken aback that the man was passing himself off as a potato seller in the streets outside his shop. The young man explained that he was not, in truth, a street hawker, but rather a self-identified grafter. His targets were the women who maintained refined homes in the city and yet were still in search of a good bargain. He would travel door to door on the pretense of selling potatoes in order to gauge the household's income. Upon ascertaining its wealth, the man would then produce a set of spectacles that had the appearance of gold but were in fact made of a near worthless metal alloy that would turn green with oxidization after a few days. Claiming to have found them in the street and that they were of no use to him, he would offer to sell his new client the spectacles at below the market cost for gold. He would vary the asking price based on the woman's outer appearance and their conversation about the sale of potatoes. By the time the woman detected the deception, the grafter had moved on to another neighborhood to ply his trade. The young man found that what he could make through the graft greatly exceeded the paltry amount of his former fixed salary. Furthermore, he explained that his rich father had recently passed away and that the estate had become so entangled that its legal fees were greater than its worth.[1]

Gilded Age authors like Johnston used the figure of the confidence man in developing a psychology to explain changes they perceived in both commercial and political life. For his part, Johnston presented stories about grafters as moral tales revealing the potentials and pitfalls of modern commercial life. For example, when recounting this particular story he noted that the young man would ultimately come to spend time in a penitentiary for his dishonesty. Yet a certain admiration for the swindler's great aptitude for moneymaking and self-transformation remained. Carefully noting the grafter's deceptive outward appearance, Johnston appreciated his honed ability to successfully play as many roles as his circumstances demanded. Morally dubious, the confidence man remained a self-reliant individual who successfully geared his trade and his own self-image toward consumer desires. Confidence men played this dual purpose because, as historian Scott Sandage has argued, in the nineteenth century "Americans swapped liberty for ambition."[2] Sandage emphasizes how, in an increasingly secular world, one's financial success became equated with one's inner moral virtue. At the same time that Social Darwinian commentators were trumpeting the moral superiority of the wealthy, a wide range of American writers argued that the act of moneymaking itself was at best an amoral and often times an immoral endeavor. Multisited discussions of confidence men, swindles, and political graft served as opportunities for reconciling this seeming contradiction. During the Gilded Age, jurists, journalists, ethnographers, and confessing criminals all came to stress that beneath the surface of the market's seemingly innocent victims brewed the deceitful passions of self-interest. Their writing brought to the fore how unethical commercial behavior toyed with one's affect and how emotion compromised one's individual freedom.[3]

Both advocates and critics described the confidence game as an activity that hinged on what they called "mass psychology."[4] Those who worked these schemes boasted of their superior mastery over other people's hidden mental machinery. As E. G. Redmond noted in his exposé *The Frauds of America* (1896), what he termed the "bunco-steerer" is "an almost infallible judge of human nature and rarely makes a mistake in selecting his subject." His account of the swindler reflected the decade's fascination with hypnotic phenomena as made popular through the figure of Svengali in George du Maurier's best-selling novel *Trilby*. Due to a powerful force of personality, Redmond's confidence man was capable of projecting his will onto his victim, who was "amazed at the temporary paralysis of volition that afflicted him."[5] Through personal magnetism and knowledge of human desire, the confidence man was capable of usurping an individual's will. He preyed on virtuous citizens and played upon their repressed and base natures, their hidden avarice and desire, drawing these emotions out into the public sphere. Beyond immediate financial losses, this literature also portrayed the risk possessed by confidence games as enslaving otherwise good liberal men to their basest passions, resulting in a loss of self-control, autonomy, and independence.

The confidence man as a distinct social type first appeared in the mid-nineteenth century, as part of the American middle class's emphasis on the importance of sincerity and transparency in one's self-representation. Techniques that enabled the honest to distinguish themselves from the disreputable permeated the period's cultural productions, from advice manuals to the arrangement of homes to women's fashion. The psychological model of the individual who informed this culture drew upon an understanding of the mind as a malleable, blank slate over which the confidence man exercised an external mastery.[6] Ideally, such a self-made man sought his livelihood through regular adherence to self-discipline, thrift, and a faith in the resulting providential rewards. This normative vision defined white manliness as "self-reliant, strong, resolute, courageous, honest."[7] In contrast to this rational and self-possessed male, women were depicted as emotional and deceitful. Yet the confidence man, with his ability to bend the will of others, was a masculine type who undercut both the self-reliance and self-mastery at the heart of the manly ideal of liberal governance.[8] Thus, he anticipated the aggressive masculinity that eclipsed the refined Victorian vision of restrained manliness. Indeed, the domain of the market economy served as an important site for the expression of this code of gendered conduct.[9] "Confidence man" became a signifier for a host of troubling economic activities, namely financial speculating, advertising of goods, and engaging in political graft.

Depictions of the confidence man are primarily found in a rather inglorious literary tradition of confession and exposé. These works fell into a number of genres: short, ephemeral pamphlets aimed at visitors to urban centers, local businessmen, or farmers; longer exposés published by police officers, mail inspectors, or agricultural newspapers; amusing and colorful recollections produced by confidence men or detectives at the end of their careers. Many of these works were published by small-time publishers, far from metropolitan centers and tailored to local concerns, problems, or crime sprees. The memoirs of confidence men, written frequently after a self-declared retirement from the trade, often served as a further opportunity at self-aggrandizement, a final cashing-in on their ingenious schemes by having the reader pay to learn from their experiences. These texts are rich catalogues of the everyday practices of grafting, but despite the availability of these sources, the confidence man remains a particularly difficult historical subject to pin down. His identity is intertwined with a refined insincerity. By his very profession, the swindler makes an unreliable witness. He literally manufactures a fictitious past in order to pursue his latest commercial exploit. Thus, despite the wealth of first-hand accounts, the reader is never certain which events depicted in the text actually took place, were exaggerated, or simply fabricated in order to spin an impressive tale.[10] To further complicate matters, journalists would often write under the guise of a swindler, executing their own literary confidence game. For these reasons, this chapter focuses on how these texts represented experiences of the marketplace and how they worked to frame expectations about commercial life.[11] These books instructed people facing a confidence man to expect a charismatic individual, who was capable of shifting his outward appearance and who embodied the aggressive, aggrandizing aspects of commercial self-making.

This chapter charts the ebb and flow of attempts to demarcate the confidence man as somehow different in kind from the denizens of what contemporaries called "the upper world." The confidence man was a powerful cultural symbol precisely because he brought to the fore the seeming omnipresence of insincerity and deception in modern social life. A wide array of Americans drew the moral equivalences between the confidence game and corporate enterprise and urban politics. In 1908 Chicago police detective Clifton Wooldridge identified the municipal graft as "the worst form of despotism."[12] Not only did this corruption impede the proper functioning of government, but seeking fraudulent deals also became the focal point of all of the crooked politician's strivings. Graft became an all-consuming and undeniable habit that dominated his behavior. Much like its effect on the individual soul, for Wooldridge the presence of the graft risked enslaving liberal democracy to the will of the criminal class. For many of his contemporaries who shared this conviction, stock market speculation, salesmanship, and municipal graft did not seem so different from the criminal swindle.

The Psychology of Caveat Emptor

As a term to describe a particular kind of criminal, "confidence man" first appeared in the newspaper coverage of the 1849 trial of William Thompson. He would approach well-to-do passersby in Manhattan, inquiring whether they had "confidence" in their fellow citizens. Upon receiving an affirmative response, the well-dressed Thompson offered an opportunity to invest in a potentially lucrative business venture. Before disclosing the details, however, he would first ask them to prove the sincerity of their statement by allowing him to borrow their pocket watch, promising to return it at the same spot the next day. He would then disappear, absconding with the watch but leaving his victims with a tidy—and, one might imagine, embarrassing—lesson on the perils of trust.[13] Thompson's scheme manipulated expectations that linked inner moral character to outward visual signifiers. He became notorious because his crime exposed the weaknesses of established mechanisms for guaranteeing trust in metropolitan interactions and crowded, mobile spaces like trains and riverboats.[14] From the outset, certain newspapers suggested the moral and methodological equivalences between this petty grafter and the Wall Street financiers. An editorial in James Gordon Bennett's *New York Herald* questioned why Thompson lived on moldy bread in the city's Tombs while his brethren enjoyed the comforts of their mansions. After all, both profited from the deceitful manipulation of other people's beliefs for their personal gain. From the very beginning, then, this criminal figure provided a vocabulary for articulating moral concerns about finance as an economic activity. For some observers, the ups and downs of speculation seemed very much like a fictitious confidence game.[15]

Historian Karen Halttunen has documented how the exposés and advice manuals produced during this period emphasized the innocence of the confidence man's victims. These books typically portrayed "marks" as rural rubes new to the illegible landscapes of America's major cities. Contributors to this genre soon identified a number of other schemes that similarly manipulated a person's misplaced trust. Bunco, gold brick, and green goods were the three earliest recognized forms of what were later called "short-cons." Bunco derived from games of chance such as card tricks and the wheel of fortune. It required one swindler to run the game while an accomplice acted the part of another player. The victim would witness his competitor's tremendous luck and significant winnings, illustrating the easiness of the game and its profitability. All along the game had been fixed and, once the mark started placing big bets, the wheel's fortune would quickly turn against him.[16]

Despite the presumption of innocence, the actual descriptions of these confidence games displayed a very different relationship between the grafter and the mark. Such was the case with the "gold brick" schemes that arose out the world of mining. The swindler would pass himself off as an itinerant prospector who had chanced upon a rich mineral find. This "miner" would offer to sell his gold below market rates, explaining his desperation for immediate cash and his lack of resources to bring the gold into town. Often the swindler would traffic in racialized notions of vulnerability and take on the appearance of a Hispanic or Native American operating on the margins of the economy. The greedy mark would realize the potential profit and happily purchase the gold, whereupon the swindler would then swap a real gold bar for one made of near-valueless ore. When the victim came to sell his prize, he would discover his costly error.[17]

The green goods scheme similarly played upon the mark's desires to take advantage of others. In its simplest form, the swindler offered to sell a set of counterfeit bills below their denominational value. In a more complex version, the confidence man, claiming to possess a printing press that could produce bills indistinguishable from genuine ones, offered to partner with his mark. Inevitably the victim would discover that the bills in his possession were distinguishable, rendering them worthless.[18] The success of the green goods scheme hinged upon the relative novelty of a paper currency backed by the federal government. With state banks responsible for the issuing of notes in the absence of nationally standardized bills, the counterfeiting of currency was endemic in the early republic. By mid-century, a national network of locally organized counterfeiting rings meant that as much as half the circulated specie was fraudulent. The Civil War forged a more cohesive nation-state, including the policing of one of its most visible symbols, a standardized national currency. Anticounterfeiting measures became a major locus of the federal government's police powers. Enforcing the Legal Tender Act (1862) led to the creation of the Secret Service, making the policing of fraud one of the earliest pillars of the formation of a more interventionist state.[19]

Contract law did not give much support to the advice book genre's attempt to portray victims as innocents. Legal decisions encouraged people to assume the omnipresence of deceit brewing below appearances of respectability and honesty and to refrain from confiding in their fellow Americans. Courts did not require that individuals subordinate their own selfish impulses in the name of any obligation to others. Instead, they outlined a series of rules for navigating the deceitfulness thought to be inherent in market activity. The legal historian Morton Horwitz has argued that the demands of the market economy led to the legal distinction between "facts" and "opinion" in the jurisprudence addressing fraud. Discussing the legal transformation of the early republic, Horwitz contended that "since opinions, estimates, or interpretations were regarded as subjective—and part of a legitimate distinction of individual talents and attitudes—the only forms of knowledge which the market system could justly protect against misrepresentation were thought to be bare statements of 'facts.'"[20] The widely cited Supreme Court case *Laidlaw v. Organ* (1817) defined the deceit necessary to prove fraud as consisting of deliberate acts of commission rather than silent omissions of relevant information. In that case, a tobacco trader who had knowledge of the declaration of peace in the War of 1812 was not held responsible for revealing this fact to an ignorant vendor, even though these new circumstances greatly improved the value of the commodity. Legally speaking, fraud required proof of intentionally conveying untrue statements as opposed to committing deceitful implicatures.[21] The net effect of this legal formalism was a constraint upon the avenues through which the courts could intervene in commercial transactions, and a stricter delineation of the terms under which an individual could claim to have been swindled.

The decision in *Rockafellow v. Baker* (1861) captures how courts at mid-century conceptualized the market, the nature of its participants, and their responsibility to one another. Echoing the language that identified Thompson's crime, Woodward insisted

> There is no confidence between buyer and seller, unless a warranty be demanded and given. They deal at arms' length. They use not each other's eyes, but each his own. The seller is allowed to express freely his opinions of the value of his wares—the buyer is at equal liberty to answer that it is naught. If there be an intentional concealment or suppression by either party of material facts which he is bound to communicate to the other, there is fraud; but neither party is bound to communicate that which is equally accessible to both. The state of the markets, the present and prospective value of a particular commodity, are among the things which are alike open to both buyer and seller, and neither is bound to instruct the other.[22]

In the absence of an outright promise otherwise, the assumption was that people will likely try to deceive one another. This was a tolerable condition because Woodward presumed that each party had equal access to the same knowledge; in his words, each party had its own "eyes." His decision posited the market as a transparent and local entity, wholly comprehensible through an individual's commonsense reflection upon the experiences of the senses. The common-law rule of caveat emptor served a crucial ideological function. In the words of economist Walton Hamilton, it was believed that "caveat emptor sharpened wits, taught self-reliance, made a man—an economic man—out of the buyer." In sum, the era's jurisprudence laid greater stress upon an individual's responsibility for self-protection and minimized any legal duty to make sincere representations.[23]

Unwary Marks and Knowledgeable Confederates

During the Gilded Age, antebellum attempts to maintain a clear distinction between the deceitful swindler and his transparent victim became increasingly untenable.[24] This period's confessional and advice literature, alongside its legal decisions, stressed instead the danger posed by the often-hidden complicity of the mark. A person's own profiteering intentions led to financial danger: inside every victim brewed both a deceivable and a deceitful self. Authors of these treatises noted that in all three of the classic confidence tricks, the victim originally attempted to take advantage of the swindler. In bunco the mark cannot believe his luck at coming across a faulty wheel that appears to consistently pay off; in gold brick he is pleased to pay less than the market value to a miner desperate for cash; and in green goods he is willing to swindle both his neighbors and the state with fraudulent currency. The active participation of the mark in the illegal scheme also insured that he was less likely to go to the authorities once he realized the swindle. To press charges against the unknown swindler required that victims confess their own frequently illegal activities. As William Moreau, the so-called King of Fakirs, argued, "even in the face of all the warnings that they have to beware of them they make themselves the accomplices of the vampires, and should be classed with them." The confidence man may be a vicious predator, but Moreau insisted that his victims collaborate in their fleecing.[25]

According to these writers, confidence games would not be profitable if they did not appeal to greed lying below the surface of seemingly virtuous citizens. As the anonymous author of the exposé of urban crime, *The Swindlers of America* (1875), warned the reader, "all swindlers are dishonest, but it is also a fact, not so generally known, that these swindlers could not exist unless there were persons of dishonest leanings, to practice their art upon."[26] The successful swindler required the active participation of his victims in schemes they knew were fraudulent yet profitable. Furthermore, the late nineteenth-century literature dissolved certain of the distinctions so important to antebellum morality, such as the virtuous countryside versus the sinful city and the agrarian rube versus the urban sophisticate. In *The Eye Opener* (1892), David H. Leeper consoled farmers: "Remember, no class of people are exempt from the sharp brains and oily tongues of the swindler. The honest, hard-working farmer, the business man, the legislator, and even the speculators and bankers, all fall victim to them." What legal historian Lawrence Friedman has aptly called "crimes of mobility" became more geographically diffuse, encroaching on a wider array of vocations.[27]

Out of this confusion over who was truly deceitful in a confidence game, a very different legal interpretation of this brand of fraud emerged in the 1870s. Like the sensational trial of William Thompson, the case that brought the issue to the fore, *People v. McCord* (1871), unraveled in New York City. In this later case, the court made the question of the *defrauded* individual's knowledge and intentions paramount in deciding whether a criminal offense had even occurred. The decision in *McCord* recognized that those defrauded in confidence schemes were often not unwary marks, but in fact the swindler's confederate. While dissolving this boundary, the court's decision still attempted to demarcate those involved in such swindles from the rest of society by denying the victims the legal protections typically available in such instances.

The police had charged Henry McCord with falsely presenting himself as a detective carrying out an arrest warrant on behalf of a local justice. He approached his victim, Charles Miller, claiming that he had been promised a $200 fee in order to secure his arrest. Believing the false representation, Miller offered a watch and diamond ring to avoid arrest. He voluntarily gave away his property to ensure that McCord would violate his supposed legal responsibility. In other words, Miller was defrauded in an attempt to convince someone else to commit a crime. The court initially found McCord guilty of defrauding Miller, but the Court of Appeals overturned the decision. Because he had parted with his property for the sole purpose of inducing an officer "to violate the law and his duties," Miller was guilty of committing a criminal offense. In the most widely discussed passage of the decision, the majority held that "neither the law or public policy designs the protection of rogues in their dealings with each other, or to insure fair dealing and truthfulness, as between each other, in their dishonest practices." In other words, the court would not enforce the proper execution of a contract that was inherently unlawful. "The design of the law is to protect those who, for some honest purpose are induced, upon false and fraudulent representations, to give credit or part with their property to another, and not to protect those who, for unworthy or illegal purposes, part with their goods."[28] According to this line of thinking, courts expected criminals to be deceitful in their interactions with one another. The law was reserved for the protection of persons with wholly honest purposes in mind.

The legal community did not see *McCord* as advancing a new doctrine. As recorded in a critical notice for the *Albany Law Journal* at the time of the decision, the case "is one of those cases that pass unquestioned when they pass unnoticed."[29] Commonsense judgments about the qualities and character of the participants' minds formed part of the everyday practice of the law. Despite this initial lack of fanfare, the decision soon became a controversial reference point for arbitrating cases involving confidence games. The legal community had to decide whether the law ought to be used to ensure honesty among thieves or whether to make the distinction between the unwary and the deceitful. This interpretation of confidence schemes bound together knowledge, intent, and responsibility. The victim's state of mind and intentions rather than those of the accused were what counted in determining whether a criminal offense had occurred.

People v. Williams (1842), a case cited as a precedent in the decision, actually underscored *McCord*'s novelty. In the earlier case, the defendant had convinced one Van Guilder to sign over the deed to his property, claiming that a third party intended to sue him. Fearing losing his place of residence, Van Guilder signed the deed over to Williams to hide this asset during the supposed lawsuit. In actuality, the threat of a lawsuit was a ruse to obtain the title to the land. When the court heard the case, it found Williams's mark ineligible for protection under the law. The justices offered two different possibilities for why he had the right to a new trial. Since Van Guilder signed over his deed in order to avoid the collection of a fictitious debt, he was simply defrauded in the process of trying to defraud another and not entitled to legal remedy. But, the judges made clear, this was not the primary reason given for dismissing the case. Such reasoning was appropriate to a civil suit, not a criminal trial. The more salient reason was the presumed mentality of the person defrauded. In principle, prosecutable frauds involving false tokens "should be such as might deceive persons of ordinary prudence." The judges found it would be "impossible to sustain this indictment without extending the statute to every false pretense, however absurd or irrational on the face of it."[30] To insulate oneself from potential fraud when conducting financial transactions, each party involved must suspend his credulity and exhibit a certain degree of caginess. In *Williams*, the judges primarily cared about prudence and the assumption of personal responsibility for ordinary risks, while *McCord* hinged on the recognition of the complicity of the swindler's victims.

Although the argument in *McCord* never became the definitive national standard, its influence flowed through later decisions. Because of the precedent it set, the New York justices in *People v. Livingston* (1900) felt compelled to find the proprietors of a green goods scheme not guilty. As they noted in their decision, "We very much regret being compelled to reverse this conviction. Even if the prosecutor [the scheme's victim] intended to deal in the counterfeit money, it is no reason why the appellant [the swindler] should go unwhipped of justice."[31] A similar dilemma faced the same court in *People v. Tompkins* (1906), where the accused used his position as an operator for the Western Union Telegraph Company to run a betting scam known as "the wire." Tompkins claimed that his position provided him with advanced information about the outcome of horse races. He would provide false results to prospective marks and recommend a nearby poolroom where a bet could be placed; the victim was assured that victory was a foregone conclusion. Unbeknownst to his victims, Tompkins was allied with these recommended poolrooms and the mark never had a chance at winning. The problem that Tompkins posed for the justice system was that all his victims planned on using illegally acquired information. The court found that it could not prosecute Tompkins since the victims of his scheme were only malicious conspirators.[32] Although *McCord*'s authority was strongest within New York, the Supreme Court of Wisconsin enrolled it into their decision concerning a green goods scam in *State v. Crowley* (1876).

While the Wisconsin court extended the geographic reach of *McCord*'s logic, New York justices protested the lack of legislation to prohibit its deployment. In both *Livingston* and *Tompkins*, the court implored the state legislature to change the statute concerning complicity in such frauds to prevent defense attorneys from using the loophole. In his dissenting opinion in the original case, Rufus Wheeler Peckham Sr. had argued that McCord should still face prosecution regardless of Miller's intentions or actions. Peckham argued that it was a mistake to see the criminal law as primarily for the protection of the defrauded individual. Rather, its purpose was the punishment for public offenses and the prevention of future frauds. On this view, the identity and intentions of the victim in this particular circumstance were irrelevant in light of the fact that McCord had committed a crime. In the instance at hand, the mark was a fellow swindler with dubious intentions, but the next victim could just as easily be a truly innocent person.[33] Critics of the *McCord* decision, such as Peckham, held that the court had incorrectly blurred the distinction between civil liability and the enforcement of the criminal code.

Caught between enforcing contracts designed to violate the law and setting swindlers free, this era's jurisprudence settled into the legal management of confidence games. Only in cases where the victims were deemed to be fully unaware would the state take action to detain the offender. The stance taken in New York can be understood as management rather than prosecution because, as critics of the *McCord* decision were quick to point out, it contributed to the proliferation of confidence men and an underground economy constituted outside the boundary of the law. While the majority opinion in *McCord* held that the decision prevented the law from sanctioning and encouraging criminal activity, the opposite effect often resulted. Because the majority of victims in confidence games often had some inkling of wrongdoing, swindlers could take advantage of the law for their protection.[34] *McCord* allowed for the growth of what became known as the "big con" in the early twentieth century, long-term schemes where the victim needed to be convinced that he was operating as the swindler's inside confederate. As one critic opined, the decision in *McCord* "grants virtual immunity to the confidence man provided the scheme by which his victim is fleeced involved the victim's own turpitude."[35]

Furthermore, the legal logic in *McCord* illuminated the problem with the predominant Gilded Age typology of the rogue, especially as popularized in New York police detective Thomas Byrnes's *Professional Criminals in America* (1886).[36] The court suggested that criminals formed a different class than citizens who, because they violated the law, lived beyond its purview and were not subject to the protections it offered. Decisions like *McCord* hinged on the assumption that the honest and dishonest could easily be disentangled and reserved legal protection for only the former. In contrast to the idea of the born criminal distinguished by a recognizable physiognomy, confidence games actually highlighted the continuum from respectable to criminal. What characterized the vast majority of swindlers was their capacity to pass unnoticed as trustworthy members of respectable society. Dishonesty characterized particular situations and social relationships rather than being written into the body or psyche in a fixed way.

The Confidence Man in the Age of Emancipation

The cultural stakes for confidence games were particularly high during the Gilded Age, a period when the ability to make contracts took on particular importance as the marker of one's political freedom and independence. Legal concerns about the confidence man intersected with key components of liberal ideology in an era marked by the experience of emancipation, the dramatic expansion of the market economy, and the rise of a salaried managerial workforce. For much of the nineteenth century, labor as a concept had broader meanings than denoting the condition of the working class. The entrepreneurial society of the antebellum North, with its politically cherished independent shops and farms, blurred the distinction between capitalist and worker. Ideal occupations were those closely linked to the production of tangible commodities under conditions of autonomous self-employment. Financial independence was intimately braided with political rights, and individuals existing in some form of economic unfreedom—whether grounded in gender, enslavement, wage labor, or debt—were defined as incomplete political subjects. Economic and political changes following the Civil War necessitated a redefinition of this republican conception of the political order. As historian Olivier Zunz argues, greater latitude was given to salaried employment for large corporations as a politically dignified kind of work.[37] Moreover, the ability—and responsibility—to make contracts involving one's own body and property emerged as the new legal standard for political freedom.[38]

Liberal regimes, which center on the democratization of sovereignty and the reining in of the excessive powers of the state, require that disciplinary power be transferred onto the individual citizen. These processes open up an autonomous realm where the subject can achieve individual freedom as long as he exercises personal responsibility in the form of containing his basest passions.[39] It was precisely a lack of self-restraint and prudence that the swindler seemed to unleash in the commercial realm. The Gilded Age confidence men threatened to unravel the balance of internalized self-control that was the hallmark of liberal individualism. By unlocking passions and promoting vice, the swindler risked making deceptive frauds into a form of habitual behavior, etched deeply into the liberal soul.[40]

This political context suggests some of the implications of portraying the confidence man as a magnetic personality that usurped the will of his victims through mastery of their psychological failings. Indeed, a psychological idiom became central to descriptions of confidence games during this period. In praising his own ability to craft fraudulent advertisements that attracted the public's attention, self-professed "grafter" Ben Kerns credited his "knowledge of human nature." In his exposé, A. J. Greiner likewise stressed the importance of the mental preparation of the mark, for when the "bunco man can seize the psychological moment to 'make his play,' his victim falls, be he a bank president, hotel man, farmer or merchant." Even the Rochester Chamber of Commerce credited the confidence man with knowledge of "human nature long before the psychologists unlimbered their guns upon it." The police detectives Eldridge and Watts noted the importance of emotion in the swindler's trade. They "have every device at their finger ends for exciting, deluding and drawing on their victims until their pockets are emptied." In *Confessions of a Confidence Man* (1923), crime reporter Edward H. Smith best captured the psychology of deception that governed the swindle. He saw it as predicated on "the artful preparation of the victim, the winning of his confidence, the dulling of his good sense, the allaying of his suspicions and the excitation of his natural avarice." In recounting the history of the profession, Smith contended that confidence games took "advantage from the beginning of the public foibles, of what is now termed mass psychology."[41]

Reflective swindlers and their critics agreed: the only certain way to avoid becoming the victim of a confidence game was by maintaining a high level of personal integrity and honesty.[42] Authors portrayed this self-policing as particularly important for men who might otherwise be carried away by commercial ventures. For example, Alson Secor, editor of the Iowa agricultural journal *Successful Farming*, encouraged his male readership to avoid being seized by immediate desires by discussing all business decisions with their more prudent wives. After all, her "instinct is often more reliable than a man's judgment."[43] The confidence man did not prey upon the truly honest, but those who were greedy for personal wealth and who sought to make a fortune without the time and effort of industrious work. Resisting the confidence man was akin to resisting the dubious temptations of the financial speculation. A confidence man required the presence of private vices among his victims in order for his schemes to be profitable. Alternatively, if one reined in one's own greed, the grafter would have nothing to target. The moral policing of confidence games reinforced this central element of liberal self-government.

If these books depicted the confidence man as threatening the freedom of others, they defined his own identity in terms of his autonomy and self-control. Although he would often operate with confederates and occasionally serve as an apprentice, the swindler was never truly in another man's employ. These relationships, even those of apprenticeship, were arrangements of convenience, where each operator stressed his independence and an intention to maximize the profits for all those involved. As Ben Kerns noted with some amusement, his occasional confederate John Hawkins "assured me that, if I ever found him guilty of participating in a legitimate or equitable transaction, I could feel perfectly free in repudiating him as an acquaintance."[44] Neither Kern nor Hawkins owed his partner anything; each was an independent operator associated with the other simply for convenience.

Grafters prided themselves on their ability to avoid regular and regulated work. For example, "Bunco Bill" proudly proclaimed that he "never worked and never will." Instead he traveled the railroad lines challenging his fellow riders to high-stakes gambling matches. He insisted that both he and the challenger place their stakes in a common suitcase to guarantee their safety during the match. The challenger would win the game, but when he came to open the suitcase, he would realize that Bill had had a confederate switch the case with another one containing a mocking note.[45] Clearly, this scheme required considerable forethought and effort, yet that it was not "honest labor" was paramount for Bill. In his memoirs, Kerns discussed how he attended law school, but lacked the patience to build his own practice. He recalled, "I was not particularly greedy for money, but I was greedy for success. I was unwilling to wait. I wanted everything quickly—right away." Grafting offered a quick avenue to worldly gain while avoiding the requisite diligence and sacrifice proclaimed necessary in success manuals.[46]

Despite attempts to characterize the confidence man as a particular social type, many Americans acknowledged the moral equivalences between the swindle and other commercial activities. Many openly wondered whether the increasingly omnipresent market economy operated much like confidence games. The *New York Herald* started this trend in 1849 when it highlighted the parallels between Thompson and the Wall Street financier. One occupation in particular frequently drew comparisons with the confidence man: the salesman. As historian Walter Friedman has documented, fewer vocations were viewed with greater suspicion, despite the salesman's crucial role in the distribution of consumer goods. Part of this distrust derived from the amorphous parameters of the trade. It encompassed the evangelical preacher, the peddler, and the seller of books. Furthermore, the majority of these men—unlike the shopkeeper—were not geographically fixed, but traveled extensively. Offering a constant stream of novel goods, the salesman's very trade would vary from appearance to appearance.[47]

The confessional literature depicted the confidence man in a similar manner, as an individual who constantly tailored his activities to meet the current demands of consumers. Even moralists who denounced swindling sometimes struck an almost admiring note. For example, Chicago police detective Clifton Wooldridge suggested that the confidence man "lives strictly by his wits and he can truthfully be said to be a witty and a hard customer to handle. He is inventive and constantly bringing out new swindles." S. James Weldon's confessional *Twenty Years a Fakir* (1899) demonstrated the blurriness of the distinction between confidence man and salesperson. Weldon defined a fakir as a traveling salesman who used puffery and fraud to move a variety of products. He specialized in no particular trade, but shifted freely from one scheme to the next based on the potential wealth it could generate. For Weldon, this included selling patent medicines, ineffective vaccines for livestock, and encyclopedias, as well as running a wheel of fortune and other gambling devices. Writing in retirement, having married and discovered religion, he still praised his earlier unsavory enterprises as a legitimate style of commercial self-making. He demanded to be shown "the man who has not the ability to draw customers to him, or sense to employ business tact and trickery, and I will show you a man who will never amount to much in the world. His brethren in the trade are using them every day, and they are the ones who succeed." Weldon defended his fellow fakirs from moral rebuke, arguing that it was better to engage in puffery than succumb to financial failure. Furthermore, he maintained that it was his fellow dubious entrepreneurs who pioneered the psychology of sales that all businesses required.[48]

With the ability to recover from financial reversals or exposure, the confidence man was a lot of things, but he was not a passive, economic failure, at least not for long.[49] According to these writers, the confidence man possessed a great aptitude for managing his outward appearance to meet the demands of the situation at hand. This dynamism is evident in the following description of an itinerate grafter operating at the end of the century:

> He was a handsome man, with big blue eyes and a striking personality; he was a fluent talker, always smooth shaven and precisely attired; he was one of the rare men who could play any part in any man's game, and his open countenance always won the esteem of those with whom he came in contact; he could preach a sermon on a moment's notice and from any text, and could pass off for a young minister with as much ease and ceremony as he could dispose of a bottle of Hawkins Pain Allayer to the most credulous for twenty-five cents.[50]

Rather than a rogue whose physical degeneracy mirrored his moral corruption, this grafter was a charismatic and sexually attractive male. Able to performance multiple roles for different audiences, he remained self-possessed in every situation.

As Weldon suggested in his memoir, the rogue confidence man laid the foundation for this new style of commercial masculinity. The fakir was the salesman par excellence as he "must move with the tide, and shift his operations from day to day. The business of this week will be the reminiscence of next. New fields, new customers, new fakes; for these he must be constantly on the alert, and work them to the most extreme limit." He calibrated his life to the unstable dynamics of the marketplace, and as an emblem of his manliness he would boldly follow the dangers of its course. When faced with a financial reversal, he always had the next scheme at the ready. Weldon was acutely aware of the disdain for his profession, yet he refused to capitulate to it. "Call me an unvarnished liar if you will, a dissembler, a hypocrite, a cheat, a dead-beat, what you like," he boldly proclaimed. "To the untutored masses a successful fakir may seem to be all of these. You think his occupation is simply skinning the public. I know that his largest triumphs are in giving every man the full value for his money, and yet securing good profits for himself. Reconcile the two if you can; I did it long ago."[51] Likewise, the salesman's behavior embodied a distinct mode of corporate masculinity, one that required a greater emotional sensitivity than nineteenth-century norms of manliness permitted. The successful agent received training to balance a sense of empathy for the customer's needs with the aggressive attitude required to best his competition. His trade, consisting of an unsentimental search for personal profit, required a heightened awareness of the emotions of others.[52] As historian Roland Marchand has argued, advertising professionals understood themselves as missionaries for modern culture, creating bridges linking the rational world of the corporation with the emotions of the feminized, consuming masses.[53]

"The Literature of Exposure" as Urban Ethnography

How was one to come to know something as elusive as commercial puffery or political graft? Starting in the 1890s, deception in the form of trickery and theatrics appeared as the most viable strategy for investigating such activities. Perhaps no group more closely identified American confidence games with the everyday operation of society than the turn-of-the-century journalists who became known as muckrakers. Theodore Roosevelt coined the term *muckraker* in a 1906 speech specifically to condemn investigative journalism that, in his view, put "a premium upon knavery untruthfully to attack an honest man, or even with hysterical exaggeration to assail a bad man with untruth."[54] The president contended that the muckrakers' fascination with the unseemly side of business and politics, combined with their sensational writing style, degraded the moral tenor of American civic life. These journalists argued that the worlds of municipal politics and high finance constituted thinly disguised swindles in desperate need of exposure. Later deployed against America's captains of industry, this form of artful deception was first developed by journalists and social scientists investigating the lives of those they defined as on the margins of American society: the mad, the homeless, and the migrant laborer. Nelly Bly was widely credited for first developing this deceptive mode of investigation with her 1887 exposé of the Women's Lunatic Asylum on Blackwell's Island. To conduct her reportage for Joseph Pulitzer's *New York World*, she feigned mental illness in order to gain admission to the asylum as a patient. In the resulting articles and book, Bly transformed her own bodily experiences into a "vehicle of publicity." By emphasizing her own physical vulnerability, she grounded the horrors facing the impoverished women confined at the asylum in her own corporeal experiences.[55] "Stunt reporting" became an eagerly sought-after genre of journalism over the next decade. Although it was begun by women trying to gain a foothold in the profession, male journalists soon co-opted the technique.

Where Bly's work blurred the boundary between madness and sanity, her male successors focused on their ability to perform different class identities. For example, Josiah Flynt attained considerable celebrity by adopting the technique to explore the homosocial world of the tramp. Josiah Flynt was the pen name of Josiah Willard, the ill-fated scion of a wealthy Midwestern family whose aunt, Frances E. Willard, founded the Woman's Christian Temperance Union. Yet Flynt's authority derived not from his genteel birth, but from the ways in which he depicted his lived experiences, downplaying his privileged connections and listing his credentials as a "Past master in the art of tramping."[56] Such claims were neither entirely unfounded nor entirely sincere. He had attended the University of Berlin in the early 1890s, but gained fame for his sketches of the hobo lifestyle, collected as *Tramping among the Tramps* (1899). As the title indicated, his method consisted of donning the garb of the migratory laborer while living, traveling, and working among them. He distinguished his own enterprise from scholars with their rarified view of human existence. According to Flynt, adhering to academic decorum would mean that the criminal would remain incomprehensible. He insisted that the methodology of so-called criminal anthropologists was misguided because it focused on studying the criminal once in confinement rather than in his natural habitat. Because prisoners inevitably falsified their behavior once they were aware of being studied, such accounts were inherently unreliable. In contrast, Flynt claimed to study "the criminal in the open" by secretly observing his everyday activities while posing as a fellow tramp.[57]

Flynt's means of studying swindlers emerged out his participation in a broader movement to understand the frequently conflated working and criminal classes through a disguised immersion into their milieu.[58] Building on Bly's precedent, reform-minded social scientists, journalists, and novelists began relating firsthand observations of impoverished conditions as they voluntarily experienced poverty for a fixed period of time. Through such outings, reporters could relate not only their observations of living conditions, but also their own bodily experiences as a surrogate for the laborer's subjectivity. These deceptive performances ultimately worked to bolster the previously held class identity of both observer and reader.[59] In discussing the trend of guides offering sensationalized tours of "fake opium joints" in Chinatown, for example, Flynt noted that "respectable people like to emphasize their respectability by bringing it into close, if temporary, contact with its antithesis. A shudderful joy results, no small part of which arises from the conviction that we are not like unto the other men."[60]

The idiom of science was an important element in this genre. Flynt drew the parallel between the laboratory-based medical investigations of parasites he witnessed during his education in Berlin and his own examination of criminals as a species of social parasite.[61] These writers frequently used the language of "experiment" to describe their research, whether it be Stephen Crane's "Experiment in Misery" (1894) or, in perhaps the most celebrated example of the genre, Walter A. Wyckoff's *The Workers: An Experiment in Reality* (1897). Unlike Flynt, who had a palpable contempt for academic life, Wyckoff held the position of assistant professor of political economy at Princeton when he temporarily cast off the genteel comforts of academia to live and work as a manual laborer. He argued that he could best contribute to the science of political economy through the experience of hoboing, associating with prostitutes and anarchists, and laboring alongside migrant workers. These authors did not define experimentation in terms of controlling variables within a confined space, but rather in terms of the collecting of their own experiences. As a kind of idiosyncratic accounting, this methodology did not consistently secure support. In 1899 a skeptical reviewer scoffed at their brand of empiricism, pointing out "the fallacy that one can know all of a thing by sharing in the doing of it." A. M. Day, an instructor of political economy at Columbia University, condemned Wyckoff for his overly theatrical approach. Day rejected the assumption that "a stranger's study (usually of necessity limited in time and scope) of the outward circumstances of the life of a class can reveal the *essentials* of that life." A reviewer for the *New York Tribune* insisted that Wyckoff "had little to do with the real tramp," with his essays clearly demonstrating "his unfamiliarity with the genus." According to these critics, the authors' earnest intentions were insufficient guarantors of the resulting observations' validity. Dramaturgy only gave one access to the surface appearances of class differences rather than providing a means of scaling their depths.[62]

For these writers, it was imperative to track their quarry in its natural habitat. Accordingly, they offered their readers a kind of taxonomic "natural history" of the margins of commercial life—a counterpoint to the naturalist's account of animals.[63] Theirs was a novel approach within the human sciences. With the exception of Frank Hamilton Cushing's studies in the Southwest, anthropologists in this period still focused almost exclusively on the study of manufactured artifacts, classical texts, and bodily measurement.[64] Even when the "Chicago school" of sociology later embraced the label of participant-observer starting in the 1920s, the majority of its practitioners derived their information from social workers rather than from their own immersion into the working-class milieu.[65] Decades before participant-observation became a highly regarded methodology, these investigators offered a form of covert intermingling with their research subjects as a means of obtaining social scientific facts.[66]

The interplay between deceit and truthfulness animated these ethnographic reports of the tramping life. Validity hinged upon the author's execution of a convincing performance. For Flynt, language acquisition posed the greatest difficulty. "On starting out on my first trip among hoboes, I thought that I had provided myself with a sufficient number of words and phrases to converse with them more or less as one of their own kind; but I soon discovered how little I knew of their language."[67] This initial lack of dexterity with their particular argot was a major impediment to being accepted as a genuine hobo. The stress on a skillful performance contrasted with the authority these narratives gained from the author's "earnestness." As a sympathetic reviewer of Flynt's work noted, "He did not conceal a pocketful of money and send a trunkful of good clothes from place to place ahead, so that he might be a gentleman on the sly."[68] These reporters needed to remain fully committed to their adopted roles while conducting their investigations. To have secretly lived as gentlemen during this process would have rendered their findings inauthentic.

In particular, physical suffering provided a crucial means of guaranteeing the authenticity of the experiences. In 1902 the *New York Sun* reported on Wyckoff's own exposure during an expedition in Colorado. Despite his "worn and haggard" appearance and severe stomach cramps, he refused to disclose his true identity until a letter accidentally slipped from his pocket. The newspaper also stressed how he resumed his sociological tramping after only a few hours of relief from the rain and the cold.[69] A true investigator could not compromise his tramping experience through access to hidden luxury items. Flynt never returned to the genteel lifestyle of his family, preferring to translate his knowledge about the behavior of tramps into work as a railroad detective while continuing his investigative enterprise. While Wyckoff ultimately passed away prematurely at his Princeton home, Flynt succumbed to pneumonia in a Chicago hotel room during the course of a covert investigation into the illicit gambling operations of the poolrooms.[70] Generating a convincing account required the embrace of such risks. In this, their self-fashioning mirrored the ethic of manly self-sacrifice central to contemporaneous narratives of scientific exploration in harsh environments.[71]

The Despotism of Graft

Despite this mixed reception, the work of these investigators provided both a methodology and a vocabulary for discussing commercial deception and its impact on political life over the next decade. After his experiences as a pseudo-tramp in the 1890s, Flynt turned to investigating the connections between the country's politicians and confidence men. As a synonym for the corruption he uncovered in municipal politics, he introduced the term "graft" into the wider American lexicon with his 1901 exposé, *The World of Graft*.[72] Formerly, the word had been part of the exclusive parlance of the nation's commercial swindlers. As journalist Will Irwin's confiding confidence man noted in 1909, "The word 'grafter' has been pulled into politics, and its original sense is lost. On my side of the police fence, we mean by it any one who uses skin games as a vehicle for stalling through life."[73] Flynt gave the term wider circulation and a new meaning, making it a keyword to signify the major political challenges facing cities in the first decade of the twentieth century.

Originally published as a series of articles in *McClure's Magazine*, Flynt's first-hand observations relayed political life in three cities. He argued that the aims of municipal reform could never be achieved by simply listening to the voices of concerned dwellers of the nation's "Upper World." Rather, the social reformer needed to enter into the subterranean system and solicit the perspectives of its inhabitants. *The World of Graft* was replete with the opinions of numerous hucksters and swindlers whom Flynt claimed to have befriended during such sojourns. He declared that he wrote from the perspective of the con man and thief, offering their particular "voice" on political questions for mass consumption. The focal point of his 1901 narrative was the practices of the graft that bound together criminals, police officers, and local politicians. Drawing on the lexicon of the tramps he had studied for years, Flynt introduced graft as "a generic slang term for all kinds of theft and illegal practices generally." On this particular "voyage of discovery" he highlighted especially the omnipresence of bribery in the everyday workings of cities. He profiled Boston, New York City, and Chicago, suggesting that in each a different form of the graft operated. For example, Chicago was known as an "honest town" in Flynt's account precisely because it openly acknowledged the politician's conspiracies with swindlers and grafters.[74]

The publication of Flynt's exposé in serial form fueled the existing political controversy over municipal patronage in Chicago. Local papers attacked the mayor's office by reprinting the national celebrity's claim about the city's reputation for being "open" to the criminal classes. The immediate controversy centered on the mayor's policy of opening police stations at night to serve as shelters for the city's homeless population. Why would his office issue such an order? Was the city providing official refuge for the criminal class from which they could prey upon the virtuous citizens of Chicago? The son of the former mayor and a future five-term incumbent, Carter Harrison Jr. skillfully managed the implications that his office and the police constituted a node in the criminal underground. In his published defense, Harrison fashioned himself as a champion of the city's poor and attacked Flynt as an outsider tainted by superficiality. Harrison argued that the reporter misunderstood and distorted the policy due to his pursuit of sensationalism.[75] Each side enveloped itself in the cause of reform. Flynt accused the mayor of an insatiable hunger for the money provided by thieves for official protection, whereas Harrison stressed his sympathy for the plight of the poor.[76]

With the publication of *The Shame of the Cities* (1904), Lincoln Steffens became the leading advocate of the claim that political graft represented the systemic breakdown of the polity. He contended that graft occurred due to the lack of legitimate mechanisms through which private power could exercise its influence publicly.[77] In making these claims, Steffens rejected the racial argument that corruption in American politics resulted from the presence of despotically minded immigrants unschooled in democracy. Political corruption resulted from arrangements particular to the American scene. He argued that in Britain politics was a gentleman's sport while in Germany it constituted an austere profession. In contrast, politics in the United States was in trouble because too many people saw it as simply another kind of business. A "commercial spirit" of personal profit possessed these "political merchants" and compromised their patriotism and loyalty to the people's prosperity.[78] The existence of the political boss who profited from graft demonstrated a fundamental weakness in the American national character.[79] Like the railroad oligarch, he was "the product of a freed people that have not the spirit to be free." He endorsed Theodore Roosevelt's solution—one, he noted, that many sneered at as naïve—that a restoration of democracy and manly freedom required "good conduct in the individual, simple honesty, courage, and efficiency."[80]

Steffens's contemporary, Chicago police detective Clifton Wooldridge, also argued that graft operated simultaneously in number of registers: the moral, the psychological, and the political. Not only did the graft cost cities financially, "but it became a certainty that it was costing something even more valuable than money. Graft became the one object of the political seeker after office."[81] Because of moral weaknesses inherent in human nature, graft threatened the democratic political system in its entirety. When a politician persistently committed criminal acts of fraud, deceit colonized his very soul and became his habit. Graft made the democrat a slave to his commercial passions and subverted the entire system.

Not only were corrupt political officials identified with swindlers, but practices of the marginal confidence man and the methods of captains of industry also became increasingly difficult to distinguish. For a number of their contemporaries, the capitalist enterprise writ large, especially in the form of the stock market and futures trading, operated like a kind of confidence game that threatened the political order. By the 1870s the standardized grading system for grain had led to the creation of futures, instruments that allowed trading on the promise of grain to be delivered. This innovation worked to detach commodities trading from the material goods involved; members of the Chicago Board of Trade were able to sell a greater number of bushels than the actual amount produced in a given year. In the 1890s the agrarian Populists, especially those in the Prairie West, focused on the fictionalized nature of such exchanges when they criticized stock exchanges. Drawing on an established moral critique of gambling, these farmers condemned the speculative nature of trading futures. Farmers complained that, through the selling of excess futures, the urban traders artificially diluted the worth of the goods that they physically toiled to produce. Farmers saw such ventures as deceptive because they were hidden from public scrutiny and seemed to bestow an arbitrary value upon the goods they produced.[82] They were not alone in holding this opinion. Flynt and fellow reformer Alfred Hodder claimed that the denizens of the upper and lower worlds "differ more in their circumstance than in their psychology." From a psychological point of view, the respected entrepreneur, "who assume[s] great risks on the chance of great returns," was identical to the gambler and thief. Both "obtain their success by a combination of enterprise, intelligence, unscrupulousness, diligence, and sheer, rude power of will."[83] Tellingly, Flynt and Hodder argued that large-scale speculation and illegal swindling coalesced in the common psychology, animating each.

The muckraking journalism that flourished in the first decade of the twentieth century shared a similar standpoint. Prefigured in earlier exposé work, this movement emerged in the January 1903 issue of *McClure's Magazine*, which contained the investigations of Steffens on the corruption within municipal governments, Ida Tarbell's history of the Standard Oil Company, and Ray Stannard Baker's exposé of labor union practices. Each taking a different object of inquiry, these longitudinal investigations collectively emphasized the political ramifications of fraud in their respective arenas. Moreover, the reports revealed each of these institutions to be actively deceitful rather than accidentally deceptive. The editorial in the issue that brought together their findings made clear the connections among commercial fraud, democracy, and the specter of political despotism: "We forget that we are the people; that while each of us in his group can shove off on the rest the bill of today, the debt is only postponed; the rest is passing it on back to us. We have to pay in the end, every one of us. And in the end the sum total of the debt will be our liberty."[84] Like most reformers who identified with the Progressive banner, Steffens held that business ought to be considered an important part of the republic, but that it was not sacred and its imperative could not go unquestioned. He argued, "if bribery is treason, if the corrupt politician is a traitor, then the corrupting business man is an enemy of the republic." Tarbell also emphasized the relationship among commerce, deceit, and despotism, arguing that modern enterprise was a manifestation of Machiavellianism. For Tarbell, the very essence of commercial enterprise, as exemplified by Standard Oil, was "lying." Industry had cultivated this tactic so that it had become "a science as perfect as the militarism of the nations," delivering America into a new "age of despotism."[85]

Likewise, the confidence man genre also came to identify large-scale businesses with illegal swindling. In *Swindling Exposed: From the Diary of William B. Moreau, King of Fakirs* (1907), contempt for contemporary business practice lends an almost elegiac, nostalgic tone to this account of a Gilded Age style of confidence man who made his fortune through get-rich schemes and puffery. When the publisher, J. B. Costello, edited the deceased's "confessions," he interweaved his own commentary in the narrative. While the king of fakirs had described himself as a "vampire" who financially suckled off the willing, his editor acerbically noted, "The sins of the Moreau, and, we may add, thousands of other crooks combined, did not squeeze as much out of the people of the United States as the unscrupulous octopus known as the Standard Oil Company, dominated by John D. Rockefeller."[86] It was evident that the real vampire was certainly not the small-time fakir.

Perhaps what was most striking was not that this equivalence was consistently drawn, but rather that traders enjoyed relative success in convincing people that what they did constituted a form of service. Defenders of stock market speculation and futures trading dismissed the claim that those involved simply engaged in a form of immoral and dishonest gambling with other people's money and livelihoods. They argued that speculation constituted a valid form of labor, analogous to the physical work of the virtuous farmer. When John Hill Jr. wrote of *The Gold Bricks of Speculation* in 1904, he was not asserting that futures exchange constituted a genre of confidence game. Rather he was exposing those institutions, namely the unlicensed bucket shops, which counterfeited the appearance of legitimate boards of trade.[87] For members of the Chicago Board of Trade like Hill, the bucket shop that allowed individuals to place bets on anticipated fluctuations in the daily market was the truly dishonest institution. Exchanges required a refined knowledge of how the economy worked and their trades rationally governed the market, but the bucket shops promoted idle speculation and outright gambling. The nation's Supreme Court bolstered this view in 1905 when it decided that stock quotes did represent a trade secret, rendering them a kind of property entitled to legal protection.[88]

Yet maintaining clear lines of demarcation between the transparent reporter dedicated to communicating truthful information and the deceitful force of enterprise proved difficult at best. The career of the stock market speculator Thomas W. Lawson manifested the oscillation between sincere exposure and deceitful swindle. In 1899 he had helped orchestrate Standard Oil's notorious takeover of the Amalgamated Copper Mining Company through the manipulation of information about its stock. Five years later, he joined the exposé racket when he began publishing a series for *Everybody's Magazine*. Drawing on his own experiences as an industry insider, he promised his readers to reveal the secrets of what he called the financier's "System." Ironically, Lawson's own dubious history and his as a confessing high-stakes confidence man endowed the monthly columns with much of their authenticity. He stressed how corporations presided over "a perfected system for the false moulding of public opinion for the purpose of making more easy the plundering of the people."[89] Lawson drew accusations of charlatanry because, at the same time as he condemned the system's manipulation of the public, he continued to manage a financial news bureau and spent thousands on advertisements to promote investment in his latest speculations. Moreover, following the financial Panic of 1907, Lawson publicly declared his retirement from exposé in order to devote his full energies to his own business of "stock gambling."[90] His career dramatically illustrated how even the mantle of the muckraker sometimes only served as another disguise for the modern confidence man.

Conclusion

The proliferation of confidence schemes in an age of emancipation complicated an image of the nation as a land populated by self-governing, free men, and an ethics derived from the virtues of commerce. The exposés and confessions of confidence men dramatically depicted the unfreedom unleashed by certain financial arrangements. In staging these concerns, the genres of swindler confessionals and muckraking journalism became productive sites for discussing the relationship among commercial opportunity, exploitation, and human nature. From the individual confidence game that made the mark beholden to another's will to the political graft that transformed urban democracies into despotic regimes, fraud usurped hard-won political freedoms. The 1849 trial of William Thompson provided the vocabulary for this understanding of the equivalences between the engine of the nation's material growth and its novel forms of crime. His case demonstrated the difficulties in distinguishing between the criminal manipulation of trust and the activities of the salesmen or the financier. Moreover, Thompson's promises of speculative wealth rewarded by trust in a stranger also raised the specter of complicity. Despite this possibility, the antebellum exposé literature laid greater emphasis on the theme of the victim as unwary innocent new to the untrustworthy city.

During the Gilded Age, contributors to the genre came to stress that this "enslavement" to the will of another had more to do with the deceivable self's hidden passions, desires, and greed than with chance misfortune. These writers determined that the victims of these confidence schemes were deceived because they initially sought to deceitfully take advantage of others. In short, they were ruled by their own psychological failings. The controversial ruling in *People v. McCord* represented an attempt to bring the law into line with this distinctly psychological understanding of fraud. In the first decade of the twentieth century, the word *graft* came to express these complicated relationships between public power and private vice. Originally referring to the techniques of individual swindlers, muckraking journalists adopted the term to disparage the political culture of the nation's leading metropolises. This act of translation was not without irony. Those who originally promulgated the term engaged in their own spectacular form of deception, albeit in the service of truth. Starting in the 1890s, a number of civic-minded writers claimed that one could not adequately depict the nature of fraud from afar. As a novel form of evidence they offered their own observations and experiences as they engaged in performances as members of the criminal underworld. In order to fully represent the many-tentacled monster of urban fraud, they engaged in their own public deceits.

These multisited discussions of the mass psychology of the market shared certain ideas about human nature that influenced the scientific discipline of psychology then taking root in American culture. These writers depicted the self as simultaneously deceitful and deceivable—indeed, the individual's deceitfulness ultimately led to his own deception. Moreover, the investigation of commercial and political fraud brought to the fore a conviction in the impossibility of accurate self-knowledge, that the human subject was a tricky character requiring deceit on the investigators' part to guarantee truthfulness. Both of these themes—the individual as both deceitful and deceivable and the centrality of deception as a tool in the human sciences—profoundly shaped the science that psychology became in the United States.

CHAPTER TWO

Hunting Duck-Rabbits: Illusions, Mass Culture, and the Law of Economy

His 1903 exposé of municipal graft established Lincoln Steffens as one of the nation's most prominent muckrakers, but journalism had not been his first choice of vocation. As the son of a wealthy banker, he toured Germany and France as a young man, absorbing the culture of their universities and arts scenes. In the winter of 1890 he arrived in Leipzig to attend the lectures of Wilhelm Wundt, widely recognized as the founder of the "new psychology," an experimental, laboratory-based science. Steffens planned on taking a degree at the university and serving as an assistant in the Wundt laboratory. His primary motivation for making the pilgrimage was "to see if I could find in psychology either a basis for a science of ethics or a trail through psychology to some other science that might lead on to a scientific ethics."[1] His presence at Leipzig was part of a larger quest to find a new foundation for morality in the modern age. Yet before the academic year was finished he had left, frustrated with the monotony of the laboratory. He returned to America to complete his self-fashioning as a moral beacon through the journalism that made him famous. Steffens's youthful misreading of the aims and scope of experimental psychology was indicative of how a generation looked to that particular discipline to provide a scientific groundwork for ethics in the wake of the Darwinian revolution and the advent of modern industry. The connection between the future muckraker's brief flirtation with psychology and his later career raises the question: what did a science largely concerned with perception have to do with concerns about deception in the commercial and political realm?

One connection was the psychologists' interest in how and why the human organism operated as a deceivable self, a fascination manifested especially through the study of perceptual illusions. The topic was so central to the new science that psychologist Edwin Boring called the 1890s "the decade of the illusion."[2] Although most psychologists had their own variation, a consensus on why illusions transpired soon emerged. This understanding drew upon numerous aspects of nineteenth-century scientific naturalism, including evolutionary thinking, statistical probabilities, psychophysics, and the conservation of energy. An illusion occurs because natural selection bestowed a strict "law of economy" upon the operation of the human mind. Humans live in a world of probable expectations. They misconceive because the majority of sensations do not receive conscious reflection, but rather are channeled down preexisting mental pathways etched by habits and evolutionary history. Deception results because expectations and interests direct one's attention when observing immediate circumstances.[3] Understood as inevitable owing to the physiological organization of the human organism, deception was nevertheless distasteful to these psychologists. Thus, they valorized conscious decisions as the guarantor of personal freedom in a universe governed by both the necessity of physical laws and the uncertainty of chance.[4] Often pleasurable, the seductiveness of the comforting state of deception made it particularly dangerous. Deception exemplified the epistemological and moral costs of humanity's finite mental resources.

Typically avoiding grand pronouncements about human nature in their laboratory reports, early psychologists expressed this aspect of their thinking in public lectures, articles for mass-circulation magazines, and textbooks. Often seen as derivative and second-rate, these communicative ventures helped secure a visible place for the discipline within the wider culture.[5] This kind of writing was an acceptable if not entirely honorable income-generating activity for scientists at a time when career trajectories and sources of funding were unclear.[6] In these works, psychologists, especially Joseph Jastrow, offered guidance on how ordinary Americans ought to manage their lives to minimize the chances for deception. Others, most notably Walter Dill Scott and Hugo Münsterberg, mobilized their knowledge about this deceivable self to receive remuneration as professional consultants to businesses.

At the turn of the century, psychologists undertook a concerted project of remaking the public's perceptual habits. Jastrow's career exemplified this trend as he "assumed the life of a vaudevillian," performing his science in the public sphere in the hopes of transforming his fellow citizens.[7] Among academic psychologists, he wrote most frequently about what he called "the psychology of deception," which encompassed both scientific and more literary investigations of those mental tendencies that led to the misinterpretation of sensations and resulted in sustaining credulous belief. In this regard, Jastrow wrote to the *New York Times* near the end of his career to express his displeasure with a book review that implied he agreed with the showman P. T. Barnum's contention that deception was essentially harmless. Instead, he argued that "indulgence in such beliefs, far from being innocent, makes for a flabby, uncritical mentality, and, carried far enough, rots the mind."[8] Unlike Barnum, who reveled in and profited from the ambiguous deceptions of the marketplace, Jastrow understood his science as a counterforce to the mentalities generated in the commercial realm. This brand of popular science changed the venues and means of communication through which people could debate the nature of deception. Yet, as Jastrow's reviewer suggested, this first generation of experimental psychologists also served as the somewhat reluctant and ambivalent heirs to Barnum's project.

The stance taken by perceptual psychologists like Jastrow contrasts sharply with the economist Simon Patten's contemporaneous call for an intellectual revolution in "mental habits" to meet the opportunities provided by the new economy of abundance.[9] In part, Patten's revolution entailed understanding the individual as a "social self," defined in relationship to others and embedded in large institutional structures.[10] While some psychologists embraced this vision, many held that their science furnished support for a more individualistic understanding of the self that was focused on prudence, restraint, and scarcity. Illusions, as a product of the law of economy, were thought to offer empirical evidence of this frugal self.

Pursuing this research also involved incorporating elements of the deceptive visual and material culture of the Gilded Age. The final two decades of the nineteenth century witnessed the increased availability of comparatively inexpensive forms of public leisure. These amusements expanded upon Barnum's museum and included the exhibitions of *trompe l'oeil* paintings, five- and ten-cent vaudeville theaters, and the midways of state and world fairs.[11] These venues prominently featured spectacles such as stage magic, midway rides, and early cinema, which captivated audiences through their artful management of expectations.[12] What insights did scientists derive from reading middlebrow magazines, watching a stage conjuror's show, or attending a fair? Psychologists, especially when dealing with the issue of deception, were as much consumers of the nascent mass culture as one of its producers. To understand the science of illusions, this chapter tracks the circulation of deceptive things between these leisurely settings and the experiments and writings of psychologists.

Making Deception Scientific

Concerns about deception had accompanied the study of perception for centuries. In the seventeenth and eighteenth centuries, European natural philosophers and showmen manipulated optics, light, and images to produce fantastical yet edifying spectacles in the form of the camera obscura, the magic lantern, and the phantasmagoria.[13] These inquirers largely confined their experimental interventions to nonhuman materials, reserving the human subject for speculation. This situation changed in the 1820s and 1830s. Inspired by Romantic interest in self-understanding, German-speaking physicians Jan Evangelista Prukinje and Johannes Müller began to study the mechanisms of visual perception. Their physiological studies aimed at bridging the Kantian gulf between the natural order and the experiencing subject by investigating precisely how the individual observes nature. Across the North Sea in Britain, Charles Wheatstone and David Brewster focused on the optics of light, a central platform of Newtonian natural philosophy. Both traditions held that the study of perception revealed the unavoidability of optical illusions. Vision, once understood as a universal experience common to all, became increasingly defined as an idiosyncratic process intimately tied to the bodily functions of the individual. Brewster, in particular, derived moral lessons from the study of vision. The presence of optical illusions served as a reminder of the unreliability of depending on one's own judgments and of being overly confident about one's own abilities. Yet Brewster did not end with such a defeatist argument; he stressed the social utility of his science. He concluded, in the words of historian Jutta Schickore, on a note of "epistemological optimism." Because natural philosophers were increasing their understanding of the mechanisms of how the subject was tricked, by following their way of thinking individuals could avoid the pitfalls of their natural endowment.[14]

Despite these antecedents, it was the reception of Wundt's experimental psychology that framed the scientific study of deception in the United States. Starting in the 1850s, Hermann von Helmholtz dominated the study of human perception. He grounded his approach in the careful measurement of the varying reaction times of living organisms to stimuli. He also developed many of the instruments for measuring bodily reactions later used in psychology. As one of Helmholtz's students, Wundt appropriated his teacher's research methods. Rather than focusing on the passing of nerve impulses through the body, however, Wundt emphasized the importance of intervening mental activity as a factor in reaction times. While Helmholtz's organic physics located subjectivity in the physiology of perception, experimental psychology shifted the locus to the mind. Indeed, Wundt emphasized a parallel relationship between mind and body rather than a strictly causal one. His research program for laboratory psychology focused on carefully facilitated acts of self-observation under standardized conditions.[15]

When Americans traveled to Europe to study psychology, they invariably chose Germany as a destination, but other nations also had programs of research that shaped the psychology of deception in particular. In Britain, Darwin's cousin Francis Galton championed a program of anthropometrics, or the measurement of bodily traits and mental abilities, that profoundly influenced American psychology. In France, psychology was closely associated with clinical medicine, especially the treatment of hysteria. During the 1880s and 1890s, Jean-Martin Charcot in Paris and Hippolyte Bernheim in Nancy debated the nature of hypnotic phenomena. Where Charcot argued that the ability to be hypnotized indicated a constitutional failing characteristic of hysterics, Bernheim claimed it was a much more common trait. Their debate brought to the fore issues of suggestion, influence, and consciousness that had ramifications on how American psychologists understood deception.[16]

In the context of nineteenth-century science, the selection of deception as an object of study may seem an odd choice for these psychologists. Scientists of Helmholtz's generation insisted that there was no such thing as a true illusion in terms of the deception of the senses. A central tenet of Helmholtz's organic physics was that whatever a healthy eye experienced was considered to be an optical fact.[17] Why did this next generation invest so much energy in reclaiming and reanimating the concept of deception if it had been banished from the scientific realm by physiological optics? In his *Illusions* (1881), British psychologist James Sully suggested the answer: what are "sometimes called deceptions of the senses" were inaccurately labeled since it was not the senses that were in error, but how the mind classified and interpreted the nervous impulses they furnished.[18]

Sully's volume testified to the resurgent interest in illusions among late nineteenth-century scientists, a resurgence that would profoundly shape psychology as a discipline in the United States. Deception did not point toward defective sensory organs, but to the vagaries of the sensorium.

The American study of illusions combined Brewster's "epistemological optimism" with regards to scientific practice with a cultural pessimism when it came to speak of the capacities of ordinary persons. Psychologists were quite confident about the ability of their fellow "men of science" to overcome the obstacles of deceptive subjectivity through the rigors of proper training. In contrast, they worried about a public that they saw as a largely undifferentiated mass of people with a severely hampered capacity for judgment. They feared that new forms of social organization posed a serious challenge to cherished forms of autonomy and self-reliance. In particular, they viewed the self-possessed man of science as caught in pincers composed of the philistine captains of industries and the masses of workers. In part, these psychologists hoped that the consumption of their brand of popular psychology might combat and keep in check these dangerous forces.

"Creatures of the Average" and the Deceptions of a Probable World

The American psychology of deception first took shape in the mid-1880s at the Johns Hopkins University. Founded in 1876 as a university dedicated to research and graduate-level study, it soon became, albeit briefly, the most important center for psychology in North America. The teaching staff and graduate students of its philosophy department included a number of individuals who were interested in what they called the "new psychology." In 1879 Charles Sanders Peirce was hired on a part-time, contractual basis to teach logic at the new institution. The son of the nation's leading mathematician, Peirce spent his early career as a contributor to the U.S. Geological Survey. He had also written extensively on the philosophy of science, statistics, and a new branch of knowledge he called semiotics.[19] In addition to these teaching duties, Peirce established a Metaphysical Club, a university-sponsored group for those with a common interest in science, philosophy, and, increasingly, psychology. He modeled the Baltimore club on a group of the same name that had convened earlier in Cambridge, Massachusetts, with Peirce, Chauncey Wright, and William James, among others. The Cambridge circle did much to translate the implications of Darwinian natural selection into a new American philosophy, which later became known as pragmatism.[20] In 1883 G. Stanley Hall joined Peirce at Hopkins. He had received his doctorate from Harvard University in 1878 for a thesis on "the muscular perception of space" and had continued his studies in Germany. Hall came to the university's attention due to the enthusiastic response to a series of lectures he had delivered on the emerging science of pedagogy.[21] The earliest students at Hopkins included Jastrow, James McKeen Cattell, the future editor of *Science*, and John Dewey, then largely a devotee of Hegel.

Jastrow entered the doctoral program in philosophy at Hopkins in 1882 and soon focused on psychological research. Hall served as his official dissertation advisor and the two shared a common understanding of the human mind as the product of evolutionary forces. During these years, Hall was formulating a psychology of adolescence that outlined how the developing mind of the individual child recapitulated the evolution of the species.[22] While he would come to repudiate his mentor's particular interpretation of evolutionary development, Jastrow's psychology of deception similarly dealt with vestigial forms of "primitive" mental behavior operating in an industrial world. Although Peirce was soon forced to leave Hopkins under a cloud of personal scandal due to rumors of marital infidelity, Jastrow later insisted that the logician's influence was paramount in his intellectual development. Jastrow embraced the importance of logic, not in the form of mathematical machines for problem solving but as the "investigation of the nature of the thought process." Most important, Peirce also introduced him to the investigation of the mind through experiments.[23]

Starting in 1883, the two executed what was widely cited as the first psychology experiment performed and published in North America. It was a psychophysical examination of a person's capacity to judge differences in sensation and the observer's confidence in said judgments. Their aim was to determine how reliably an individual could detect small changes in pressure when differing weights were pressed against his index finger. The apparatus was simple: a post office scale, a one-kilogram weight, and smaller weights that could be added or removed from the scale during the course of the experiment. The personnel involved were also minimal, with Peirce and Jastrow fulfilling the role of experimenter and observer, each exchanging roles in turn.[24]

Peirce and Jastrow insisted that obtaining valid results required maintaining a certain level of ignorance in the observer. At the time, Wundt had built experimental psychology upon a particular relationship between the experimenter and the participant, described as "the observer." Wundt held that unaided self-observation risked exposing the scientist "to the grossest self-deception," but laboratory apparatus and proper training allowed for certain individuals to accurately record their own mental experiences when presented with carefully measured stimuli by the experimenter.[25] Psychology demanded honest, transparent observers who acted as faithful recording instruments of their own consciousness. In contrast to Wundt, Peirce and Jastrow modeled their experimental observer not on the self-possessed, gentlemanly introspector, but on the unsavory gambler trying to best a game of chance. They separated the observer from the experimenter by a screen, making him unaware of whether additional weight had been added, removed, or maintained. Furthermore, even the experimenter was left unsure about the sequence of stimuli. Peirce and Jastrow achieved this goal through the randomization provided by a shuffled deck of playing cards. The color of the card determined whether to diminish or increase the weight. Randomization ensured that the observer could not discern a regular pattern.[26] The observer's ignorance did not require deceit on the part of the experimenter: the chance generated by the shuffled deck left both men uncertain about each trial's course.[27]

Probability governed not only the design of this experiment, but also the deceivable self it materialized. Based on their findings, Peirce and Jastrow argued against Gustav Fechner's notion that there existed a "*Schwelle*," a threshold beyond which the difference in sensations could no longer be detected. No such definitive point existed; rather, the observer operated along a continuum of misperception. The experiment offered a view of the mind as an uncertain judge, one that tended toward likelihood on what they called a "subliminal" level. In their conclusion, Peirce and Jastrow speculated that the existence of subliminal judgment had the potential to explain contemporaneous telepathic phenomena and maybe even the mysterious "insight of females."[28] Following these experiments, Jastrow continued cultivating his interest in the nature of misperception. Where his mentor Peirce understood the significance of living in a statistical age in terms of inhabiting a cosmos governed by chance, similar resources drew Jastrow to the conviction that existing mental habits had the tendency to induce deception.[29]

Such an attitude was already evident in Jastrow's earliest individually published article. "Some Particularities in the Age Statistics of the United States" (1885) considered those mental habits that expressed themselves as behavioral trends in large populations. He analyzed the findings of the 1880 national census and found a disproportionate number of individuals who stated their age as ending with zero compared to those with a nine or one. For example, there were far more people claiming to be fifty than either forty-nine or fifty-one. This statistical aggregate of ordinary Americans had a propensity to record its age in round numbers such as thirty, forty, and fifty. His interest in the census was psychological. This "10 exaggeration" illustrated a mental predisposition, a distorting judgment. He cautioned that future census takers had "to appreciate how enormous the attraction towards round numbers really is."[30] His explanation did not concern the circumstances that led individuals to erroneously record their age, but rather identified tendencies within populations. After all, the rise of statistical thinking had reconstituted the study of individuals away from particular motivations in favor of discerning propensities and patterns in the aggregate.

Jastrow next turned to a consideration of the role of racial and sexual differences in the phenomena. His findings reaffirmed the political hierarchies of his day with white, American-born males the most accurate in their reporting followed by male immigrants, then women. The population most prone to exaggeration and hence deceiving census officials were "colored peoples," a category that included immigrants from Asia, but was mainly populated by the recently emancipated slaves. Jastrow argued that he had uncovered the geography of mendacity. He suggested that contact with racial minorities served as a kind of mental contagion that affected white males from New Mexico, whose numbers included Hispanics. The denizens of the southern states were found similarly wanting as living in "too close intimacy with the 'round-number loving' Negro seems to be dangerous to statistical accuracy." These regions were in stark contrast to the more reliable and accurate zones of New England and the Midwest. In these passages, Jastrow identified age exaggeration as a kind of "mendacity" expressed as a psychological propensity to err. Furthermore, he directly correlated this tendency to individuals aggregated into sexed and raced populations. Arguing that such deeply engrained tendencies were impossible to root out of the population, he suggested that future census takers ought to reformulate their questions and approach individuals with greater skepticism in order to produce a more accurate final product.[31]

The psychology program at Hopkins quickly disintegrated, with those involved becoming major figures in the field at other universities. In 1889 Hall left when asked to become the founding president of Clark

University in Worcester, Massachusetts. From there he published his pioneering *American Journal of Psychology* and continued to train numerous psychologists. Among Jastrow's peers, Cattell quickly left after being denied a fellowship for which he felt entitled. He received his doctorate under Wundt in 1886. He then briefly studied medicine at Cambridge University, where he corresponded with Galton about anthropometry. Upon his return to the United States, he founded the laboratories at the University of Pennsylvania and Columbia University. Dewey's interests shifted from German idealism to pedagogy. He was an influential professor at the University of Michigan, then Chicago, finally also settling at Columbia in 1904.

When Jastrow graduated in 1886, no jobs in psychology materialized until his appointment two years later at the University of Wisconsin.[32] He began his career as a freelance science writer, contributing not only to the recently established *Science*, but also to genteel periodicals like *Popular Science Monthly* and *Harper's*.[33] He quickly became psychology's most prolific publicist. Pursuing his vocational interests as a scientific authority on matters of the mind required that Jastrow straddle the worlds of academia and popular publishing, tailoring his scientific output to the demands of middlebrow publishers.[34] Even after his appointment at Wisconsin and the creation of discipline-specific journals like the *American Journal of Psychology* (1887) and the *Psychological Review* (1894), he continued to contribute to a wide variety of periodicals.[35]

Jastrow utilized these commercial ventures in popularization to advance his understanding of the mind as the locus of deception in a probabilistic cosmos. In an 1888 address delivered before the Anthropological Section of the American Association for the Advancement of Science and published in *Popular Science Monthly*, Jastrow first proposed his "psychology of deception."[36] While acknowledging that pronouncements about the capacity for appearances to deceive went back to ancient times, Jastrow argued that only scientific psychology could properly illuminate the problem of human deception. In what would become a hallmark maneuver, he formulated his argument in opposition to "common sense" and its notion that it was the *senses* that deceived.[37] As a psychologist, he argued that it was not human physiology that was primarily to blame for deception, but rather the mind's interpretation of the sensory data. He stressed that every perception had two natures: the nature of the object perceived and the nature of the observer. The former was related to the external conditions of the physical world while the latter was concerned with the mechanisms of perception, one's reliance on inferences, and the power of affect: "Not only will the nature of the impression change with the interests of the observer, but even more, *whether or not* an object will be *perceived at all* will depend upon the same cause." "Expectant attention" framed what the individual observer was capable of perceiving.[38] With this turn of phrase, Jastrow signaled his intellectual debts not only to German experimental psychology, but especially to clinical explanations of supposed mesmeric or occult phenomena in terms of unconscious physiological actions.[39]

Jastrow located the root of deception in those predispositions that led the observer to misinterpret sensory experience. Certain objects (such as optical illusions) or persons (such as the spirit medium or pickpocket) were particularly prone to inducing misjudgment, but this was because they preyed upon a preexisting mental organization. These pitfalls were somewhat avoidable, but the prevention of deception required self-discipline on the part of the individual observer. The easy road of quick conclusions and pleasurable beliefs must be eschewed in favor of rigorous brainwork. As "creatures of the average," the deceivable selves resulted from routinized habits of mind that led to taking the appearance of things at face value. After all, as humans "we are adjusted for the most probable event; our organism has acquired the habits impressed upon it by the most frequent experiences; and this has induced an inherent logical necessity to interpret a new experience by the old, an unfamiliar by the familiar."[40] The possibility of deception derived from the very anticipation of the expected that allowed humans as an intelligent species to operate in an uncertain but probable universe. While this mental machinery as a whole was a beneficial attribute bestowed upon humanity as a population through natural selection, this well-worn path also led into superstitious credulity.

Jastrow's account of deception resonated with Peirce's pragmatist contemplation of the nature of belief in a scientific age. Both men stressed the role of habits in sustaining certainty. In 1877 Peirce published his influential essay "The Fixation of Belief" in *Popular Science Monthly*. Peirce later described this essay as his attempt "to gain the ear and the interest of a large public, and, at the same time, to place the rationale of reasoning in so clear a light as to make it really serviceable in reference to loving questions of science."[41] At the time, he embraced a psycho-physiological understanding of the mind to explain the relationship between knowledge and belief. Belief was analogous to nervous association, the process whereby the mouth waters, in Peirce's example, at the scent of a peach. In contrast to pleasurable and comforting beliefs, doubt is expressed by the irritation of a nerve. This produces further investigation as a reflex action. Peirce prized the mental state of irritation above all others since it alone provided the proper spur to true inquiry. Pleasure could never serve as the impetus to knowledge since it merely reinforced preexisting beliefs.[42]

There were significant differences between Peirce and his student. Peirce's focus in these essays soon shifted to the elucidation of the proper course of scientific investigation, but Jastrow's interest remained on the psychological conditions of belief. Peirce insisted that logic was the shared product of a community of knowers that "must extend to all races of beings with whom we can come into immediate or mediate intellectual relation. It must reach, however vaguely, beyond this geological epoch, beyond all bounds." In contrast, Jastrow, along with most of his fellow psychologists, wanted to inscribe reasonable thinking and right-minded action in the judgment of the individual person. They remained resistant to Peirce's vision of logic as "rooted in the social principle."[43] Peirce later distanced himself from his early formulation of pragmatism, but this articulation had the most influence on psychologists. Jastrow pursued his mentor's passing interest in making deception itself into an object of scientific inquiry.

These pragmatist motifs illustrate how the psychology of deception emerged from what Jastrow called a "Darwinized outlook."[44] This was a stance shared by many of his contemporaries. In a 1900 dissertation written under Hall's direction, Norman Triplett laid out the evolutionary dimension of deception. His aim was to explain the psychological principles that allowed the stage magician to deceive his audience. According to Triplett, the conjuror's tricks originated in "a universal instinct to deception—a biological tendency appearing throughout the animal world from simple forms to the highest orders, which acts as a constant force in the process of natural selection—as a means of preserving the self or species." Triplett argued that a connection existed between the simplest deceptive organisms and their modern counterparts. Commercial fraud descended from the organism's instinctive mimicry in order to survive either by eluding predators or trapping prey through trickery. "Our commercial life is redolent of fraud from our gigantic infant industries with their specious pleas of inability to compete with foreigners, to the small grocer who is made by Puck to ask the clerk if he has sanded the sugar, larded the butter, and graveled the coffee." He claimed that susceptibility to such behavior represented "atavistic tendencies" in the species. According to Triplett, as primitive man sought to wrest himself from the state of nature his animistic religious beliefs sustained credulity in the religious conjuror. These shamans consciously maintained their power by deceiving their credulous adherents about the nature of their powers.[45]

Triplett held that this deceivable self resulted from the mind's governance by the "law of economy."[46] According to this principle, new sensations travel the easiest, most economical route and therefore flow along pathways established through habits. Humans have the tendency to categorize new sensations in terms of what they have already experienced or what they desire to see. In *Talks to Teachers* (1899), William James formulated the law as follows:

> In admitting a new body of experience, we instinctively seek to disturb as little as possible our pre-existing stock of ideas. We always try to name a new experience in some way which will assimilate it to what we already know. We hate anything absolutely new, anything without any name, and for which a new name must be forged. So we take the nearest name, even though it be inappropriate.[47]

Natural selection granted humans the unique attribute of a decisionmaking consciousness, but they still operated in an economy of scarce mental resources in need of conservation. Consciousness made no use of novel phenomena unless they were of immediate utility.[48]

A number of historians have connected James's thought with the economic developments he witnessed. Lewis Mumford noted the significance of James's "persistent use of financial metaphors." Mumford argued this language indicated the extent to which "James's thought was permeated with the smell of the Gilded Age: one feels in it the compromises, the evasions, the desire for a comfortable resting place."[49] Later historians have focused on how James's "cash-value" theory of passing truths like unchallenged currency resonated with the financially expansive practices of turn-of-the-century futures trading and credit lending.[50] In contrast, his "law of economy" points toward a more frugal understanding of the self and the world, one harmonious with the central tenant of nineteenth-century scientific naturalism. Indeed, an understanding of nature as a finite economy was central to both natural selection and the conservation of energy. Early advocates of extending this economization of nature to include the human mind included Herbert Spencer, Ernst Mach, and the American philologist William D. Whitney. In the case of James's psychology, habits facilitate preservation by alleviating the burden of having to constantly form conscious choices while simultaneously consigning the individual to the comfortable pathway of the most probable.[51] This permitted the human organism to operate efficiently without having to fully draw upon the mental resources of consciousness, but it also risked leading one into self-deception due to an unjustified classification of sensations in a novel situation. The law of economy turned the mind into a Malthusian world with a paucity of resources in need of conservation, not Patten's optimistic "new civilization" of abundant consumer goods and expansive credit.[52]

Jastrow connected this economization of the mind to racial development. He understood both credulity and mendacity (both as mental traits and moral qualities) as biological variants prominent among "primitive" populations but recapitulated or transmitted in troublesome ways in more civilized individuals. In his own day, Jastrow was certainly not the leading advocate of a scientific basis for racial hierarchies. Yet, in the 1880s and 1890s, he displayed a fairly consistent willingness to utilize the tropes of scientific racism in order to promote his science and ingratiate his own writings in the world of middlebrow publishing.[53] Jastrow used this anthropological idiom in his earliest attack on psychical research published in a middlebrow periodical. In an 1889 essay for *Harper's Monthly*, he argued for an "evolutionistic" perspective on the problem of psychical phenomena. Citing the authority of the British anthropologist E. B. Tylor, Jastrow held that the credulous belief in psychical phenomena represented "reversions to a more rudimentary state of thought." Much like the "child" and the "savages," the faithful adherent of spiritualism inhabits a world of fearful mysteries and invokes the otherworldly to explain the perfectly ordinary things he or she could not account for otherwise.[54] At a time when spiritualism was gaining an increasingly respectable hold in the culture of America's upper classes, in the pages of a genteel magazine Jastrow performed his own calculated sleight-of-hand trick by associating spiritualism with savagery and racial recapitulation, warning his readers that such pursuits were not appropriate for the refined tastes of their class. He argued that relying on primitive expectations did not provide an appropriate exit from the stresses of industrial life, but led merely to self-deception.[55] To adhere to psychical research's doctrines was to compromise white privilege and the achievements of civilization.

Jastrow was not alone in linking deception to evolutionary narratives of racial progress. In his 1890 survey of "children's lies," Hall stressed the functionality of deception as a tactic in the struggle for existence that persisted in unwanted ways among the civilized: "Truth for our friends and lies for our enemies is a practical, though not distinctly conscious rule widely current with children, as with uncivilized, and indeed, even civilized races." The order of the sentence captured Hall's recapitulationist logic. He first described a behavior as childlike, which he then identified with the uncivilized. Only with this connection made did he insist that such mental traits also appeared among the civilized. Through these chains of associations, Hall sought to demonstrate how evolutionary history bestowed upon the civilized certain seemingly undesirable qualities. A similar viewpoint characterized much of Jastrow's thought. Even in the 1920s, after rejecting Hall's "appealing" recapitulationist theory, Jastrow still argued, "We do not completely escape or emerge from the childlike or the primitive; our past pursues us. The problems of life, social and personal, industrial and political, cultural and recreational, are due to the strong tendency to revert to childish patterns of behavior."[56]

This understanding of deception, which braided together epistemology and ethics, served as the foundation for Jastrow's popular psychology: a program that aimed to instruct how to wisely navigate the world of appearances. Jastrow's rather vague solution required attending to the irritating and discomforting itch of doubt and the cultivation of a critical attitude to combat habits that directed perception onto the easy path of self-deception. Incredulity led to intentional choices and actions. In articulating these concerns, Jastrow's response remained voluntaristic. Science popularization gave the individual reader the techniques to detect deception, but it was the individual's personal responsibility to acquire such habits in order to operate prudently.[57]

For these psychologists, living in an uncertain cosmos governed by chance demanded a probabilistic mindset. The individual was compelled to act based on his or her judgment of what was most likely to be true. History and experience had etched associations that facilitated quick judgment into human beings as a population. The law of economy, inscribed as "a condition of progress in the individual and the race" according to Triplett, rendered the majority of mental functions automatic, easing the perceptual burden.[58] If reliance on these habits was the evolutionary inheritance of the civilized races in the aggregate, it was the free individual's obligation to struggle against comforting mental habits.[59]

Stage Magic and the Theater of the Laboratory

Despite the importance of the topic of deception to the new psychology, these scientists came to the study of illusions comparatively late. The commercial landscape of the Gilded Age featured a wide array of amusements whose pleasures hinged upon their ability to toy with an audience's attention, expectations, and beliefs. Between 1860 and 1910, this "magic business" was at its peak and encompassed not only conjuring shows on the stage but the practice of sleight-of-hand tricks in the domestic parlor. These magic shows revolved around card tricks, sleight-of-hand, and mechanical devices designed to make objects disappear. These performances resonated with the psychology of deception as they hinged upon controlling the audience's attention and expectations. Stage magic fascinated psychologists because conjuring was a deceitful profession whose performers were honest that they were deceitful. Audiences came to these performances willingly anticipating the likelihood of trickery within the show's clearly defined boundaries. The pleasure derived from stage conjuring depended upon the audience members' simultaneous captivation with the magician's skillfully executed deceptions and their desire to understand the secret mechanism behind the trick. The experience involved both the inducement of deception and the cultivation of incredulity.[60] This was a secularized form of magic that naturalized the phenomena under consideration while concealing the modus operandi as a trade secret. The magic show was thus a deeply felt but carefully circumscribed experience of deception. This sensibility extended into the midways of fairs, which featured wondrous and deceptive landscapes of rollercoasters and mazes of mirrors that solicited puzzlement and delight.[61] One of the most striking aspects of the new psychology was the manner in which its scientists made explicit links between their laboratory practice and these mass amusements.

Indeed, a highpoint of Jastrow's early attempts to popularize psychology was his involvement with the 1893 World's Columbian Exposition, the definitive cultural event of the final decade of the nineteenth century. Centering on the nation's recent commercial and technological achievements, the exotic pavilions and electrified boulevards of the Chicago exhibition grounds thrilled fairgoers. The exposition was intended as a hymn to Americans' capabilities of conquering the continent, its indigenous peoples, and natural resources. It was also mired in controversy; African American attendees were segregated and black intellectuals were excluded from organizing displays, while Native Americans were transformed into exhibited objects.[62]

The fair also provided the venue where Jastrow coordinated the first major public exhibit presenting psychology as a science to American fairgoers. It appeared under the auspices of Frederic Ward Putnam's Department of Ethnology. Jastrow displayed a number of laboratory instruments acquired from colleagues with the aim of demonstrating certain psychological principles.[63] Indeed, he used the fair as an opportunity to introduce a new instrument, the aesthesiometer, into psychological practice. The device used two movable pins to determine the distance on the forefinger at which the observer could detect the two pricks as distinct sensations.[64] The most celebrated individual Jastrow tested was Helen Keller, the thirteen-year-old deaf and blind girl then creating a national sensation. Keller's examination demonstrated the variety of senses investigated under the rubric of anthropometry: not only sight and hearing, but also touch.[65] Jastrow used a series of recently developed psychological instruments to probe Keller's observational capacities, to comprehend the reality she experienced. In his published report, he went to great lengths to emphasize her abilities, "the remarkable alertness and receptivity of mind displayed by her."[66] As important as these material artifacts from the psychologist's toolbox were, they did not garner the most interest. The exhibit's centerpiece consisted of the psychologist taking mental and anthropometric measurements of the attendees in collaboration with the anthropologist Franz Boas. Jastrow hoped to collect the mental measurements of a large number of Americans as they completed a series of tasks.[67]

Yet psychologists and replicas of their laboratories did not monopolize the discussion of psychological principles emerging from the late-nineteenth-century fairgrounds. Amusements capitalizing on the titillating effects of illusion engaged scientists and fairgoers alike in contemplating the nature of perception. The Chicago fair introduced the midway as a space for leisurely amusements distinct from the more serious exhibition halls. When the Midwinter Fair opened in San Francisco in early 1894 as a West Coast response to the highly successful Columbian Exposition, it featured an attraction that apparently gave "one the most peculiar sensation ever felt." Known as the "Haunted Swing," it was an illusion that individuals not only perceived but also briefly inhabited. Fairgoers would enter a large room furnished to resemble a parlor. In its center hung an enormous swing suspended from a horizontal bar. Once seated, from the perspective of the rider, the swing appeared to rotate to such a degree that the rider ought to be tossed out. In actuality, the swing itself remained stationary throughout the experience. The entire room, which the fairgoer expected to be fixed to the ground, was suspended on the bar and could be rotated by the ride's proprietors. Debuting in San Francisco, the "mysterious

FIGURE 2.1 Exposure of "the true position of the Haunted Swing." Albert Allis Hopkins, *Magic: Stage Illusions and Scientific Diversions, Including Trick Photography* (New York, Munn and Co., 1897).

and wonderful" swing toured the country over the next decade, becoming a fixture at Coney Island in 1904.[68]

The Haunted Swing fascinated scientists, with physicist Robert Wood repeatedly visiting the fair because of its allure. In the *Psychological Review*, he reported on "the curious sensations" he experienced as a fairgoer riding the swing. What particularly intrigued Wood was "that even though the action was fully understood, as it was in my case, it was impossible to quench the sensations." He found that his superior knowledge of the physical laws involved provided little protection from the force of his expectations. He consistently failed as he "tried to suppress it and reason against it."[69] Wood's account of his experiences was consonant with the intentions of the swing's promoters. Advertisements directed the audience toward the presence of an illusion without revealing how the deceptive effect was accomplished. The financial success of amusements like the swing hinged not on seamlessly deceiving the senses, but in drawing people's attention to the presence of an illusion and getting them to talk about how the trick occurred.

The divergent experiences of Jastrow and Wood illustrate the different ways that scientists engaged with the exhibitory culture of the Gilded Age. A number of historians have documented the presence of scientists as producers of the fair's contents. This was especially true of those anthropologists who helped design the ethnographic displays.[70] Jastrow certainly fulfilled this role in Chicago. Neglected in such accounts is Wood's role as an enchanted consumer of a particular midway ride's deceptive effects. He offered a glimpse of the man of science repeatedly purchasing tickets and never quite overcoming the disorienting sensations he experienced on the midway. The scientists interested in the mind could happily learn important lessons from the designers and proprietors of the Gilded Age's deceptive amusements.

The psychologists' captivation with stage magicians further demonstrated the role of scientists as avid consumers of mass culture. Jastrow once more led the way, building relationships with a number of celebrated magicians, the professional performers who materialized deception on the stage much as he felt he did in his laboratory. Antagonistic toward spirit mediums, he depicted stage conjurors with considerable sympathy. In 1897 he pronounced, "The student of the psychology of deception takes his place with the audience and observes how readily their attention is diverted at critical moments, how easily they overlook the apparently insignificant but really essential settings of the trick, how the bewilderment increases and the critical faculties lapse as one bit of sleight-of-hand succeeds another."[71] Passages such as this illustrate the ways in which the learned professor transformed himself into the conjuror's pupil. Jastrow depicted the psychologist as part of the audience rather than as a privileged backstage observer. At the same time, he distinguished himself from the credulous spectator due to equal attention he dedicated to the events on stage and the audience's reactions.

A number of Jastrow's contemporaries shared his captivation with stage magic. From Berlin, Max Dessoir expounded on "the psychology of legerdemain" derived from the memoirs of famed conjurors. In 1894 Alfred Binet invited five conjurors to visit his Laboratoire Psychologique at the Sorbonne in Paris, where he observed, interviewed, and photographed them. American psychologists were even more fascinated by magicians. When Yale psychologist E. W. Scripture condemned his peers for their insufficient attention to accuracy when making their observations, he held up the French magician Robert-Houdin as a paragon of accurate observation conducted with a mere glance. Over a number of years, Hall went about purchasing the readily available parlor versions of the tricks developed by stage conjurors. In 1908 he boasted of his ability to successfully execute a full performance as an amateur magician. It was at Hall's suggestion that Triplett undertook a classification of various tricks and their psychological foundations. Despite this widely shared interest, Triplett granted that Jastrow was "the first to enter upon this field" with his 1888 *Popular Science Monthly* article.[72]

In 1896 Jastrow orchestrated a series of experiments that transformed the magician into an experimental object. He invited Alexander Herrmann and Harry Kellar, the country's most popular conjurors, to visit his Wisconsin laboratory during one of their western tours. Both Herrmann and Kellar shared Jastrow's interest in providing a naturalistic explanation of their abilities. Modern magic had become a secular art largely due to the promotional strategies embraced by its practitioners. Stage conjurors adopted the tuxedo as a form of bourgeois dress and practiced the art of exposé, targeting the methods of the spiritualists, the pickpocket, and their fellow magicians.[73] As part of a conscious project to make their trade more amenable to middle-class sensibilities, these Gilded Age magicians—at least in most circumstances—abandoned the air of occultism that was traditionally linked to their trade. Where the Davenport Brothers, under whom Kellar had apprenticed, blurred the boundary between stagecraft and authentic spiritualism, Kellar took the techniques he learned from them to conduct exposés.[74]

Competition among performers abetted the disenchantment of magical phenomena. Promotion involved both the cultivation and deflation of the deceivable self as conjurors boosted their own shows by claiming that their rival's mysterious feats actually consisted of easily replicable tricks.

The stage performance involved the execution of deceptive tricks and then the revelation that an illusion, rather than a supernatural event, had just occurred. Herrmann's show emphasized his own manual dexterity and practical intelligence in accomplishing his feats.[75] Because the minutia of his craft received such attention from the audience, Herrmann noted that the magician's techniques and devices were constantly copied. To maintain one's reputation required that the magician "must be an inventor, mechanical and scientific" who persisted in creating new tricks through "time, thought, and weary labor." He also discussed the "natural aptitude" for accuracy demanded of the stage conjuror. A successful career in the field required that the person be "alert in body and mind" and "cool and calculating to the movement of a muscle under all circumstances."[76] Herrmann's celebration of the virtue of accuracy did not mean that he also adopted the ideal of sincerity. While willing to detail the effort undertaken by magicians to ensure a precise performance, he unsurprisingly kept secret the actual workings of his profession's tricks.

Jastrow invited Herrmann and Kellar to his laboratory because he wanted to compare the magicians' results on a series of mental tests with those of other experimental subjects to determine how they compared against the average observer. He sought to determine whether the conjuror, either through natural endowment or professional training, had acquired superior or unique perceptual skills. Similar to his earlier tests on Helen Keller, the measurement of touch was of prime importance in defining individual ability. The experiments began by testing the magician's tactile sensibility using the aesthesiometer. He found that the magicians actually possessed a coarser sensibility than the ordinary person, who could detect the two sensations at two millimeters. Herrmann perceived the difference at three and a half millimeters and Kellar at two and a half. Such findings were somewhat surprising considering Herrmann possessed the superior reputation in the realm of sleight-of-hand whereas Kellar was better known for the management of technological wonders. Furthermore, although both conjurors had trained themselves to be ambidextrous as a necessity of their profession, under experimental conditions they performed poorly in a test that required them to move their two hands equally from a common starting point.[77]

Other experiments affirmed Jastrow's central premise that although these gentlemen might possess perceptual skills superior to the average individual, their mental capabilities fell within the range of the normal. For example, he wanted to test the visual acuity of the magicians to determine whether they could take in vast amounts of information more quickly than the ordinary individual, as Robert-Houdin claimed. Accordingly, Jastrow presented a series of flash cards, each with a different patch of color, for half a second to determine whether Herrmann could detect the color at a mere glance. The magician responded correctly in half the instances, but executed the test perfectly when the colors were linked to differently shaped patches. Jastrow also conducted an alphabet test wherein twenty-five letters were repeated in random combinations in a series of two hundred and twenty-five. The experimental subject had to pick out a select letter in a period of ninety seconds. Herrmann marked off ten correctly, but had nineteen incorrect responses in his first attempt. The more disciplined Kellar fared better, making seven correct identifications with no errors in his first attempt and eleven in the second. Such results placed the conjurors within the range of the average subject, who succeeded in marking eight correct responses.[78]

Although the experiments successfully demonstrated that the magicians did not possess any extraordinary powers, Jastrow noted some dissatisfaction with the experience. In a concluding note, he discussed how he felt that Herrmann, in particular, could have performed better in the laboratory. The problem was that the conjuror acted too quickly, intuitively responding to tests without giving the situation adequate thought. He completed the tests in half the time they usually took, but his results were usually below the norm.[79] In short, the prestidigitator lacked the prerequisite self-discipline required of a truly reliable laboratory subject. This was undoubtedly due to the very different sensibilities that the experimental psychologist and the entertaining magician brought to the situation. There was a major tension within the new psychology between its claim to reveal universal mental processes common to all and the conviction that only properly trained individuals could accurately observe their own consciousness in a laboratory setting. Producing reliable responses required certain sensibilities and training from both parties; it was a cultivated rather than a natural attitude.

Popular illusionism appealed to psychologists because it relied upon the simultaneous cultivation and disclosure of the deceivable self's perceptual vulnerabilities. By introducing magicians into his Wisconsin laboratory, Jastrow hoped to make apparent how talented individuals succeeded in deceiving spectators. This research focused on the abilities of the deceiver rather than the limits of the deceived. In other instances, however, he and a number of his fellow psychologists pursued the question of how the staging of attention, expectations, and beliefs shaped observation in such ways that led ordinary persons into becoming the deceivable self. Despite Herrmann's disappointing performance, the psychologists who toyed with deceptive instruments came to embrace his ideal of accuracy in the absence of sincerity.

Toying with Weights and Subjects

The "size-weight illusion" that captured the interest of a number of psychologists for much of the 1890s illustrates how the experiments that materialized the deceivable self built upon the legacy of stage magic. These investigations both incorporated objects taken from the wider commercial culture and cast the psychologist as an insincere performer. In 1891 the French physician Augustin Charpentier claimed to have uncovered a new aspect of the psychophysics of touch. He found that when he presented subjects with objects of equal weight but of differing size, they invariably described the larger object as feeling heavier. Charpentier offered two explanations for the illusion: that individuals were influenced by the psychology of expectancy based on visual cues or, alternatively, that the neurophysiology of the skin's contact with the object produced the illusionary effect. European investigators pursued both lines of interpretation, but American scientists embraced only the former framework.[80] Cultural assumptions that bounded the discipline in the United States made the illusion thinkable only as the product of psychological suggestion.

Between 1893 and 1898, work on the size-weight illusion became the centerpiece of the psychology laboratory at Yale University. In 1892 E. W. Scripture, another Wundt-trained American, joined George Trumbull Ladd as a member of the university's philosophy department. Although Ladd had penned an 1887 textbook sympathetically surveying experimental psychology, he remained committed to a more philosophical study of the mind. In contrast, Scripture insisted that the only way to comprehend mental life was through experimentation, the tinkering with apparatus, and, especially, rigorous measurement. This became known as his "try it" approach. He combined these convictions with a passion for popularizing the new science and forging for it novel, practical applications.[81] Yale hired him with the specific task of running the laboratory. During these years, Scripture and his students investigated a wide array of illusionary experiences from the sensing of expected tones and tastes to the effects of psychoactive drugs on consciousness.[82] Within the decade, the conflict between Yale's two psychologists led to their mutual dismissal, but initially the presence of both men attracted a number of students to conduct experimental programs under their guidance.[83]

Among these early recruits, Carl Seashore worked extensively on the size-weight illusion as part of his attempt to measure the presence of "illusions and hallucinations in normal life."[84] According to Seashore, hallucinations were commonly but incorrectly identified exclusively with mental abnormality or weakness. Instead, he found that they "may be experimentally produced in mild forms in fairly normal states, for the purpose of studying their nature." This finding was important for it meant that illusions were not necessarily the product of a somatic pathology. In the case of the size-weight illusion, Seashore found that its strength increased with perceived differences in size among the test weights and argued that this was due to the influence of past experiences. The suggestion made by regular visual associations of size with weight overwhelmed the fleeting impression of the moment. Like Jastrow, Seashore saw deception as deriving from a compromised attention: the individual's disappointed expectations based upon past associations and deeply felt interests. Even made aware of the size-weight illusion, the observer continued to fall prey to it because his "interpretative consciousness" persisted in ignoring the momentary aberration in favor of the associations built over time through habit. Both psychologists understood deception as the absence of willed action due to its circumvention by a habituated response.[85]

Some of Seashore's contemporaries suggested that the omnipresence of illusions within the human mental economy offered evidence for the apparition of otherworldly entities. For example, in the midst of the Yale research, the British Society for Psychical Research published the findings of its 1889 census on hallucinations. The census tabulated the frequency of hallucinations among its respondents and its authors argued that this provided evidence for the existence of paranormal phenomena.[86] Seashore disagreed. Because they followed a regular pattern, the presence of illusions demonstrated the fundamental lawfulness of the mind. Indeed, Scripture formalized the findings of the Yale laboratory into a mathematical "law of size-weight suggestion" in 1897.[87] Seashore's study carved a space where hallucinations belonged neither to the somatic medicine of the neurologist nor to the unexplainable otherworldliness of the psychical researcher. Instead, hallucinations constituted lawful reaction manifested in certain situations due to instilled habits.

The end results of the Yale studies of illusions included not only a lawful explanation of the mind, but also a new scientific apparatus that had deceit built into its design. Describing the Yale program for measuring hallucinations, Scripture stressed that those "experimented on did not know they were deceived." By and large, tools rather than persons accomplished this task. The psychologists intended for the appearance of their apparatus to deceive participants schooled to expect greater weight to correlate with larger size. With the assistance of the university's laboratory mechanic, the psychologists built two sets of blocks, one of uniform size but varying weight, and the other of constant weight but differing size. Seashore described how deception itself was inscribed into the apparatus: "that the appearance of the surface should not suggest any definite material, the blocks were all painted a dull, smooth black with optical varnish"[88] No longer did the psychologist depend upon the apothecary's standardized weights, but rather used a novel apparatus designed in light of psychological principles for psychological ends.[89] Scripture boasted of building cubical wooden boxes filled with lead to successfully restage the experiments "before large audiences."[90] Such performances made the deceptive weights into icons of the new psychology's experiments.

The Yale experiments on the size-weight illusion altered the relationship between experimenter and subject. In the earlier Peirce-Jastrow experiment, a screen had kept the subject unaware of the changing visual appearance of the weights. In contrast, the success of the Yale experiments hinged on presenting the subject with a stimulus designed to confound expectations and intentionally deceive. Scripture and Seashore introduced deceit into the everyday practice of laboratory psychology. Tellingly, Seashore revealed in his autobiography that his original test subjects consisted entirely of divinity students. Originally, he thought that individuals pursuing such a degree would constitute a particularly suggestible population, even though his experiments ultimately demonstrated that these individuals fell well within the norm.[91] His selection defined the subject as inferior to an experimenter bestowed with superior knowledge, authority, and incredulity. Moreover, as the selection of divinity students indicated, a major appeal of the size-weight illusion for these scientists was its potential to offer an explanation of belief in the form of the psychology of suggestive expectation. The Yale psychologists placed the deceivable self into a narrative of human development. Another of Scripture's students examined the size-weight illusion as an aspect of his examination of the mental development of school-age children. He concluded that children gradually grew into the illusion, being most susceptible between the ages of six and nine, with its effect slowly waning during the teenage years.[92] Scripture and his students offered a developmental account of credulity where naïve belief was associated with a normal but childlike mentality. These broader commitments help explain why in the United States the illusion was interpreted wholly in terms of the psychology of suggestion, a mental mistake caused by unwarranted expectations. The size-weight illusion offered an account of their contemporaries' captivation with wondrous, deceptive distractions, one that bolstered the psychologist's sense of epistemological superiority.

Alongside this materialization of the deceivable self, experiments on the size-weight illusion subtly altered the role of the psychologist in the laboratory. Armed with deceptive apparatus and providing misdirection, the experimenter became something of a deceitful performer. Seashore argued that commercial tricksters, whether the stage conjuror or the medium, used techniques similar to his own in order to delude their audiences. Likewise Scripture introduced his gloss of these experiments in *Thinking, Feeling, Doing* with an anecdote about his favorite magician Robert-Houdin.[93] As they learned from magicians, the study of the deceivable self sometimes required calculated insincerity on the investigator's part.

The Global Traffic in Pseudoptics

In spite of the technical prowess of experiments such as Scripture's, psychology maintained a porous border with mass culture. Psychologists took a keen interest in another element of the era's deceptive visual culture: ambiguous figures. These images regularly appeared in both popular illustrated magazines and psychological journals. No other object better illuminates how deception as a scientific concern was produced in relationship to both laboratory experimentation and the leisurely amusements of the new mass culture. Illustrators working for mass-circulation periodicals designed the majority of these visual tricks to amuse subscribers. Scientists would borrow and refine these images into paper tools for use in laboratories. They later accompanied anthropologists on expeditions in the South Pacific, where they were used to test the mental abilities of the indigenous population, and were further deployed by psychologists to communicate the principles of their science to leaders of the advertising industry. Just as psychologists used the bodies and performances of stage conjurors to make sense of the processes of deception, these visual tools helped materialize the self promoted by the new psychology. Psychologists increasingly used visual illusions circulated through middlebrow publishing to provide a medium for communicating the particular knowledge of their discipline.

The example of Hugo Münsterberg, head of the Harvard psychological laboratory, is illuminating.[94] Soon after his arrival from Germany, Münsterberg collected a set of images under the collective title of *Pseudoptics* (1894). Published by the board-game manufacturer the Milton Bradley Company, the set consisted of a series of optical illusions promoted as an amusing family game. In his correspondence with the publisher, Münsterberg expressed anxiety about having his name associated with what could be interpreted as a commercial venture and asked that the collection of puzzles not be branded with his own name. Bradley wrote that he had "not felt that it was necessary for you to conceal the fact that you furnished the brain for the thing but still we have carefully followed your instructions in this respect." Despite such reticence, Münsterberg was deeply concerned with how the product was priced and marketed. He discussed with Bradley strategies for encouraging psychologists and other educators to procure the product.[95] The two had some success. For example, Cornell's E. B. Titchener recommended using the set as a standard tool for optical research in his influential article detailing the apparatus necessary for a properly equipped psychology laboratory.[96] Later, the set was also used on the anthropological expedition to the islands of the Torres Strait in 1898, probably the first use of experimental psychology instruments in field research.

Psychologists enrolled ambiguous figures into their programs of research because these images brought to the fore how habituated expectations shaped what people were capable of perceiving. What made these ambiguous figures particularly interesting as scientific objects was that no single individual, institution, or discipline had exclusive rights over their design, production, and interpretation. In present-day parlance, illusions were an open-source technology. While individual illusions were frequently branded with a particular psychologist's name, they were never subject to proprietary rights. When Frank Angell reviewed *Thinking, Feeling, Doing* in the journal *Mind*, he accused Scripture of numerous instances of plagiarism. The Yale psychologist's response avoided the main charge of stealing words directly from Wundt and focused instead on the fact that most of the illustrations "were commonplaces like the optical illusions that are 'borrowed' by every book on psychology."[97] These origins made ambiguous figures distinct from other psychological instruments. For example, by the time of their publication in 1921 Hermann Rorschach had standardized his famous inkblots into uniform sets in order to produce a reliable object whose varying interpretation would materialize the psychiatric patient's innermost subjectivity. The Swiss psychiatrist and his anointed successors closely managed the design, manufacture, and subsequent circulation of these nonfigurative blotches. They argued that giving up control of the images would compromise the instrument's validity.[98] In contrast, the optical illusions of an earlier era were diffuse in both their origin and production. Unlike the tightly regulated Rorschach images, the details of ambiguous figures were invariably altered by their various users to heighten certain intended effects that the individual author desired.

When Jastrow introduced the "duck-rabbit" into psychology, he was borrowing from existing print culture. The image first appeared in an 1892 issue of Germany's oldest nationally circulated illustrated paper, *Fliegende Blätter*. Jastrow first noticed it in an 1893 *Harper's Weekly* reproduction.[99] Originally, the image was not understood as an illusion. Instead, the journals framed the image by emphasizing the artist's ability to depict the similitude of two distinct animals, and its diffusion was based on its ability to bewilder and amuse readers. It had no connection to academic psychology and was bereft of edifying or pedagogical goals. When Jastrow incorporated the mass-produced image into his writings, he altered it slightly to reinforce his psychological argument. Due to the angle of its head, the original duck-rabbit was not as ambiguous an image as Jastrow's subsequent one. In the original, the image of the rabbit predominates and the viewer needs to tilt his or her head to be able to see the duck clearly. With the bill-ears pointing horizontally, as in Jastrow's version, the two visions could coexist more easily. More significantly, Jastrow altered the text that accompanied the illustration, subtly reframing how the observer ought to appreciate the image. Initially, the caption asked, "Which animals resemble one another most?" The question oriented the viewer to consider the ingenuity of the illustrator to be able to capture the appearance of two animals in a single illustration. In contrast, Jastrow asked, "Do you see a duck or a rabbit, or either?" Where before the image embodied the impression of two creatures, Jastrow's formulation suggested that one animal would be more or less prominent depending upon the perceptual habits of the individual viewer. This shifted the significance of the illusion

FIGURE 2.2 Jastrow's version of the duck-rabbit. Joseph Jastrow, *Fact and Fable in Psychology* (New York: Macmillan, 1900).

away from the representation on the page toward the mental processes in the mind. Such an arrangement bolstered Jastrow's claim that perception was more than simply the reception of sensory experience. It depended upon the mental work of interpretation. One perceives not only with the senses, but, more important, also with what he called in this essay "the mind's eye."[100]

The ease with which ambiguous images traveled from the popular to the scientific press mirrored their ability to travel in the physical world. Among laboratory apparatus, the ambiguous figure printed on the page was both considerably less expensive and more compact than other tools for studying perception, qualities that held great appeal and enabled these pseudoptical devices to travel the world. Such ease of transport and communication served, however, to reveal the cultural embeddedness of certain illusions. For example, the illusions collected in Münsterberg's *Pseudoptics* proved of limited value when they accompanied W.H.R. Rivers on Cambridge University's anthropological expedition to the Torres Strait. Trained as a neurologist, Rivers was responsible for carrying out psychological and physiological investigations of the indigenous population of the region. This project gained considerable notice in scientific circles since it was widely credited with the novelty of exporting the equipment of the modern psychological laboratory into the field in order to perform mental experiments on individuals perceived to be barely exposed to civilization. Yet in many instances, Rivers and his associates encountered difficulties in engaging their test subjects with the experimental apparatus. The use of illusions as research materials required the subject's active engagement in the test. Rivers argued that the lack of receptivity pointed toward racial differences in the capacity to perceive and interpret certain visual cues. For example, he speculated that the Western subject matter made it difficult for his informants to be tricked by the illusion that the figures of a man and a boy were of the same height even though visual cues included in the image generally induced the observer to believe otherwise. The Western signifiers of the bourgeois top hat (for adult male) set against the schoolboy cap (indicating youth) had no resonance among the Papuans, who subsequently registered no difference in height between the figures.[101]

Despite these setbacks, the optical illusions did play a crucial function in how Rivers interpreted the mentality of the islanders. Rivers found that the Torres Strait inhabitants were not particularly prone to the deceptions of optical illusions. He reported that many of his informants were less susceptible to these visual tricks and "were certainly quite as good observers as the average European." Despite these results, the data from the expedition was still used to illustrate the existence of natural hierarchies of race. He explained his experimental subjects' immunity to the visual illusions that misled European minds in terms of their lack of mental development. Living so close to a state of nature, they were more prone to perceive infinitesimal details rather than being able to visualize the images as a whole. To survive as primitives, the inhabitants had to be more acutely aware of nature's minutiae while the more civilized European's perception was capable of taking in a wider scope of objects and details. For example, he explained their inability to be fooled by the Müller-Lyer illusion by the fact that they were more likely to direct their attention to a comparison between the two line segments while the European would recognize the need to understand the figure as a whole. Jastrow concurred with this interpretation when he reviewed the findings of the expedition for the journal *Science*. If the islanders were less susceptible to deception compared with peoples of European descent, it had to do with their propensity to focus on subtle details and an inability to focus on the more general picture.[102]

On such a reading, deception, as governed by the law of economy, was best understood as a normal and necessary facet of advanced, "civilized" life, and, indeed, these same images were used in one of the most self-consciously "modern" of fields: the advertising industry.[103] Advertisers boasted of their role in enhancing consumer choice by connecting purchasers to the goods they actually desired. At the same time, advertisers emphasized how this freedom of choice was frequently compromised by the individual purchaser's cognitive vulnerabilities. In the various editions of his *Theory of Advertising* (1903), applied psychologist Walter Dill Scott used a similar array of optical illusions as Jastrow and Rivers to explicate how the mind of the consumer functioned. For Scott, among the most pressing psychological problems facing advertisers were "errors of expectancy." Scott used Jastrow's duck-rabbit image to emphasize that such errors resulted from how the mind categorized sensory experience. According to Scott, poorly crafted advertisements unintentionally deceived the purchasing public with their ambiguous visual cues. In discussing "illusions of apperception," Scott initially analyzed, through anecdotal evidence, the graphic organization of unsuccessful advertisements. For example, he described two competing groceries in the Chicago area, Winter's Grocery and Robinson Brothers, whose inserts in the morning paper were virtually indistinguishable. Scott related the experience of an acquaintance whose wife asked him to purchase a large order from one of the outlets that was having a sale. Not realizing the difference, the man purchased the goods from the wrong grocer and, when scolded by his wife, he produced what he sincerely believed was the advertisement for the correct store, but which of course was not. Scott's other examples of misleading advertising were even more visual in orientation, and he gave examples of distinct manufacturers sharing advertising space on the same page of a newspaper and magazine, resulting in customers soliciting business from the unintended one. From these examples, Scott moved to a series of ambiguous images from the psychologist's repertoire, each illustrating how one's preconceptions affect what one perceives.[104]

Scott's use of optical illusions to explain psychological principles to advertisers completed the circle that linked experimental psychology and commercial culture. Objects that originated as visual images meant to entertain the purchasers of mass-circulation periodicals came to explain how the minds of consumers operate. These minds could be conceptualized more clearly through the use of these images. Scott drew upon these visual tools to explicate the mental mechanism through which advertisements mislead and inadvertently deceive the consumer. These images "illustrate the same principles of illusions of apperception, but they make it clearer than any confusion of concrete advertisements can possibly do." Where Jastrow utilized the image to convince observers to operate prudently, Scott used the same visual tool for his own ends, to rationalize advertisements.[105]

Optical illusions were visual images that could recreate the uncertainties of observation in a laboratory, a field expedition, or the text of a popular exposition. Their representational ambiguity captured in a way that words could not the process through which the mind could miscategorize sensations based on expectation. As such, they were an important resource in the articulation of a psychology of deception. They were particularly apt tools because as images they had originated as entertaining illustrations in mass-circulation periodicals. They were artifacts of the popular culture whose perceptual habits were under scrutiny by turn-of-the-century psychologists, making them ripe for appropriation.

Where Scott used ambiguous figures to educate his corporate patrons about how to rationalize their advertisements through visual seduction, others utilized these same images to urge individual observers to operate prudently. From the 1890s onwards, a number of perceptual psychologists abandoned their laboratory experiments to speak to a wider public. William James, undoubtedly the most prominent and best remembered, was far from an isolated individual. Certain persons like Münsterberg and Scott embraced the possibilities of corporate patronage, but others were ambivalent. Jastrow was among those who favored an edifying popular psychology that would make people more aware of the costs of habituated expectations. They explored the social consequences of attentive action and, more important, the dangers of its absence.

Contemporaneous with these debates, William James advanced the idea in his highly influential *Principles of Psychology* (1890) that there existed a "social self." According to James, a fundamental feature of human identity was the recognition of one's existence by others. He wrote, "No more fiendish punishment could be devised, were such a thing physically possible, than that one should be turned loose in society and remain absolutely unnoticed by the members thereof." The "young intellectuals" of the early twentieth century expanded this brief discussion in the *Principles* into a new social philosophy. For George Herbert Mead, perceptual psychology could not provide an adequate account of the self since it excluded the inherently social nature of the individual. Although critical of the current practices of the commercial trusts, these intellectuals saw the potential of achieving true democracy by emphasizing the interdependence of individuals, a reality that the structure of the corporation seemed to make manifest. As discussed by historian James Livingston, these intellectuals saw pragmatism and the social self as offering a means of conceiving the dissolution of an older, "sovereign" individuality in favor of the new sociality. For their part, perceptual psychologists largely rejected such attempts to locate truth within the population (the community of knowers) rather than the individual. In both their science and their professional organizations, they reaffirmed the primacy of the sovereign agent as the most reliable knower. Talking about deception served as a means for discussing their regrets and worries about the large forces they saw surrounding them.[106]

In spite of his prominent role in articulating the idea of "social self," James was ambivalent about the trend toward large-scale organizations that he witnessed. He embraced a muted form of "anarchism." His concerns about the growth of an American empire fed his opposition to bigness for its own sake. James emphasized instead the need for political pluralism and the imperative to protect individual choice against homogenizing entities like the large trusts. Never an overt revolutionary, he saw his writing and lecturing as political contributions, encouraging others to think independently.[107]

Among perceptual psychologists, George Stratton most clearly drew lessons about social organization specifically from his studies of perception. The University of California psychologist had trained with Wundt in the 1890s, conducting a series of highly regarded experiments on the relationship between vision and touch. Employing a radical form of self-experimentation, he lived for several days at a time looking exclusively through lenses that turned the retinal image upright. Seeing through the lenses placed Stratton in an inverted world where his sense of sight did not correspond to that of touch. During the course of several days, he found that his sense of touch and vision gradually came back into harmony.[108] In a 1903 popular book, Stratton interpreted his own experimental work and that of others on illusions to explain to the public the true import of psychology. What the study of illusions demonstrated, he argued, was the inappropriateness of the "socialistic direction" prevalent in the United States. By this he did not refer to a specific political ideology, but rather to a trend that privileged the importance of social conditions and social embeddedness as explanations over "the inevitable inner development of the individual." Perceptual psychology illustrated the fallacy of overemphasizing collective or social experience. Instead, each individual has within him a standard by which to detect even those illusions that best the race. He can judge his particular impressions in the light of the system of his experience; and "reality" is what emerges from a kind of interaction and self-checking of his various sense-perceptions. The distinction between reality and illusion thus is not an exclusively *social* distinction.[109]

Stratton championed decisive willfulness according to one's inner standard as the counter to the pernicious influence of social suggestion. Individual freedom resided in the action of deliberately "self-checking" one's perceptions and avoiding the pitfalls of deception.

Many psychologists articulated similar lessons about the nature of deception. They advocated a voluntaristic form of mental self-help, one predicated on securing the autonomy of the individual. Their conclusions resonated with the natural philosopher's epistemological optimism of an earlier era, provided that the reader listened to the psychologist's guidance. Seashore concluded his study of hallucinations on the optimistic note of self-management through the new knowledge gained: "We may now make an approximate allowance for the illusions of weight and for all other illusions, due to intellectualized feelings, as they become recognized." When Triplett detailed the threats posed by pickpockets, fortune tellers, and the "harebrained financial schemes being exploited to relieve the credulous of their means," he concluded with psychology's practical utility of demonstrating "how perfectly simple in reality are some apparently wonderful things."[110] Despite the motifs of autonomy and self-reliance, these scientists depicted unassisted, individual discernment as prone to error and requiring the support provided by articulated knowledge of psychological principles.

Conclusion

In 1926 Jastrow wrote to his editor at Houghton Mifflin, promising a sequel to his earlier books on deception, provisionally entitled "Excursions of a Psychologist."[111] Although the manuscript went uncompleted, the proposed title provides a telling clue about the approach to psychology that he advocated. Whether rooted in a narrative of progress or declension, historians have largely concurred that the establishment of psychology as a scientific discipline has been the story of a particular space: the laboratory. Certainly these scientists privileged the creation of these spaces in the late nineteenth century as key to delineating their discipline. Yet from the outset, the methods and approaches of psychology were more heterogeneous than the image of a laboratory revolution would indicate. The activities of Jastrow and his compatriots illustrate how the establishment of psychology as a discipline also involved the scientist's frequent excursions into the realm of commercial culture. This is not to suggest that they thought that psychology should be a full-fledged field science, but rather to highlight the traffic of objects and persons that flowed in and out of the laboratory, and the manner in which that permeability was productive of both the subjects and methods of the new psychology.

These scientists were both producers and consumers of the deceptive visual and material culture of the Gilded Age. Crafting psychological tools from illustrations borrowed from respectable periodicals, scientists also produced popular science books intended to educate the purchasing public about its perceptual habits. They attended the fairs not only as exhibitors, but also as enchanted audience members. These excursions had much to do with the psychologists' fascination with human deception in its many forms. They shared this concern with a wide variety of contemporaries, from the entertaining stage magician to the muckraking journalist. Indeed, one of the most influential products of these excursions was the psychologist's encounter with the magician. Jastrow originally used these professionals in a limited fashion to demonstrate how the mind obeyed natural laws. Yet their influence on the discipline was more significant, as psychologists came to embrace the magician's acumen for deceptive stagecraft as a model for designing scientific investigations. In this sense, psychology's fascination with deception was polyvalent, ranging from edifying amusements to the scandalously fraudulent. The end result was that the psychologists came to embed their scientific practice firmly within Gilded Age traffic in deceptive visual and material culture, a positioning that helps explain the particular cultural power of the new discipline.

The resulting psychology also expressed some of the ambivalences these scientists felt about the new mass culture. What emerged from this encounter was a vision of the human as a profoundly deceivable self. One of the recurrent ways of explaining why humans were deceived was due to the mind's governance by a law of economy. According to this logic, the mind is an entity with limited resources at its disposal, coping with an abundance of novel sensory experiences that threaten to overwhelm it. For these scientists, the new economy left the self-reliant individual vulnerable and in need of guidance.

CHAPTER THREE

"Not Our Houses but Our Brains Are Haunted": The Arts of Exposure at the Boundaries of Credulity

On Halloween evening 1906, chaos reigned as the Los Angeles Anti-Faker Society descended upon a special séance held by local medium Elsie Reynolds. Members of the group, accompanied by two plainclothes police detectives, infiltrated the popular event by posing as the Hadleys, a credulous couple from Kansas. At the height of the proceedings, "Claude Hadley" rushed from his seat to grab the spirit at the center of the séance, revealing instead a cheesecloth-covered Reynolds. The exposure soon devolved into a physical brawl as supporters and detractors fought to free or constrain the medium's body.[1] The next morning, Hadley's "wife," Bessie Beatty, published her account of Reynolds's tricks, emphasizing the air of credulity that prevailed at the séance. In the muckraking series that followed in subsequent Sunday editions of the *Los Angeles Herald*, "business mediums" and "commercial spiritualism" came under particular scrutiny. At this time, the city's mediums specialized less in communicating with departed loved ones than offering advice about real estate purchases, marital choices, lawsuits, and other financial matters.[2] In her newspaper series, Beatty relied on her own deceitful performances to reveal the spiritualists' fraudulent trade. Dressed in costumes and assuming false identities, the journalist presented herself as an unwary innocent soliciting advice from the city's mediums. She argued that the medium's commercial-mindedness made the charlatan particularly vulnerable to exposure. Because of their greed, these spooksters actually made "easy marks."[3]

This chapter delves into the contemporaneous muckraking narratives of medium exposure written by scientists. It examines four different wonder-workers between 1909 and 1924, with particular attention given to the encounter between psychologists and the Italian medium Eusapia Palladino. A number of themes central to Beatty's journalism also animated these exposés. First, the subject's commercial status invariably played a constitutive role in defining whether she possessed any scientific worth. Second, much like Beatty, these psychologists found deceit an invaluable tool for besting individuals they considered deceitful. Exposures were a contest over who had the moral authority to deceive. These narratives also toyed with the question of who was truly the deceivable self: the paying public, fellow scientists, or, as Beatty suggested, the mediums themselves. Finally, just as Beatty learned important lessons from another profession, in her case the stage actor, psychologists also looked to different fields—detective work and stage magic—for their epistemological and moral models.

Especially during the Progressive Era, psychologists wrote about those whom the magician Harry Houdini called the "miracle-mongers" as the analog in the realm of mental life to those financial parasites plaguing the nation.[4] Usually these occult powers manifested themselves in performances for a fee in darkened rooms or at fairs and carnivals, in other words, in the grubby world of commerce. Initially, America's "men of science" ignored the spiritualist movement; the American Association for the Advancement of Science refused to discuss such matters at its meetings. Although early enthusiasm soon waned, the situation became more complicated in the 1860s as spiritualism began to attract adherents with considerable reputations in other areas of learning, giving its claims added credence. A new strategy of confrontation and exposure emerged in the 1870s. Physicians, especially neurologists, began to actively campaign against the healing regimes proffered by spiritualists and clairvoyants. This skeptical attitude was largely continued by those psychologists who studied perception in the laboratory and sought to demonstrate that, in the words of the famed neurologist George M. Beard, "not our houses but our brains are haunted."[5]

The sustained enthusiasm for spiritualism has drawn considerable interest from historians grappling with the place of "unchurched" religiosity in the modern world.[6] Recent historical scholarship has largely rejected the low opinion of the paranormal held by turn-of-the-century skeptics. These historians have traced how various strains of occultism contributed to the modernist articulation of a "deep" self ruled by the unconscious.[7] They have documented also how psychical research, which strove to subject paranormal experiences to scientific methods, separated from popular spiritualism in the 1880s. Moreover, while many psychologists used public pronouncements against these movements as a means of demarcating their science in the public eye, they also borrowed certain elements from their opponents' investigative repertoire. For example, both the growing sophistication of the randomization of experimental trials and the questionnaire owe much to the innovations introduced by psychical researchers.[8]

Building on these insights, this chapter outlines how psychology's encounter with these potentially deceptive wonder-workers further transformed the scientific practices and understanding of the self prevalent within the discipline. What these performances drove home was that the honesty of the examiner was no guarantee of the veracity of the final results.[9] Early twentieth-century representations of psychical research were awash in talk about "scruples." Scientists complained that spiritualists were unscrupulous frauds profiteering from a credulous public willing to pay their fees. It was the scientist's civic duty to remove these predators from the marketplace. The other arena where the language of scruples figured prominently was in describing the scientist's own behavior. These men of science spoke readily of the need to sacrifice existing moral codes of conduct to achieve the goal of obtaining the truth. Psychical researchers and psychologists cherished differing virtues in their pursuit of the truth behind human wonders. Psychical researchers particularly prized accuracy, seeking to subject spiritualists' phenomena to quantification and instrumentation.[10] When it came to debunking such phenomena by psychologists, the question of sincerity was paramount. In these encounters, sincerity was unevenly distributed among participants. It was a quality that the incredulous scientist demanded of every authentic spiritual medium. She had to be pliant and transparent about her activity, and she needed to operate independently of the marketplace. For their part, the scientists involved largely abstained from the virtue of sincerity. They contended that adhering to its demands would compromise their service to truthfulness since full disclosure would render impossible honest observation of persons known for their trickery.

When conducting these investigations, psychologists self-consciously modeled their approach on the deceptive activities of the stage magician and the detective. From the former, they learned how to orchestrate elaborate tricks through stage management and sleight-of-hand. The model of the detective involved the public suppression of affect and the embrace of an amoral, hard-boiled masculinity. Although hard-boiled masculinity became central to the genre of detective fiction between 1920 and 1945, earlier traces—namely the embraced deceit and the use of physical force in order to best serve the truth—featured prominently in Allan Pinkerton's fictionalized accounts of his life, such as *The Detective and the Somnambulist* (1875).[11] As in Pinkerton's stories, divining the perpetrator's true identity was not the focal point in the psychologists' narrative.[12] Rather their "cases" involved the revelation of an already known suspect's secret modus operandi. Like detective fiction, the psychologists' exposés trafficked in the tropes of the corrupting influence of money, the prevalence of faked identities, and the perils of the double-cross. The genteel codes of mutual respect and understanding that governed early relations in the laboratory dissolved in these particular investigations. These popular science narratives featured both tough male investigators passing judgment on an immoral world as well as mysterious female antagonists whose seductive appeal required resistance. Psychologists saw their subjects as deceitful selves in need of careful and often secretive handling. To fulfill their moral responsibility to police fraud in the realm of mental life, psychologists embraced a cynical amorality as a public stance, but this cultivation of intentional insincerity generated difficulties in their relationships with psychics, the press, and their fellow scientists.[13]

Mental Muckraking

In 1879 William James described George Beard as "on the war path about spiritualism."[14] Beard had gained national renown ten years earlier when he coined the term *neurasthenia* to explain the mental exhaustion that arose owing to the taxing sensory demands of civilization. While the product of cultural circumstances and a disease that flourished particularly in the United States, neurasthenia, Beard insisted, had somatic causes.[15] Not everyone adhered to his materialist interpretation of such illnesses, however, and in the wake of the Civil War a regular trade developed in healing through animal magnetism, communication with the dead, and clairvoyance about the future. Beard became involved in the attack on spiritualism because he felt that these unlicensed practitioners solicited his clients with their alternative methods. Perhaps most troubling to men like Beard was the class dimension of spiritualism: clairvoyants had managed to secure the patronage of the rich, refined, and educated.[16] Certainly many adherents could be dismissed as female hysterics according to the logic of the time, but such a strategy could not account for all believers. Intimating notions of racial degeneration, Beard suggested that "outside the negroes and spiritualists, there are probably not a thousand persons in this country who have even a lingering faith in witchcraft."[17] An expert in mental phenomena, Beard deemed himself particularly well equipped to provide an explanation for the apparent abilities of spiritualists. He fashioned himself as a debunker working in the public interest and he took the project on as a kind of scientific crusade.

At the heart of Beard's interpretation of spiritualism was the trance, a concept first advanced by Mesmerists but redefined by Beard to serve as an important element of his neurological vision of the mind. A trance involved the concentration of one's cerebral activity within a small portion of the brain. The remainder of brain activity was suspended, rendering the individual an automaton open to suggestion. According to Beard, trances were fairly commonplace mental phenomena and the popular appeal of spiritualists resulted from widespread susceptibility to such states. Trances operated like an easily transmitted mental contagion, for they "may spread like flames in a dry forest" within groups.[18] He contended that the widespread positive testimony from refined, cultured participants on behalf of spiritual manifestations were the product of this commonly experienced neurological state. Beard theorized that trances were not a marker of mental pathology but rather afflicted the "normal" population. As he noted, "if a belief in spiritism, in animal magnetism, in clairvoyance, in mind-reading, in the evidence of the senses, and allied delusions is proof of insanity, then this United States of America is but one vast asylum."[19] Under these conditions, the unscrupulous medium could play upon the wishes of the deceivable self, convincing him or her of the reality of the spirits. Beard's account was wrought with a paradox: while supposedly lacking volition as mere automata, these individuals still willfully perceived what they desired.

Asked to help craft a law prohibiting the clairvoyant trade, Beard expressed skepticism about the potential success of such a measure. Because the American legal system was grounded in the commonsense evaluation of individual testimony, he felt that it would prove near impossible to prosecute spiritualists since they could produce many witnesses enthusiastically testifying to the reality of the phenomena they observed. What was required was a transformation in the way eyewitness testimony was evaluated, away from the lay assessments of lawyers and juries in favor of the neurologist's expert insights into how the human machine functioned. In other words, the attacks on spiritualists were meant to protect the realm of physicians, but they could also carve out a greater space for neurologists in the legal sphere.[20]

In an 1879 article for *Popular Science Monthly*, Beard proposed certain reforms to experiments involving living human beings based on his experiences with spiritualists. He mocked those scientists credulous toward spiritualism for assuming "that if the subjects were honest, the results of the experiments must be accepted by science." Beard insisted that an individual's education, culture, and social standing must be disregarded, as moral character was no guarantor of trustworthiness. Instead, the scientist had to assume deceit or unconscious deception in every human encounter. He introduced a maxim that he felt should govern all subsequent research. Simply put, "deception, whether voluntary or involuntary, can only be scientifically met by deception; it must be beaten with its own weapons. No experiment of this kind in which the results depend in any way on the honesty of the subjects experimented on can be of any value in science." He tried to disentangle a necessary connection between the honesty of the scientist and the resulting truthfulness of the science. To best serve the truth, scientists had to take recourse to deception.[21]

A number of factors limited the immediate impact of Beard's recommendations. While he suggested that this disposition should govern all forms of experimentation on living human beings, his own examples made clear at the time that this was primarily a problem particular to spiritualism. This focus fit well with the editorial vision of Edward Youmans, who made *Popular Science Monthly* into a leading antispiritualist organ. Read in this context, the article seemed most relevant to the exposé of spiritualists rather than the human sciences writ large. Furthermore, Beard was a highly visible clinician dealing with a host of nervous ailments rather than an experimentalist. Finally, his death a mere four years later, while only in his mid-forties, minimized Beard's direct influence on experimental psychology.

Despite Beard's efforts, in the last decades of the century the spirit trade continued to proliferate and even gained some measure of scientific respectability. A number of the day's most prominent scientists were willing to at least give advocates of spiritualism the opportunity to prove their claims. In Britain, the movement was bolstered by the endorsement of the naturalist Alfred Russel Wallace, the chemist William Crookes, and the physicist Oliver Lodge. Such interest culminated in the founding of a scientific body assembled for the purpose of testing claims of paranormal capabilities in 1882. The Society for Psychical Research sought to carve a path distinct from the commercial world of spiritualism by advocating a critical yet ultimately sympathetic stance toward the existence of the extraordinary powers. With regard to causation, the group as a whole took an agnostic stance, willing to endorse either naturalistic or supernatural explanations on a case-by-case basis. Although often disparaged for its chosen object of study, society membership included an international array of respected scientists. Sustained scientific interest in the paranormal was bolstered in the United States with the creation of the American Society for Psychical Research (ASPR) in 1885. During its early years the ASPR benefited from the dedicated leadership of Richard Hodgson, who emigrated from Britain for this purpose in 1887. Upon Hodgson's sudden death in 1905, James Hyslop, a former professor of philosophy at Columbia University, succeeded him. These men held out hope that communication between the living and the departed could be achieved. Both also conceded, however, that certain phenomena were best explained in terms of unusual yet wholly natural abilities, the delusions of an abnormal psychology, or deliberate fraud.

Early membership of the ASPR included many of the psychologists interested in perceptual illusions, prominent among them William James, Joseph Jastrow, and G. Stanley Hall. Their presence was far from marginal, with Hall serving as the ASPR's first president in 1885. Born in 1844, Hall later recalled how he "was brought up among spiritists and always kept more or less tabs upon the whole matter." He expressed a boyhood captivation "with table-tipping and levitation, slate-writing, inspirational speaking, and all the phenomena of séances," despite his parents' objections. Notwithstanding encounters with many disappointing charlatans, Hall continued to follow the claims of spiritualists throughout his life.[22] Jastrow and George Nuttall's 1886 study of the "existence of a magnetic sense" exemplified the initial trend of subjecting psychical phenomena to experimentation. Tellingly, they published their findings in both the *Proceedings of the American Society for Psychical Research* and the equally new journal *Science*. The two insisted that they were unable to manifest any such phenomena under proper laboratory controls. The same year, Hall praised Jastrow for his "rigid feeling" on the matter.[23] Jastrow quickly emerged as the leading spokesperson on the topic within the psychological community. In contrast, James remained convinced that a fundamentally new understanding of the self and the natural world could emerge from psychical research and he condemned both "the credulous spiritist" and "dogmatic scientist" in equal measure.[24]

The relationship between the spiritualist and scientific communities was quite different in the United States compared with Britain, France, and Germany. In each European country, numerous natural scientists with esteemed credentials in other fields remained passionate adherents of psychical research throughout their lifetimes. In the United States, the popular spiritualist movement was larger than in any other country, but most members of the nascent scientific community distanced themselves from the formal investigation of psychical phenomena.[25] Although James remained a dedicated and vocal affiliate of the organization until his death in 1910, the other psychologists soon abandoned the ASPR, disappointed with the quality of phenomena they were asked to evaluate. Outside the association, however, these psychologists continued to associate with genteel psychical researchers as well as expose those whom they deemed to be frauds.

Due to its popular appeal, spirit mediumship had an ambiguous legal status at the beginning of the twentieth century. Spirit mediums were frequently charged in criminal cases with obtaining money under false pretenses and Progressive Era legislative bodies passed new regulations seeking to curtail the business of mediumship. For example, a 1907 city ordinance made it illegal to advertise one's ability to predict the future within Chicago's limits. Fortune-telling was prohibited since city officials determined that its primary purpose was "profiting by the credulity of the public."[26] At the same time, civil courts were reluctant to define spiritualism as necessarily fraudulent, especially when it came to the tricky issue of probating wills. In *Keeler v. Keeler* (1889), the court determined that belief in spiritualism was so widespread it constituted an unremarkable set of religious beliefs. The courts only invalidated a will if the judge determined that the testator's "free agency is destroyed" due to undue influence. This required the medium's direct intervention in the drafting of the will as a legal document. The precedent setting case was *Robinson v. Adams* (1870), where the testatrix believed that it was the spirit of her dead husband as channeled by a medium that dictated the content of her will.[27] In *McClary v. Stull* (1895), when a Nebraska widow, Elizabeth Handley, left her property to a charitable organization, her siblings contested the will on the grounds that she exhibited the delusions of the insane. Their primary evidence of her disturbed mental state was her conviction that she could communicate with her dead husband through the mediumship of a planchette, an early version of the Ouija board. In its decision, the court emphasized Handley's ability to manage her financial affairs for ten years following her husband's death and concluded that using mediumship to speak to him was the legal equivalent of consulting with one's living spouse.[28] Courts did not want to entertain debates concerning the veracity of privately held religious convictions. In general, judges focused on the issue of whether testators had freely chosen how to dispense with their property or whether they seemed unduly influenced by another party. The authenticity of spiritualist beliefs was a marginal issue when it came to the court's decision.[29] As a result, the muckraking narratives written by journalists and especially scientists served as a more of the primary means of policing the actions of spiritualists.

Eusapia Palladino: The Commercial Trickster versus the Deceptive Scientist

Although notorious for indulging in deceitful trickery when given the opportunity, Palladino was among the world's most famous spirit mediums, attracting enthusiastic endorsements from criminologist Cesare Lombroso in her native Italy and later the physicist Oliver Lodge.[30] In a number of experiments conducted in Naples, Paris, and Warsaw, she exhibited remarkable powers, and, through the mediumship of the spirit "John King," mysterious raps were sounded, tables were lifted, and a ghostly hand appeared. A mysterious wind would billow through the thick black curtains fronting a cabinet during the séance even though the windows were closed. In the wake of such performances, an eerie breeze would emanate from a scar on Palladino's forehead. Although some heralded Palladino as the most powerful medium of the day, even sympathetic psychical researchers criticized her propensity to augment her powers for the sake of a good show. A 1908 examination of her powers in Naples had produced overwhelmingly enthusiastic endorsements from witnesses, but a series of tests in Cambridge, England, in 1895 had generated incredulity as the scientific observers repeatedly caught her cheating. Because of this reputation for fraud, the ASPR's leader James Hyslop worried about what her arrival in the United States in late 1909 would bring. Despite demonstrating some of the most remarkable psychical abilities on record, the woman could be a severe liability for the precarious movement's attempts to ground psychical phenomena in scientific evidence. Skeptics, on the other hand, relished the opportunity that Palladino presented: to publicly debunk one of their best-known opponents through scientific tests.

Between October 1909 and April 1910, both camps attempted to stage séances that would benefit their own needs. Hereward Carrington, a young British psychical researcher and former stage conjuror, managed this American tour. Confident of her abilities in the wake of the Naples demonstrations, Carrington was convinced he could now persuade the Americans of her unique value. He invited leading psychologists to act as witnesses to Palladino's remarkable powers. According to this plan, stage conjurors—acknowledged experts in deceiving performances—would first evaluate Palladino to determine whether she was using trickery. He explained to the scientists that should the magicians "endorse her phenomena as genuine (as I feel they will) I hope that scientific men will take up her case for careful study, and I desire to know whether you would join such a committee—composed solely of scientific men."[31] Both Hall and Hugo Münsterberg expressed initial interest in participating in sittings with the medium, on the condition that the membership of the scientific committee that was asked to evaluate her met their expectations and that they could implement their own tests.

Two developments brought discord to the originally civil relationship between Carrington and the psychologists. A few weeks after his first contact, Carrington sent them a generic form intended as preparation for Palladino's visitors.[32] Addressed as mere "sitters," they were cast as passive witnesses to her spectacular powers, with their ability to intervene in the proceedings severely curtailed. The accusation that Carrington was using the names of scientific men to promote Palladino as a commercial spectacle further disrupted the cordial relationship. In November 1909 Charles Dana, professor of nervous diseases at Cornell and also a member of the scientific committee, penned an urgent series of letters to his fellow investigators, urging them to withdraw from their association with Carrington. It was not the prescribed conditions for the investigation that worried Dana, but rather that, as he wrote Hall, "it seems to me as though he [Carrington] were using the names of gentlemen of scientific prominence to exploit his wares."[33] Hall withdrew from the committee immediately after receiving Dana's letter about commercial exploitation, almost a month after he received Carrington's note for sitters.[34] Hall objected to the constraints laid upon him in the earlier instructions, but it was this affront to his status as a disinterested, noncommercial gentleman of science that moved him to withdraw.

Others remained committed to completing the task of defeating Palladino. For example, Dickinson Miller, chair of philosophy at Columbia

University, refused to cede ground to the Italian medium, to permit her to ply her trade unchallenged on American soil. As he wrote to Hall in late November, "it is, in my belief, amply worthwhile that men of science should investigate the case. I should add that my hope is to see it smashed by American men of science."[35] Miller's repeated use in his correspondence of the somewhat outmoded term "men of science," rather than the more modern neologism "scientists," was telling. A "man of science" was a broader category, simultaneously a moral figure rather than just a skilled and well-trained technician. It could be used to encompass allies based outside of the university system yet within the boundaries of respectable opposition. The man of science was also a deeply gendered identity.[36] National pride and masculine guile came together in Miller's vision of a definitive breaking of Palladino on American soil. Similarly nationalist sentiments, namely the desire to prove the superiority of American science and scientists on the world stage by bettering their European peers, seemed to be an important motivation for convincing others to pursue the project despite qualms about the publicity. In a later interview with the *New York Times*, Dana boasted, "The Americans showed her up where the Europeans failed."[37] Despite the absence of psychologists, a scientific committee was convened with Dana, Miller, and a number of physicians and natural scientists.

This first committee conducted their investigation in William Hallock's physics laboratory at Columbia University. Its task was to evaluate the manifestation of a number of phenomena with special attention paid to Palladino's ability to levitate the séance table and the appearance of a ghostly hand in a cabinet during the course of the sitting. The committee members quickly became convinced that she was merely repeating the fraudulent behavior exhibited during the 1895 investigation, but there was some concern about their inability to explain how she managed to levitate the table. After six sittings, Palladino and her manager ended the laboratory investigation without the American scientists having definitively bested her on all fronts.[38]

Miller was unwilling to let her leave America with even part of her reputation intact. In April he convened a second scientific committee, distinct in membership from the first, and developed a new approach. This group concluded that exposing the medium as an outright fraud would require deception on the part of skeptical researchers. In part, the conditions imposed by Carrington necessitated their use of deception. Among the eleven prescriptions in his "Notes for Sitters" was the stipulation that "whatever the attitude of the sitter toward the medium may be, no suspicion be OPENLY manifested at the sittings, as this is liable to spoil phenomena."[39] Palladino's proponents argued that an openly hostile demeanor would ruin the conditions under which her unique abilities manifested themselves. Hence, investigators needed to feign conviction in her powers during the actual séance while secretly testing her natural abilities without her knowledge. Carrington had originally proposed two sequences of investigations: one executed by stage conjurors, the subsequent one by scientists. The plan of the second committee combined the two, with magicians collaborating with psychologists without the knowledge of Carrington or Palladino.

Perhaps surprisingly, considering his own vocation, Joseph Jastrow (now a member of the committee) felt that scientific men were not necessarily the best kind of witnesses to such events. In May 1910 he published his report on Palladino, narrated as a legal trial, in Norman Hapgood's muckraking weekly *Collier's*. There Jastrow insisted, "Men who approach a problem by the aid of their scientific training have a proper dislike, and many of them a pronounced unfitness, for detective work. They are skilled in setting up apparatus and wrestling the secrets of nature; but it is not quite in the line of their business to set traps."[40] His argument was given a populist twist a few months later by a confessing former medium who claimed that "scientists are the easiest people in the world to deceive. It is far easier to fool half a dozen of them than six ordinary working men."[41] Jastrow's intention was far from presenting scientists as enfeebled, clueless fops, more easily duped than the hardy manual laborer. Instead his argument allowed him to make sense of how otherwise credible and respected scientific men like James and his European counterparts had been convinced by various mediums over the years: their training had made them trustful. After all, Jastrow noted, unlike the medium, "Nature is puzzling and evasive to a degree, but she plays fair." When discussing the potential existence of psychical phenomena a year earlier, James had asserted that "nature is brutal enough, Heaven knows; but no one yet has held her non-human side to be *dishonest*."[42] James was fully aware of the methodological "snares" of rendering psychology into natural science, but he remained convinced that spiritualism belonged to the natural order and hence adhered to nature's law of honesty.

For Jastrow, like George Beard thirty years earlier, there was a fundamental cleavage between nature writ large and human nature. The methodologies applied to one were not suited for testing the other. The great failing of the first committee, why its investigation ended inconclusively, was that it remained too married to the regularity of the laboratory and its machinery. To subject the spiritualist to rigorous examination required drawing upon ways of knowing derived not from the natural sciences, but from detective work and stage conjuring. These occupations demanded from their members both a commitment to accuracy and a rejection of sincerity. Jastrow publicly argued that Palladino's critics had to embrace the epistemology of the private investigator rather than take the approach of a laboratory technician, asserting that "the detective attitude is uncongenial rather than inaccessible to the scientist."[43] He held that professional magicians were the experts best equipped to confront the spiritualist because their knowledge of sleight-of-hand and the practical management of an individual's attention were invaluable resources.

Jastrow's 1910 article not only represented a stark reversal in his own published opinion, but it also marked a change in the dominant claims previously made by scientists about the prerequisites for a legitimate investigation into the paranormal. Jastrow had earlier insisted that psychical phenomena could only be properly understood "when they are investigated by the same methods and in the same spirit as are other psychological problems."[44] He proclaimed repeatedly that the supposed powers of the medium could be debunked only when subjected to the rigors of laboratory experiments and complained when spiritualists refused to submit themselves to such tests. Jastrow's opinions were widely shared. In 1871 the British physiologist W. B. Carpenter advanced the influential argument that amateur scientists made improper witnesses at the séance table due to their lack of professional training. While these men might be attentive observers, they lacked the specific skills bestowed by an education in experimental inquiry that was necessary to uncover the spiritualist's true modus operandi. Carpenter made these arguments because of the prominent role of sympathetic stage magicians and respected scientists without university affiliations in the British debates of the 1870s. His stance belonged to the movement to demarcate the sciences of mind from spiritualism in order to professionalize the former.[45] Almost forty years later, the first grantee of a doctorate in psychology from an American university was making the opposite case: laboratory training schooled the scientist in the reliable laws of nature, making him particularly and overly trustful. Jastrow revived Beard's claim that deception and objectivity ought to operate in concert to achieve reliable results when dealing with living human subjects.

After the dissatisfying conclusion of the physics laboratory investigations in January, the later sittings were characterized by changes both in venue and personnel. In April 1910 Palladino's opponents staged what Miller called "two finally decisive" séances a week apart in the house of Columbia professor H. G. Lord. In correspondence with his colleagues, Miller insisted that they keep the very "*existence of the project* as a secret."[46] Among those in attendance were university professors, but also established psychical researchers, society women, and stage conjurors. Each group had its own particular role to perform. Following standardized procedure, two men were assigned positions to the immediate left and right of Palladino to hold her arms and restrain her feet, in order to guarantee the genuineness of her phenomena. The committee selected for this task James L. Kellogg and W. S. Davis, conjurors with experience in psychical investigations. A gentleman with similar credentials, John W. Sargent, was assigned the task of sitting opposite Palladino and observing the scene as a whole. Davis, the secretary of New York's Metropolitan Psychical Society, was also a reformed medium who bolstered his reputation as an infallible debunker by emphasizing his insider knowledge of the trade. This experience risked proving a liability should Palladino recognize these seasoned observers, and Davis noted in his own report, "Kellogg, Sargent, and I were introduced by our middle names so that she would not know us."[47] Misdirection was meant to ensure that Palladino acted as she would at any other séance.

The examination was notable for the absence of laboratory instruments as tools for evaluation. In fact, the only apparatus used during the April 17 proceedings, an electroscope, was "to serve as a psychological decoy" to mislead Palladino.[48] The stated purpose of the instrument was to detect the presence of static electricity and when they presented the device to Palladino, the assembled scientists requested that she place her hands on the globe and concentrate so that her powers could be measured. The electroscope's true function of was to distract Palladino at the outset of the séance so that two new participants could secretly enter. The plan was that Palladino, engrossed by trying to generate a measure on the electroscope, would not notice the introduction of these actors into the proceedings. While she may have been cognizant of the skeptical inquirers surrounding her at the table, she would remain ignorant of the hidden participants.

During these carefully planned moments, two men clothed in black stockings and caps quietly crept into the room and took their place under the chairs of the spectators. Their introduction had been planned well in advance and they had been waiting patiently upstairs for this

FIGURE 3.1 Palladino confronted by the unobserved but observing observers. Joseph Jastrow, "The Unmasking of Pal[l]adino," *Collier's Weekly* 45, no. 8 (May 14, 1910): 21.

choreographed opportunity. These men reproduced the dramaturgy of the famed magic trick the "Black Art," first made famous in America during the 1880s by the rival shows of Alexander Herrmann and Harry Kellar. Those magicians performed the Black Art on a dimly lit stage in a room covered in black velvet and surrounded by a ring of gas lights. The highly visible conjuror, dressed in white, managed to make a series of objects appear and disappear with the assistance of an unseen confederate. The trick required that "there is an assistant on the stage all through the act, but as he is dressed in black, with gloves on his hands and a hood over his head, made of black velvet, he is not seen by the spectators, whose sight is somewhat dazzled by the open gas jets."[49] This unobservable helper revealed and hid various objects by covering and uncovering them with additional fabric during the course of the act. In terms of their darkly dressed confederates and the deployment of distracting visual devices to conceal their presence, both the Palladino exposure and the Black Art operated on near-identical principles.

Jastrow bestowed these "unobserved but observing observers" with considerable epistemological authority.[50] The first was Warner Pyne, a Columbia student and "a trusted friend" of the host. The other was J. L.

Rinn, a produce merchant, magician, and psychical researcher, "who has long been deeply interested in the methods of mediums and has proved himself a reliable observer."[51] Their ability to take in events, to detail her subtle movements without being detected, was a point of pride for the anti-Palladinoists. There was a palpable giddiness in their success in finally fooling the commercial huckster, for the unobserved observers could see Palladino's legs move beneath the table as the phenomena were manifested. Jastrow's telling turn of phrase illuminated the multiple qualities bestowed upon these secretive witnesses. "Unobserved" indicated the importance of their capacity to witness events without their own presence being detected by the quarry. They could watch without interfering. "Observing" signified that these were attentive watchers who possessed the talents and discipline to deduce subtle clues. Although they might be

FIGURE 3.2 The modus operandi of the Black Art. Albert Allis Hopkins, *Magic: Stage Illusions and Scientific Diversions, Including Trick Photography* (New York, Munn and Co., 1897).

hidden from sight, they were not removed from events and were acutely aware of which phenomena to note. Finally, an "observer" was a clearly defined role in scientific psychology. The observer was a highly trained individual who reliably and truthfully reported his own introspective experiences when presented with stimuli in the laboratory. He had sufficient skills, responsibility, and self-control to complete the task assigned to him. The observer was also disinterested. In his *New York Times* report, Miller insisted that Rinn was impartial when it came to evaluating the veracity of Palladino's claims. He quoted the magician as stating, "I am simply an observer. I do not know what will happen. Perhaps a foot will come out from under that dress and perhaps it will not."[52] They were neither gawking and credulous spectators nor prejudiced partisans. The secretiveness and insincerity of their self-presentation bolstered the accuracy of their observations.

The ideal of an unobserved but observing observer was not the expression of a disembodied view-from-nowhere interpretation of scientific objectivity. Rather Jastrow tied this form of watching to certain persons whose authority derived from a distinct set of perceptual skills performed in specific locales. It was an epistemologically privileged vantage point and not just anyone could engage in such activity. For these were not anonymous individuals; they were repeatedly identified in public representations of the séance as a guarantee of the reliability of their findings. Relying on natural scientists was a mistake under the conditions of the séance; they made poor witnesses since they lacked both the detective's skills in disguising themselves and the magician's familiarity with the modus operandi of trickery. The proper observer in this situation combined the formal learning of the university student with the practical experience of the conjuror. Arriving at the truth of the matter required observers who could simultaneously be tricky, eluding the detection of the medium, and serve as reliable witnesses of good character.

Another important element of the debunker's apparatus was the presence of respectable society women in the audience. According to Jastrow, their primary purpose was "to allay her [Palladino's] suspicions."[53] They were intended to give the aura of an authentic séance by supplying one of its most common features: enthusiastic female participants. The committee assumed that if Palladino had been met by an exclusively male audience, she may have guessed that the invitation to perform was a trap. Women, complete with their enthusiastic reactions to the proceedings, were there to authenticate the proceedings as well as to provide an additional cloak for the scientific investigations of the conjurors and the psychologists. Yet the men of science also portrayed the presence of these women as an obstacle. Their emotional reactions, especially their credulous obedience to Palladino's instructions, risked exposing the unobserved but observing observers in a number of instances. At one point, Palladino requested that all the participants stand as part of the séance. As Davis recalled, "we had spies under our chairs and we did not want Eusapia to see them, though the two ladies sitting at our table, in their inexperience, proceeded to obey the medium's commands."[54] In contrast to these women's blind obedience, Davis boasted of his own quick wits as he faked a muscle cramp that prevented him from standing. This reminded the anonymous ladies of their purpose. Although necessary ornaments, serving much like the distracting electroscope, these women were not deemed full participants or active agents in the debunking. Significantly, no published account of the exposure was written from their perspective, although female contributions to psychical research were commonplace at the time. The staging of a debunking was an activity constituted as needing the guile, incredulity, and experience of the male unbeliever.

A week following the séance with the unobserved observers, the same scientific committee booked another sitting with Palladino. Although its members felt that they possessed sufficient evidence of her trickery, they had yet to reveal this conviction to Palladino. Prior to the second séance, certain members met at the Columbia faculty club to discuss the roles they were to play. These subtly shifted, with Kellogg and Davis still professing to be in awe of her powers while Jastrow was to perform the role of the hostile skeptic, barking orders "to check trickery" at the other men. The organizers feared that if Kellogg and Davis were to now seem skeptical of her, Palladino might insist that they be replaced as her controls. For this subsequent outing, Palladino's monitors would exercise the strictest physical controls while intently policing her movements. No worldly means would be permitted to make the spirit's presence felt on that particular evening and the resulting sitting was an absolute failure as the "performance simmered down to nothing." For her part, "[Palladino] complained of the violence of her restraint," which the investigators assured the reader "was in reality a most gentle but firm contact."[55]

The committee's report had an impact on American champions of psychical research. William James became convinced that "Eusapia, if *true*, deserves to have it *never known*" due to her repeated mendacity. In June 1910 he urged Carrington to abandon her cause in the United States, "to leave yourself in the past history of the subject without trying to connect yourself with its future history." To maintain his association with Palladino would, according to James, permanently damage the young man's reputation and the cause of psychical research more broadly. James still held out some hope for the ultimate veracity of Palladino's powers, but due to her past behavior he did not think she could ever convince those of a skeptical temperament.[56]

Palladino's own response to having been bested violated the norms of civil exchange that governed Anglo-American psychical research. According to witnesses, she became like an animal as she lost her self-control and began to wail and howl. Miller had earlier expressed a desire to see her smashed, and by the end of the séance she was described as "crushed and defeated."[57] The scientific committee had achieved its epistemological victory over Palladino and exposed her racial degradation in the process. Such interpretation is not surprising considering that skeptical psychologists viewed belief in spiritualism as a kind of racial recapitulation. For Jastrow, "its closest affiliation is with the time-worn, crude practices and beliefs of primitive people." Hall once described opposition to spiritualism as a kind of "crusade against this recrudescence of savage and barbarous concepts in our modern life."[58] With such expectations, it was not surprising that her genteel observers would interpret her behavior as a kind of savagery. Such a reaction, furthermore, reaffirmed their confidence in their moral and cultural superiority over the commercial fraud.

These reassurances were necessary as, in the face of their gleeful success in besting the notorious medium, a pronounced moral ambiguity continued to mark her exposure. As his friend John Jay Chapman wrote William James, "you have done harm too—D. Miller exposing Pal[l]adino—under sofas—newspaper talk. . . . Miller was sick with excitement & all of them together—who started the spook hunt as a scholar's recreation?"[59] From Chapman's perspective, the giddy pleasures derived from the "spook hunt" were antithetical to the proper public image of the scholarly life. "So far as lay in her power she has degraded some of her American investigators," Miller conceded in the *New York Times*, "for they were forced against their will to the only device that would eradicate superstition and terminate a scandal to science—to watch from hidden places."[60] Unmasking the charlatan required the scientific man to operate in a similarly backhanded and insincere fashion, mastering and emulating her kind of trickery.

Jastrow addressed the uncomfortable equivalence between the medium and the scientist. He admitted that the scientific man "has no choice but to set aside his scruples and meet the 'medium' on her own ground." Accordingly, "one must swallow one's distaste and enter into the spirit of the game, or dismiss it as unworthy of serious attention."[61] Jastrow pinpointed a tension at the heart of his brand of popular psychology with its spectacular demonstrations: the truthfulness of the observer did not necessarily guarantee truth, but could rather work against it. Scientific men occasionally needed to abandon traditional scruples and approach their subjects with cunning, guile, and even dishonesty. Deception was not the polar opposite of objective science; indeed, it could furnish an aspect of its methodology. To transform Palladino's own body into a transparent, truthful witness did not require the introduction of laboratory technology, but rather necessitated that she be lied to and interrogated using the ingenuity of the magician. To approach her with sincerity would simply result in being manipulated by this dishonest actor. Yet the skepticism of the psychologist-cum-detective came at a cost. In Miller's words, it left the man of science morally "degraded," even if it saved him from being duped.

Such ambiguities were not lost on the press, which tended to portray the scientific committees not as the neutral witnesses, but as representatives of a single faction, the "anti-Palladinoists." By claiming to give equal coverage to both sides in the dispute, newspapers reserved the position of impartial arbiter for themselves. As such, they did not accept the May verdict as final and eagerly published reports that continued the controversy. Their pages contained rounds of combative claims particularly after the magician Rinn offered a thousand dollars if Palladino could prove her abilities genuine. The *New York Times* went so far as to offer a room in its office tower as a neutral space for the warring groups to meet and conduct tests.[62]

Even certain members of the scientific committee publicly objected to the deceit recounted by Jastrow. Although he concurred with the scientific consensus that Palladino was a charlatan, physicist Robert Wood did not accept "the so-called exposure of her method of raising the table, published in *Collier's*."[63] Wood, a member of the January committee, had engaged in hidden observation by watching the movements within the cabinet through a hole cut in its top from the vantage point of a neighboring instrument case. He rejected the authority of the experienced conjuror hidden under the table and favored the use of another kind of unobserved observer: the relatively novel technology of X-rays. He had constructed a special cabinet with an X-ray tube hidden inside and a fluorescent screen placed on the opposite wall. Wood placed a switch to operate the coil on top of the cabinet where an observer watching events could activate the system once the "ghostly" hand manifested itself inside. The resulting recordings would determine the true nature of what took place. Because X-ray imaging could "be used without the medium's knowledge," it had all the advantages of hidden human observers without relying upon their subjective observations.[64] It took judgment out of the hands of the magician—a merely temporary ally—and returned it to the domain of natural scientific inquiry. Wood offered his solution as both a more accurate and ethical alternative to the deceits of human observers, although it was never implemented due to Palladino's unanticipated abandonment of the laboratory tests at the end of January. Wood enacted another form of boundary work in claiming that the conflict could only be definitively resolved with the tools of his own discipline in spite of a scientific consensus on the matter.

The acerbic exchange of letters between Jastrow and Münsterberg over their respective conduct during the Palladino affair further illuminated the problems associated with using insincerity and deceit to fulfill one's obligations to the truth. Tensions between the two erupted following Jastrow's anonymous review of Münsterberg's *Psychology and Social Sanity* for *The Dial* in 1914. Münsterberg had grand expectations for this particular popular work, dedicated as it was to applying "the results of scientific study of the mind" to solving social problems. Since it was one of his least technical works and explored political questions central to the Progressive Era reformers, Münsterberg had expressed hope that it would be his most commercially successful book. It also included his account of his own activities during the Palladino affair. The disagreement between the two psychologists was somewhat surprising considering they occupied basically the same epistemic ground. Both were absolutely convinced from the outset of Palladino's own insincerity and the need to expose her, although Münsterberg was somewhat more sympathetic toward the notion that she may have been an unknowing hysteric. Furthermore, they were united in the conviction that men of science may not serve as the most reliable form of witness at such spectacles. Münsterberg concurred "that scholars are especially poor witnesses in such a case. They are trained through their whole life to breathe in an atmosphere of trust." In other words, they agreed that the very mechanisms of trust that ruled in the laboratory made the scientist a vulnerable and unreliable witness outside its confines.[65]

They differed in terms of the responsibilities of proper disclosure: Jastrow felt Münsterberg's conduct worked to publicize Palladino's claims.

In February 1910, before the scientific committee had announced its conclusions, Münsterberg published an account of his encounter with Palladino in *Metropolitan Magazine*. In the article, Münsterberg described his participation in a December sitting, claiming that an unnamed gentleman had caught Palladino's foot astray in the cabinet behind her, where the spirit seemed to manifest itself during the course of the séance. Jastrow felt that the style of reporting in the *Metropolitan* article gave the impression that scientific exposure was "a hasty affair" rather than a "comprehensive one."[66] Absent from Münsterberg's December investigation was the sufficient control over the situation as represented by the follow-up séance conducted by Jastrow's group. Moreover, unlike the reports from the April séances, Münsterberg withheld the names of those who aided in trapping Palladino. Jastrow felt this decision made Münsterberg's report less a detailed exposé and more a sensational piece of publicity.[67] Certain newspaper versions of these events stated that the unnamed individual who detected the movement of Palladino's foot into the cabinet was Münsterberg's formal "assistant" or confederate.[68]

Jastrow questioned not the accuracy of Münsterberg's findings, but the transparency with which he communicated them. Where they agreed on Palladino's status as a fraud, they differed about how to publicize the procedures involved in exposing her. Jastrow's condemnation of his colleague assumed that if psychologists were to use insincerity to trap their commercial opponents, certain forms of transparency had to be in place to guarantee a trustworthy account. Anonymity and solitariness were anathemas to the means through which Jastrow publicly conveyed the Palladino affair. His popular account clearly outlined the individuals involved, the roles they performed, and the kinds of expertise they possessed. Below the surface was the suggestion that Münsterberg used the situation as an opportunity for personal aggrandizement and commercial gain, thereby confounding the boundary between the fraudulent medium and respected university psychologist.

Münsterberg was certainly affronted by this implication. He fashioned himself as "the most ardent opponent of mysticism" in the United States, insisting that "men like James are considered as spiritualistic party men" while "men like Prince and Jastrow have at least developed theoretical ideas of subconscious which seem to the public not unfavorable to spiritualistic claims."[69] As the nation's leading skeptic (an identity that would come to be challenged from a surprising corner), he did not appreciate insinuations of dishonesty from a scientist whom he considered his inferior. Münsterberg assured his Wisconsin colleague that he had no desire to encourage public credulity in the woman's abilities, deferring such a position unto others. He wrote, "I went to those Pal[l]adino meetings as a direct challenge to James, who claimed that it was my duty to examine such affairs which are personally distasteful to me." Although certain newspapers represented him as complicit in the affair, he denied the charge vehemently. After Jastrow wrote to reassure his disgruntled colleague that the review was a friendly one, Münsterberg insisted he had never been offended. Münsterberg's initial response, however, made apparent his discomfort with Jastrow's charge.[70] The psychologist's adoption of accuracy without sincerity when confronting spiritualists did not simply cause tension between Jastrow and Münsterberg. It came to shape the whole enterprise of debunking spiritualists. While garnering attention for their successful exposures, psychologists also cultivated a sense of distrust when it came to their methods.

Mrs. Piper's "Honest Living," Miss Tanner's Deceptive Profession

Unlike Palladino, who was notorious even in psychical research circles for her trickery, Leonora Piper was a medium universally acclaimed for her sincerity. This honesty and respectability made her a valued experimental object for investigators from a variety of perspectives. James first alerted psychical researchers to her existence in 1885, after he had attended a number of her séances. To assure her moral rectitude, the ASPR did not originally turn to scientific methods. Instead, Hodgson employed a private detective to trail her and her husband to determine whether she acquired information about her clients by dishonest, worldly means.[71] In the ensuing decades, Piper became the most cherished spirit medium in America. While there was considerable debate over the nature of her ability, no one impugned her character.

When Piper entered a trance state during which her own personality was completely subsumed, her body became the medium that enabled a variety of spirit controls to communicate. By 1909 one of the most powerful of these controls was the supposed spirit of Hodgson, the deceased former head of the ASPR. The society published two major reports that year that represented the culmination of years of observation: one produced by Hyslop, the other a lengthy inquiry authored by James. The Harvard psychologist suggested that Piper's trance and spirit manifestations seemed most likely to be the product of abnormal psychology, in the form of disassociated personality. Hyslop's was the more ambiguous of the two, suggesting that although it seemed unlikely that spirits actually communicated through Piper, perhaps she acquired knowledge about the dead relatives of those who attended her séances through mental telepathy.

In 1909 Hall, along with his assistant, Amy Tanner, convinced psychical researcher George Dorr to partially fund six sittings with Piper. They saw their purpose as dispelling the uncertainty about Piper's true nature. Although the Clark University psychologists did not engage secretive witnesses in the manner of the second Palladino committee, they too argued that deceit was an integral component of a successful investigation. During the course of their first sitting with Piper, Hall inquired about a series of fictitious deceased acquaintances. Most notable was a young niece named Bessie Beals, whose spirit "Hodgson" readily contacted. The psychologists noted the ease with which the control produced the "pseudo-personality," especially considering her nonexistence. In addition, Hall recounted a highly falsified account of his relationship with the living Hodgson, which his spirit counterpart enthusiastically endorsed.

In his published notes on the case, Hall discussed the moral dilemmas raised by securing scientific knowledge through deception. Reflecting on the Hodgson "personality," Hall wrote,

> And back of all in my consciousness is the marvel how he can possibly accept the absurdist gaff I can think of with such implicit and immediate faith. Surely all his life and since he must have been used to dealing with people who treat spirits with implicit honesty, and his acceptance of my involuted lie fills me with qualms of conscience. But I am a detective in quest of truth, and the end must justify the means.[72]

Once again, a psychologist made the analogy between the epistemology of the detective and that of the skeptical researcher. Moreover, he emphasized how serving the needs of science required that the individual investigator sacrifice his "qualms of conscience" in the name of a greater moral project.

In her book *Studies in Spiritism* (1910), Tanner made a telling comparison between the relationship between scientist and subject in the Piper investigations and in the psychology laboratory. She explained to her reader that the techniques used with Piper fell well within the purview of established norms. Tanner made the case that insincerity was a prerequisite for objectivity in psychological research. When studying illusions, the investigator "deliberately leads the subject astray, distracting his attention to unessentials by his remarks, arrangement of apparatus." This did not mean "that he is morally telling a lie, or that he implies any doubts of the subject's veracity." She held that "in all psychological experiments where the subject's own consciousness is concerned, it is not wise to let the subject know the nature or purpose of the experiments, because such knowledge inevitably modifies his attitude and vitiates the results."[73] In such situations, the informed experimental subject has the not entirely conscious desire to meet the experimenter's expectations. Tanner deemed this form of experimental sleight-of-hand as different from the fraud the investigator sought to expose. Yet Tanner's need to offer an explanation to the reader undercut her insistence that such techniques were ethically unremarkable within the moral economy that governed psychology experiments.[74] Indeed, Hall highlighted the novelty of his approach in his correspondence with the international community of psychical researchers. These correspondents were particularly interested in Hall's own use of deception. From England, Lodge requested that Hall share his "careful record of these sittings, including no doubt a full statement of your different experimental devices and what you call the 'tricks and deceits practiced upon the control.'"[75]

The Piper investigation presented the psychologists with a rather unique set of circumstances. Unlike the majority of spiritualists who were open to charges of fabricating otherworldly phenomena, Piper overwhelmed them with her sincerity. She did not seem to embellish her powers and shared in their desire to understand the root cause of her trances. Furthermore, the secondary personality, or the spirit control, impressed Hall and Tanner with his straightforward, credulous manner. The situation presented the psychologists with an individual who was neither dishonest nor genuine. Yet as an object of inquiry, Hall insisted on the need to examine her though the use of deception and misdirection to fully understand the nature of her trance and alternative personalities.

This choice of methodology had consequences during the course of the six sittings as Piper and "Hodgson" objected to and resisted such trickery at numerous turns. It caused severe strains in the relationship between experimenter and participant during the third sitting, when Hall tried to introduce a posthypnotic suggestion. He decided that once Piper was in her trace he would mention to the control that a gas leak had occurred, but had been fixed in an adjacent room. The control responded violently to the false suggestion, repeating, "Do not deceive me unless you wish to be deceived." The trance quickly broke off.[76] A few weeks later, Hall and Tanner decided to experiment on the extent to which the spirit Hodgson could perceive Piper's bodily sensation. They wanted to test his sensitivity toward pinpricks on the back of Piper's hand using an aesthesiometer. While they debated whether to inform Piper of the procedure prior to the trance, they "thought the tests would be purer if we did not do so." This lack of disclosure nearly brought the entire investigation to an end, as Piper, her daughter, and Dorr all objected to the physical harm that had occurred during the trance. Piper and her supporters claimed that the Clark psychologists were not forthcoming about their methods and that they secretly conducted tests to which she did not consent.[77] In order to restore the cordial relationships necessary to complete the experiments, Hall quickly assumed a position of apologetic deference. The quality that made Piper a particularly prized object of investigation, her genuineness of intent, also constrained the ways in which psychologists could use deceit to elicit an authentic response.

This resulting distrust caused great difficulties for Hall. His series of sittings had a purposeful structure and he stressed that if the final one could not be completed then the experiment would be for naught. He deemed the sixth sitting to be "the most important of all, because it enabled us to bring to a head matters for which the other sittings had been more or less preparatory."[78] The subtle deceptions over the previous weeks would come to the fore as Hall confronted the spirit control during the final trance with all the ruses he had developed. He revealed the fictitious dead relative and the falsified details of his relationship with Hodgson, each disclosure driving home the psychologist's lack of conviction in the spirit's authenticity. The experimenter confronting the research subject with his own deceptions furnished a key moment in the examination itself. At this moment, he hoped that Piper would acquiesce. As in the Palladino case later that year, psychologists insisted that they could legitimately deceive in their investigations but psychical researchers and spirit mediums did not possess this privilege. In spite of the revelations, neither "Hodgson" nor Piper yielded to Hall and Tanner. The control insisted that his memory was fallible or that he had misunderstood the psychologist's words.

Through deceit, Hall claimed to have bested the psychical researchers' most prized exemplar of spirit communication, yet he hesitated in making his findings public. Among other actions, he shifted authorship of the report onto Tanner and postponed its publication for a year. He expressed this reluctance in his letters to other scientists who shared an interest in Piper, for example, urging James not to circulate confidential correspondence about his findings with Lodge and the British society for "fear that it might affect Mrs. Piper's standing in his eyes." Hall was uneasy with the ethical burdens acquired during in his role as an active debunker. As James sardonically noted, "They become *responsible* now, whereas hitherto Hall & Münsterberg have enjoyed the *otium cum dignitate* of irresponsible little sneers & digs from a safe distance."[79] Engagement in psychical research made novel ethical demands upon psychologists and reconstituted them as moral agents.

Hall delayed exposing Piper because he concluded that "she earns an honest living for herself and her daughters." Unaware of the nature of her abilities, he deemed her innocent of commercial fraud and hence entitled to the minimal fees paid by psychical researchers. She was an honest business woman and he did not want to deny her a living. He made this position clear when, after much prodding, he finally wrote Lodge several months after completing his tests. Hall suggested that his findings "would tend to lessen Mrs. Piper's Boston patronage and income and probably give her pain. Thus, altogether we have been in no haste to print our results."[80]

The attitudes of American scientists toward Palladino and Piper were diametrically opposed. These differences were certainly marked by religion, ethnicity, and class. In Palladino they could see nothing but a deceitful Italian washerwoman attempting to fool the world's leading scientists. On the other hand, they depicted her contemporary as a respectable Boston woman trying to eke out a living for herself and her family. Equally important were their relationships to the market. Palladino aroused ire for her considerable commercial traffic, while Piper largely confined herself to sittings with credulous scientists. Because of her more restricted engagement in commerce, Piper's debunkers curtailed the hard-boiled stance so prominent in the Palladino exposure. Piper's relationship with even the most skeptical of psychologists remained governed by codes of gentlemanly civility.

From its origins in spiritualist manifestations in upstate New York, spiritualism had become closely linked to certain strains of nineteenth-century feminism in both ideology and practice. In many respects, spiritualism was a subversive movement with autonomous female entrepreneurs making profits through the embodiment of masculine spirit personalities. Yet the movement still adhered to certain Victorian gender norms insofar as the spiritual medium supposedly did not acquire power through her own agency. Rather she served as the passive channel through which the male control spoke.[81] Furthermore, men belonging to various professions closely policed the spiritualists' activities. In his account of the magician Harry Houdini's encounter with the medium Margery Crandon, the historian John Kasson argued that exposure is best understood as a performance of masculinity. Houdini, who had made a career of transforming his own body into an object of consumption, was deeply perturbed by the woman's claim to be able to materialize ectoplasm from her orifices as well as the profane and sexual utterances of her masculine control Walter. Just as he designed a number of spectacular instruments to trap his own body, only to successfully escape during the performance, as he tested Crandon's powers he produced a number of devices that constrained her motion.[82]

As the recurring language of "men of science" in both the popular press and private correspondence attests, the scientific investigation of fraudulent female mediums was also defined as a male enterprise. In such encounters, psychologists gradually shed the trappings of gentlemanly civility in favor of a more rugged, hard-boiled masculinity. Debunkings gave psychologists the opportunity to use their intellect to best these female claimants to authority over matters of the mind. They portrayed exposures as arduous, manly work that required considerable mental effort to successfully overcome the trickery of women where previous investigators had failed. Denouncing the cheats of the medium as commercial ruses, they insisted that their own deceptions served science. As a rule, women were rarely granted much agency as active participants in the process of exposing a fraudulent spiritualist; their role was, at best, one of supposed passive participation.

The gendered nature of spiritualism was at the forefront in 1913 when Münsterberg fired off an angry letter about academic freedom and psychical research to Columbia psychologist and *Science* editor James McKean Cattell. Cattell was a leading voice for the position that professors, rather than university presidents and boards of trustees, ought to control the curriculum. The hierarchical Münsterberg had long opposed Cattell's program of giving greater authority and autonomy to faculty members, but due to recent events in Cambridge he had changed his opinion. The Harvard trustees had accepted $10,000 for the establishment of the Richard Hodgson Memorial Fund to investigate psychical research and the possibility of the soul's immortality. As the university's leading psychologist, Münsterberg took offense that he had learned of the bequest through the newspapers. He also worried about how the highly publicized fund would affect his and the school's reputation. Arguing that he had "fought for twenty years against the cheap mysticism of the New England public," he felt that Harvard's financing of such a project "repudiated the whole spirit of my scientific endeavors." Münsterberg contemplated resigning over the new fund, but ultimately decided to remain silent publicly. He concluded his letter by assuring his colleague that he would "endure bravely the congratulations of hysterical women who see with gladness that I have at last turned into Hyslop's track."[83]

In this context, Tanner's authorship of a skeptical exposé altered some of the dominant gender relations that governed psychical research at the time. Having completed her dissertation on word association at the University of Chicago in 1898, she next turned to child psychology. She received an appointment at Wilson College in Pennsylvania a few years later but by 1907 found that "one is truly unfortunate, from a financial standpoint, to be a woman with a love of philosophizing!"[84] Tanner had determined that, in terms of both intellectual advancement and financial security, her position at the women's college was insufficient and she asked Hall to consider making her a fellow at Clark University. Hall, who admired her book *The Child*, made arrangements for her to join the staff at Clark. Although continually frustrated by the limited opportunities for women even at the coeducational Clark, she served various functions at the university from 1907 to 1918, ultimately being named Hall's assistant. It was in this capacity that she was involved in the extended examination of Piper. Gender inequalities, which had seriously curtailed the opportunities available to Tanner as a female psychologist, were replicated during the Piper investigation. In contrast to her being the sole author of *Studies in Spiritism*, wherein she offered her own interpretation of Piper's psychology, Tanner was designated as merely a "note-taker" during the actual sittings.[85] Much like Piper, Tanner was expected to be an honest, transparent recorder of events rather than an individual engaging in scientifically minded subterfuge. Hall was the individual responsible for artfully manipulating the situation.

Like Bessie Beatty who preceded her, Tanner identified herself as a "new woman."[86] Indeed, both were actively involved in progressive reform programs, including child welfare efforts, workers' rights, and feminism.[87] Beyond her work as a reporter, Beatty also founded "Happy Land," a summer camp for California's poor children.[88] For Tanner, activism marked not only her role as a child psychologist, but also her earlier social investigations. In 1907 she published a study of her experiences living as a waitress to gain insight into the mental effects of working conditions. In order to gain accurate results, she had disclosed neither her true identity as a psychologist with a doctorate nor the intentions of her study to those who employed her.[89] This study required that Tanner engage in a form of deceptive class-passing similar to that first made popular by Josiah Flynt and Walter Wyckoff a decade earlier.

Where an earlier strain of feminism embraced spiritualism, these women sought professional identities rooted in its rejection. Just as her exposé series transformed Reynolds and her ilk from wonders to petty frauds, so, too, did Beatty metamorphosize from a reviewer of light theatre and society gossip to hardened investigative reporter. By the end of the year, she was covering labor troubles in Nevada and, in the ensuing years, she became one of the few American reporters to witness the Russian Revolution firsthand. Tanner's *Studies in Spiritism* offered a similar narrative of transformation. In the preface, she stressed that she did not "enter upon my work with any spirit of antagonism, but rather in a spirit of doubt that inclined toward belief."[90] During the course of the book, she moved from this feminine subject position of potential credulity to embracing a kind of scientific masculinity characterized by cynical doubt. Obtaining this identity centered on the sacrifice of scruples because certain deceptive performances best served the truth.

Beulah Miller in the Shadow of the Vaudeville Stage

Münsterberg made the tensions among credulous wonder, humbuggery, and commercial showmanship the focal point of his account of the supposed abilities of ten-year-old Beulah Miller. Known briefly in the winter of 1913 as "the girl with the x-ray eyes" for her seeming ability to describe details of objects hidden from her view, Miller quickly attracted both fascination and contempt. Locals in her hometown and advocates of psychical research felt they had uncovered a genuine telepath. The hypnotist John D. Quackenbos emerged in the ensuing months as Miller's most enthusiastic supporter on the national scene. A former professor of rhetoric at Columbia University, Quackenbos had resigned his chair in 1893 under a cloud of rumors that the majority of the faculty did not support his presence. He had dedicated the subsequent years to the promotion of hypnotism as the human race's salvation. He claimed that it could easily rehabilitate criminals, increase a person's intellect, and even provide a bridge to communicate with the deceased. The unfortunately named Quackenbos was adamant that Miller possessed unambiguous telepathic abilities as well as the power to see through objects, claims that generated considerable interest in the national press. For example, the *Washington Post* sent its own reporter to her town of Warren, Rhode Island, to conduct a study of the young wonder and to solicit positive affidavits from her pastor and a local judge, among other respectable witnesses.[91]

One might suspect that Münsterberg would be quick to associate Miller with previous fraudulent marvels and keep his distance, but in February 1913 he also traveled to Warren to investigate. In a subsequent report, he emphasized how he "was as deeply startled and overcome with wonder as I was after the first night with Eusapia Palladino."[92] This was not an auspicious start. While newspaper accounts stressed how the wondrous sentiments seemed to guarantee authenticity, for Münsterberg the very aura of wonder led him to suspect that deception lay at the heart of the matter. In this instance, he invoked the fraudulent Palladino to contrast with the innocent Miller, whom he ascertained was simply unaware of the true nature of her powers. Again, economic value was central to Münsterberg's comparison of the two wonders. He emphasized his physical revulsion at entering "Palladino's squalid quarters" to encounter the "sham psychologist, sitting in a trance state at a table surrounded by spiritualistic believers who had to pay their entrance fees." In such an environment, "everything suggested fraud." This made exposing the "humbug" a relatively straightforward task. In counterpoint, the innocent Beulah Miller demonstrated her abilities "before any commercial possibility suggested itself." As he told the *New York Times*, "It is a pity how the community forces sensationalism, commercialism, and finally humbug and fraud on a simple country girl." Deception was extrinsic to Miller, a product of other people's attitudes. Münsterberg praised her upstanding New England mother, who turned down propositions of having Miller make a profit from her talents on the vaudeville stage. Her mother's denial of Miller's commercial potential allowed for the preservation of her scientific worth. Münsterberg condemned financial opportunism in those he examined even while being constantly attacked on this very issue himself.[93]

This abstention from the marketplace transformed Miller from a potential fraud into a genuine psychological object. It rendered her more sincere and transparent. Dismissing the notion that Miller was executing a commercial humbug, Münsterberg offered a naturalistic explanation of the precocious young wonder. The young girl's family indicated that she was capable of reading the minds of others. For example, if someone studied a playing card away from her view, she could determine its face value. Münsterberg determined that her talent was analogous not to the spiritist, but rather to the stage magician, a commercial actor whose abilities Jastrow and others had already made amenable to the psychologist's naturalistic explanations. Like these showmen, young Beulah had the ability to detect subtle movements in people's bodies, providing her with clues as to their thoughts. Her talent was particularly heightened when either her sister or mother was present with the tester, as she was even more capable of reading their clues. The primary difference was that, unlike conjurors who practiced and mastered this skill, Miller's was largely an unconscious process of which she was not aware. For this reason, Münsterberg was willing to include discussions of her faculties in his talks on the lecture circuit.[94]

Not all were as impressed by her psychological as opposed to psychic abilities. Speaking on behalf of the Metropolitan Psychical Society, W. S. Davis put a new twist on the argument advanced by Münsterberg and Jastrow during the Palladino affair. Where he had collaborated with psychologists in the former case, Davis now claimed that only a stage magician had the wherewithal to unmask deception. He dismissed the ability of scientific men to cope with the "chicanery" represented by Miller because they "were not skilled in the detection of trickery." For Davis, the public at large may have found it reassuring "to hear that gentlemen with titles are conducting a 'scientific' investigation of psychic phenomena, but when we find that scientists are already on record as credulous or incompetent, or both, and that this is especially true in the present instance, it would be well to find other investigators."[95] Davis went further than the claim that fraud could be detected if only Miller would subject herself to a proper examination. He boasted that he could use his theatrical acumen to produce in any other child an identical performance. He suggested that this hypothetical protégé could appear before the Brooklyn Philosophical Society, where the credulous doctors would try to detect what tricks he had trained her to produce. Despite Münsterberg's persona as the leader in the promotion of the public understanding of science, Davis dismissed the Harvard psychologist as "leading the mildly credulous." The *Washington Post* reported that Davis did not "consider a scientist a competent judge, because scientists are not schooled in legerdemain and conjury."[96] The stage conjuror could serve as the scientist's ally during these exposures, but he was an ally who played by his own rules.

Münsterberg's attempt to carve out a space for his psychology, by explaining the extraordinary in terms of abnormality rather than outright fraud, put him in a difficult public position. Deception seemed central to the revelation of Miller's true nature, but the psychologist was the magician's student in such matters. His own previous arguments led to a situation where the redeemed former conjuror laid claim to greater skepticism, leaving the man of science seemingly mired in credulity. In a March editorial, the *New York Times* backed the abilities of the stage magician over those of the scientist. Dismissing the positive affidavits provided by family members, the paper wrote the testimony

> of scientists, whatever their eminence, is little or no better when, as invariably happens, they are men utterly unaccustomed to dealing with deception, conscious or unconscious, and as ignorant as any gaping bumpkin at a country fair of the tricks, ancient and modern, which are the stock and trade of the professional prestidigitateur.[97]

No mention was made of the psychologist's special expertise when it came to such cases, but a month later the same paper praised Münsterberg for subjecting the Rhode Island wonder to "his coldly scientific methods of investigation."[98] Both editorials failed to recognize the practical lessons learned by psychologists from their extended encounters with stage magicians. When it came to dealing with wonders like Beulah Miller, Münsterberg's coldly scientific methods owed a significant debt to Davis's lively stagecraft.

Willetta Huggins and the Credulous Doctors

Silent during the Miller case, Jastrow came to the conclusion during his encounter with Willetta Huggins that the credulity of learned doctors was a serious problem. Starting in 1921, various papers began promoting the sixteen-year-old, deaf-blind Huggins as an even more wondrous successor to Helen Keller. Downplayed in later coverage, the earliest accounts suggested that Huggins possessed an "occult sense" that compensated for her lack of sight and hearing. The directors of the Wisconsin School for the Blind introduced her to the state governor and brought her before the legislature to demonstrate these talents. Later, they locked her in a dark bank vault and then asked her to identify the color of six different pieces of string sealed in envelopes.[99] On April 26, 1922, participants at a meeting of the Chicago Medical Society witnessed her remarkable spectacle. At the request of Dr. T. J. Williams, the physician responsible for presenting her to the gathering, Huggins performed a number of tasks before the assembled medical experts to illustrate her unusual powers. When given a bouquet of variously colored paper flowers, she was able to distinguish every color by simply smelling each flower in turn. Next, Williams presented her with several paper bills of various denominations and she determined their monetary value based on the texture of the money. The final demonstration involved a man sending a message along a twelve-foot tube that Huggins was asked to decipher based upon the vibrations the tube produced. According to initial reports, the young woman exhibited remarkable talents in all three tests, leading some to theorize about the alternative, compensatory powers of the deaf and blind. The published report speculated that based upon the Huggins case, the deaf and the blind had the power to cultivate sensory powers distinct from the senses of the "normal" population.[100]

As Jastrow had been present at the Chicago demonstration, the officers of the American Medical Association invited him to publish his verdict on Huggins's remarkable talents in their journal. In order to prepare his report, he was permitted to further examine the wondrous youth. Although he felt that there was insufficient evidence to prove that the Huggins case was an example of deliberate and intentional fraud, Jastrow quickly dismissed the claim that she possessed extraordinary perception. He argued that Huggins had been misdiagnosed and that her "ordinary" senses of sight and hearing were not completely useless, but rather merely severely limited. Jastrow concluded that she suffered from slit vision and, therefore, could perceive some traces of vision and that her deafness had psychological and not physiological origins.[101]

Huggins and her supporters dismissed Jastrow's report. They intimated physical abuse, claiming that his rough treatment of the subject in a dark room "filled with chemical fumes" had inhibited the manifestation of her true abilities. Northwestern psychologist Robert Gault announced that he was convinced of the genuineness of the Huggins phenomena in the pages of the *Journal of Abnormal Psychology*. Huggins found herself to be cured of blindness two years later. In a final twist in the case, she credited her faith and the healing techniques of Christian Science, an exemplar of the mental healing movements that so disturbed the Wisconsin psychologist.[102]

In the *Journal of the American Medical Association*, Jastrow explained that the Huggins case was not really about the abilities of this particular woman, but rather what he perceived to be the widespread problem of credulity among physicians. He cited the Chicago demonstration as an acute case of "the will to believe," an emotional predisposition to act credulously toward phenomena one wanted to believe were true and an overreliance on one's own immediate sensory experience. In doing so, he consciously turned on its head James's justification of faith by exposing some of the weaknesses in this fortification of belief in the modern world. Where James saw a staunch and active commitment to tentative beliefs as the only protection for religious faith in a scientific world, Jastrow worried about how such credulity could permeate and dominate other spheres of life. To James's dilemma of whether there was a greater danger in believing in too little rather than too much, Jastrow responded that the preponderance of credulity required the nurturing of skepticism as the true moral path.[103] Jastrow translated the Huggins case from a test of a particular woman's physiological capabilities into an ethical drama centering on the social "problem of deception."[104]

In passing his verdict on the case, Jastrow argued that his psychological expertise was not only relevant for the scientific evaluation of Huggins, but also ought to be directed toward his colleagues. In this case, supposed men of science had failed to apply sufficiently critical judgment. The significance of the case lay in how it illuminated the psychological predisposition to believe in such a wondrous young woman. What Huggins demonstrated was that the forms of popular enthusiasm that led to uncritical credulity were not limited to the solemn contemplation of religious truth or to the unruly masses attending a commercial spectacle. They also infused the beliefs—and hence the scientific observations—of the most learned professionals. In such instances, Jastrow reasoned, it was insufficient to simply eliminate deliberate frauds, thus permitting doctors to freely accept any given testimony. Physicians needed constant vigilance against the possibility of unconscious deceptive behavior on the part of their patients. Jastrow shifted the focus from Huggins's capabilities toward an investigation of those who observed her. While she was not a commercial fraud, certainly the medical men who professed faith in her were easily deceived.

Conclusion

In the age of muckraking journalism, psychologists produced their own variation on the genre with their exposés of mediums. In these exposés, psychologists did not construe all wonder-workers as the same. One key criterion used in evaluating these individuals' scientific merit and the psychologist's treatment of them was the medium's relationship to commercial exchange. Nurtured by the desire of scientists to police commercial fraud in matters relating to the mind, these encounters came to transform how scientists understood the nature of human experimentation. Those psychologists involved in the exposure of spiritualists adopted a hard-boiled stance where the tough scientist-detective resisted the seductions of the deceitful female medium. Performing one's duty to the truth required that the scientist sometimes operate in an unscrupulous fashion, a lesson learned primarily from stage magicians. In such instances, it was not entirely clear who precisely was being deceitful and who was being deceived. As William James suggested, participating directly in such underhanded activities bestowed upon the psychologist a new set of ethical dilemmas. The psychologist now had the responsibility of managing the deception involved in interacting with participants.

Through these exposures, psychologists worked to redefine who could legitimately participate in the study of the mind. Psychologists largely endorsed a view of the natural world as fundamentally intelligible through the human senses, but with one crucial caveat. Human nature was knowable, but immediate surface appearances were deceptive and one could not rely upon one's preliminary commonsense judgment. As a field of inquiry, psychical research brought this caveat to the forefront. Led by Jastrow, psychologists came to emphasize that disinterested, reasoned judgment was not a natural capacity among the majority of observers, even highly trained scientists and physicians. Their approach contrasted directly with the one advocated by the great showman P. T. Barnum a generation earlier. The showman tried to open up the scene of inquiry to reflect the democratic character of the market. In contrast, these scientists sought to solve the problem of credulity and conviction by removing the evaluation of potentially deceptive persons from the purview of the individual judgments made by members of the public. As George Beard contended in the late 1870s, widespread, sincere testimony could not serve as the measure of truth. Despite their claims about scientific method, psychological exposés relied more on carefully choreographed dramaturgy. In this regard, psychologists followed Barnum in claiming that certain forms of deception might serve the greater good, while condemning the deceits of others. These scientists favored the ingenious trickery characteristic of the stage magician and private detective over mechanized procedures of truth-telling.

The significance of these influences is clarified by contrasting the psychophysics of illusions in the laboratory with the art of debunking the spiritualist in the field. The former certainly required a division of labor between the experimenter and subject, but these were exchangeable roles, governed by an ethos of trust and mutual respect. Carl Seashore's autobiographical comment about the credulous theology students hinted at another potential relationship between experimenter and subject, one that came to fruition in the exposé of psychical wonders. In the psychologists' role as muckrakers, a new ethos emerged that constituted the psychologist as epistemologically superior to their subjects yet bestowed with a morally ambiguous status. During the early twentieth century, this arrangement migrated from the domain of the wonder-workers to that of the mundane laboratory subject as psychological practice would increasingly be governed by a distrust of potentially deceitful and deceivable selves.

CHAPTER FOUR

The Unwary Purchaser: Trademark Infringement, the Deceivable Self, and the Subject of Consumption

In November 1937 newly appointed Supreme Court Justice Hugo Black delivered his first opinion as a member of the nation's highest court. The case revolved around the Federal Trade Commission (FTC)'s attempt to prohibit the Standard Education Society's campaign of selling "free" encyclopedias while charging for a mandatory subscription service for the loose-leaf "supplements" that actually constituted the books' only pages. A year earlier, a lower court had dismissed the commission's worries, arguing that no one was "fatuous enough to be misled" by the company's methods.[1] In his opinion, Black took a different tack when he claimed, "There is no duty resting upon a citizen to suspect the honesty of those with whom he transacts business. Laws are made to protect the trusting, as well as the suspicious." Moreover, that "a false statement may be obviously false to those who are trained and experienced does not change its character nor take away its power to deceive others less experienced."[2]

This judge, widely known as a strict constitutionalist, seemingly rejected the long-standing legal tradition of defining the default legal subject as a prudent, reasonable, and even suspicious man. Instead, he championed the idea that the law ought to assume that such an individual was reckless and easily duped. Furthermore, he advocated his position without citing any legal precedents. When *New York Times* reporter Lewis Wood covered the case, he linked Black's opinion to a speech Franklin Delano Roosevelt had made four years earlier. The president had warned traders in investment securities that his administration would place "the burden of telling the whole truth on the seller."³ Was Black articulating a radical new doctrine whereby the New Deal state would protect the rights of innocent consumers from the manipulations of unscrupulous business interests?

In siding with the FTC, the Court's decision actually reflected an established legal tradition for dealing with cases of trademark infringement and deceptive advertising. A particular understanding of consumer perception and behavior lay at the heart of these decisions. During the same years that scientists toyed with sensory illusions and wrestled with psychics, American judges materialized their own commonsense counterpart to the psychologist's deceivable self, which they called the "unwary purchaser."⁴ In cases of trademark infringement, judges constituted the typical purchaser as an individual with little personal responsibility or cognitive ability. Vision and hearing occupied a paradoxical position in these jurists' psychological theory. On the one hand, judges held that over time these senses stamped habituated associations into the minds of purchasers that governed the decisions a person made. On the other, they understood the momentary experience of seeing or listening as an unreliable basis for willful action. Without the ability to touch an item, inspect it closely, and reflect upon it, a person might make a decision that is untrustworthy. People necessarily relied upon their senses in the fleeting moment of commercial transactions, but frequently found themselves betrayed by their own minds and led into deception.

The legal evocation of common sense signaled two things: first, the judge's attentiveness to the particular circumstances in the case at hand and, second, his reliance on a seemingly self-evident, pre-existing body of knowledge when it came to making inferences about human nature. Although sometimes questioned, this common knowledge about the psychology of consumption had very real effects on how the law defined the boundaries of the legitimate economy. In the nineteenth century, as well as today, there was no Leviathan-like state apparatus intently policing manufacturers and protecting consumers. Although largely suspended in the tradition under discussion, the rule of caveat emptor (buyer beware) remained the most powerful imperative in tort law.⁵ Yet greater awareness of deceptions produced in the minds of purchasers that threatened the profits and livelihood of other free men demanded that caveat emptor be modified in particular instances. Certainly, the notion that it was the government's responsibility to protect consumers as a particular kind of citizenry did not become institutionalized within the American state until the New Deal.⁶ Indeed, in cases involving trademark infringement, deceptive labeling, and related wrongs, judges suspended the rule of caveat emptor in order to protect the interests of established manufacturers and retailers, not those of individual consumers. This inclination became apparent during the so-called butter wars, when state intervention aimed to secure the interests of a large industry from encroachment by a novel product and competitors. A product of the big business–oriented political culture of the Gilded Age, political reformers of later eras would also find use for the idea of the unwary purchaser.

The legal embrace of the unwary purchaser as a careless consuming subject illustrates the limits of the default of "the reasonable man" in the liberal rule of law. For many judges, the "ordinary purchaser" was most likely to be an unwary one who acted quickly, on impulse, and with little reflection. For example, in a 1910 case of trademark infringement, the judge declared, "The law is not made for the protection of experts, but for the public—that vast multitude which includes the ignorant, the unthinking and the credulous, who, in making purchases, do not stop to analyze, but are governed by appearances and general impressions."[7] This consumer was vulnerable to fraudulent commercial activity such as trademark infringement and deceptive advertising deployed by the unscrupulous to impinge upon their competitors' place in the market. In order to protect the reputation and trade of "honest" businessmen, many judges enshrined the customer's right to be careless. This enabled them to prohibit certain trade practices and regulate specific consumer products in the name of preventing the deception of the purchasing public. These decisions particularly shaped the market for a host of inexpensive and semidisreputable goods, such as patent medicines and caffeinated beverages, as well as common household items like flour and soap.

Turn-of-the-century trademark cases constitute an important if largely forgotten chapter in the relationship between science and the law. Challenging the reign of the law's common sense, one of the nation's leading trademark lawyers, Edward S. Rogers, argued that the legal assessment of the consumer's psychology lacked both empirical foundation and standardization across jurisdictions. To achieve these twin goals, Rogers commissioned experimental psychologists to develop a quantitative scale to represent the confusion generated by branded products in the minds of a typical consumer. Judges, however, met this psychotechnic intervention with indifference, favoring instead their own assessment of the purchaser's psychology derived from legal precedent and their own common knowledge about human nature. Despite the absence of authoritative expert witnesses, these cases represented a psychologization of the law insofar as they required that judges reflect upon how the mind functioned within the framework of the marketplace. These cases illuminate how the law and psychology collaborated and competed to constitute the minds of the purchasing public in the "assemblage of the subject of consumption."[8] When historians have discussed questions about consumer psychology during this period, they have focused primarily on the perspective of advertisers.[9] Trademark infringement cases draw out how a legal version of consumer psychology differed from the one deployed by advertising agencies but had equally influential effects. The figure of the unwary purchaser was more than a discursive fiction, an inhabitant of legal texts. The unwary purchaser as a legal object came to occupy a node in the network through which the law constituted the realm of social life known as the "economy."[10] The judicial vision of consumption may not have always reflected the behavior of actual purchasers, but the effects of a judge's reliance on this conceptualization were real, shaping the contours of the legitimate market by adjudicating the legality of retailing and marketing practices and apportioning rights to specific styles of labeling or packages, thereby framing the behavior of both merchants and consumers.

The Legal Assemblage of the Consuming Subject

The unwary purchaser was born out of the booming commercial expansion of the Gilded Age when manufacturers and retailers in a wide variety of industries deployed numerous techniques to gain a privileged position within the marketplace. Manufacturers and merchants placed greater emphasis on the importance of branding and advertising in stimulating greater levels of consumption and rationalizing their relationship with purchasers. The American state played a key role in defining the boundary between the legitimate and the fraudulent. One of the most powerful, if contested, techniques to accomplish these goals was the new laws concerning labeling and trademarks.[11]

The 1870 federal copyright act included clauses that offered established businesses the ability to protect their recognized identity from competitors seeking to profit from their established goodwill with customers. If a manufacturer felt that a competitor infringed upon his established trade due to the similarities in name or image between their products, he had the opportunity to take the competitor to court in the hopes of having a cease-and-desist order issued. Rights to a trademark hinged upon prior use, with the public building the association between product and symbol through long-term exposure.[12] In the 1879 *Trademark Cases* decision, the Supreme Court declared that a trademark did not "depend upon novelty, invention, discovery, or any work of the brain. It requires no fancy or imagination, no genius, no laborious thought." In this regard, the law made a clear distinction between trademarks and patents. Furthermore, the 1870 act defined the standard for infringement in terms of whether an ordinary purchaser was likely to be deceived by the similarity. As the Supreme Court announced in 1893, competitors "have no right, by imitative devices, to beguile the public into buying their wares under the impression that they are buying those of their rival."[13]

In grounding their decisions about what constituted trademark infringement in the credulous nature of the ordinary purchaser, judges frequently cited *McLean v. Fleming* (1878) as a precedent. Not surprisingly, the case concerned a skirmish between two competing "merchants of health." With its secret formulae, testimonial advertising, and startling therapeutic claims, the field of proprietary medicine often furnished the grist of legal regulation.[14] In 1834 Charles McLane, a Virginia physician, began to manufacture and sell liver pills in labeled, wooden boxes. Through a series of sales and inheritances, two Pittsburgh pharmacists, John and Cochrane Fleming, acquired the exclusive right to use the trade name, "Dr. McLane's Liver Pills" in 1853. Five years later, the remedy's immense popularity led the Fleming brothers to announce that they were dedicating the entirety of their pharmaceutical business to its manufacture and sale. Following the Civil War, their advertisements were a ubiquitous feature in the nation's periodicals. Appealing in particular to farmers in the South and the West without regular access to a physician, they promised that their pills offered relief for a host of digestive or "bilious" ailments. These adverts also cautioned readers to ensure that they obtained the genuine McLane liver pill and avoided imitations. By 1876 the firm possessed thirty employees and boasted of annual sales of approximately 1.5 million pillboxes.[15]

To promote brand recognition, the Flemings gave their pillboxes a distinctive black label with fine, diagonal lines engraved into it. The company wrote the name "Dr. C. McLane's Celebrated Liver Pills" on the label in white lettering. In 1872 the surviving Fleming brother, Cochrane, filed a case to restrain James McLean from allegedly infringing upon his company's established trademark. Since 1849 McLean had manufactured and marketed a similar laxative under the trade name of "Dr. McLean's

FIGURE 4.1 McLane's Liver Pills trade card featuring a potentially unwary purchaser. Author's collection.

Universal Pills" from his base in St. Louis, Missouri. In 1866 McLean adopted a pale red label with white lettering. The court had to decide whether McLean's trade infringed upon Fleming's exclusive right to use the name "Dr. McLane's Liver Pills." In 1878 the case reached the Supreme Court of the United States.

The case centered on establishing a legal standard for the degree of similitude necessary to constitute infringement. At what point did two distinctly branded commodities come to so resemble one another that the rights of the original proprietor had been transgressed? In its decision, the Court determined that a litigant did not have to prove that an exact similitude existed. Rather, the appearance only had to be sufficiently similar so that it was "likely to mislead one in the ordinary course of purchasing the goods."[16] Such cases required that the presiding judge visually inspect the goods, contemplate the sound of the trade name, and estimate the perceptions of the purchasing public. If the judge determined that an ordinary purchaser was likely deceived by the similarity, then, legally speaking, infringement had occurred. Malicious, deceitful intentions were not necessary, only deceptive effects.

McLean v. Fleming made consumer psychology central to the arbitration of trademark cases, and later decisions determined that this ordinary purchaser was a truly unwary one. For example, *Wirtz v. Eagle Bottling Co.* (1892) raised the issue of whether the differences between the labels of two competing beer distributors would be noticeable to the eye of an ordinary purchaser. The judge grounded his decision in the "common knowledge" that the ordinary buyer was not likely to exercise particular caution when purchasing such relatively inexpensive goods. As a result, it was incumbent to determine whether the differences between the two labels could be perceived by a mere "glance" rather than by sustained contemplation.[17] In *Fairbank v. Bell* (1896), the appeals court overturned the lower court's decision that prudent customers using their eyes, ears, and common sense would be able to detect the difference between two brands of powdered soap. Judge E. Henry Lacombe noted that the ordinary purchaser had neither the intelligence nor the experience of a learned judge in these matters. Consumers tended to associate the goods they habitually purchased with their general appearance rather than with a specific name. As a result, if a retailer mingled the two goods on the shelf, unsuspecting and nondiscerning purchasers would likely be deceived. This justified the intervention of the courts. Not every judge concurred with this assessment of the purchasers' impoverished cognitive abilities. In an 1898 case involving rival tobacco brands, the judge spoke of what a person could observe by the mere glance of the eye. He decided that in this case there could be "no reasonable ground to apprehend that any man of ordinary intelligence would be misled."[18]

Despite such disagreements, the frequently cited opinion of James Jenkins in *Pillsbury v. Pillsbury* (1894) laid out most fully the unwary purchaser's presumed psychology. Jenkins's activities during the Panic of 1893 had earned him a reputation as a strong, if not notorious, defender of large-scale commercial interests. Most troubling, a Milwaukee grand jury indicted him for bank wrecking after he and his fellow directors attempted to remove funds from the insolvent Plankinton Bank.[19] In December of that year, he issued a controversial injunction forbidding the workers of the North Pacific Railroad Company from striking due to a 10–20 percent decrease in their wages after the company had been placed in receivership.[20] These combined actions led to calls for his impeachment, a fate he narrowly avoided. Instead, Grover Cleveland promoted him from the district to the circuit court that same year. The influence of Jenkins's opinion most clearly illustrates how the unwary purchaser doctrine emerged from a jurisprudence bent on advancing commercial interests with little concern given to issues of consumer protection.

The *Pillsbury* case revolved around a dispute between two Midwestern flour businesses. In 1869 Charles Pillsbury had entered into Minneapolis's burgeoning flour industry by purchasing a large share in a mill. Ideally located to establish direct contact with the farms producing the majority of the nation's wheat, Pillsbury introduced a host of technological innovations to improve the refinement of the grain. Three years later, he continued the process of vertical integration by directly promoting his flour to consumers. He branded his product as "Pillsbury's Best XXXX," claiming it constituted the highest-quality flour available on the market.[21] In 1889 an English syndicate bought his company but maintained the widely recognized Pillsbury brand. Four years later, their lawyers charged that two Illinois entrepreneurs, L. F. Pillsbury and Ephraim Hewitt, conspired to use the former's name to fraudulently profit from the goodwill that their company had built with purchasers over the course of two decades. Not only were they using the fortuitous name, but their packaging closely emulated the appearance of Pillsbury's Best.

In his decision, Jenkins highlighted the importance of psychological qualities such as attention, memory, affect, and recognition, as well as questions of personal responsibility:

> A specific article of approved excellence comes to be known by certain catchwords easily retained in memory, or by a certain picture which the eye readily recognizes. The purchaser is required only to use that care which persons ordinarily exercise under like circumstances. He is not bound to study or reflect; he acts upon the moment. He is without the opportunity of comparison. It is only when the difference is so gross that no sensible man, acting on the instant, would be deceived, that it can be said that the purchaser ought not to be protected from imposition.[22]

According to the law, the purchaser's psychology mixed deeply etched impressions with a fickle inattention and an attraction to the mere surface appearance of things. The habitual exposure to visual images, in the form of a particular brand's distinct markings, left a deeply ingrained impression on the individual psyche, like a seal pushed into hot wax.[23] Through repeated exposure to specific visual markings, an impression became ingrained in the subject's memories, making them familiar and readily recognizable to the eye and ear. If a brand was satisfactory on previous occasions, the mere intimation of its visual or auditory appearance was sufficient to entice the purchaser's desire. It was assumed that the purchaser was denied the opportunity to compare among brands. Rather, he was perceived as acting in the moment, engaging in speedy transactions that involved little to no inspection or reflection. During the act of selecting goods, this subject's attention span was indeed lessened.

The use of such language did not make judges a ready audience for the testimony of psychologists, alienists, or neurologists. These professionals found it difficult to get courts to admit their evidence to demonstrate that a criminal suspect was insane and, hence, not morally responsible.[24] No one questioned the sanity of the unwary purchaser; he was supposed to be an individual possessing ordinary mental capability and sensibilities. These laws were geared toward nonpathological persons and forms of consumption.[25] The issue was whether ordinary purchasers had the opportunity to exercise their good judgment and whether the differences between products were sufficient to be detectable to a person with average observational abilities.

As a legal subject, this consumer was relatively free of responsibility, to the point where many judges argued that he had the right to be careless. This emphasis on carelessness contrasted starkly with the dominant vision of the liberal subject in nineteenth-century jurisprudence. In most contexts, lawmakers underscored the autonomy, independence, and self-reliance of the law's reasonable man. Citizens were expected to act prudently and embrace personal responsibility for life's risks as an emblem of their freedom.[26] Though sane, the legal purchaser-subject could be inattentive, frivolous, lacking good memory, and at the mercy of the retailer, depending on the judge involved. In one case, the judge went so far as to assert that "the ordinary retail purchaser of soap powder for consumption is not usually of a high degree of intelligence."[27] These psychological qualities made the purchaser vulnerable to the frauds and deceptions of the marketplace. In order to protect the legitimate businessman's income, the court could not assert purchasers' full responsibility for their actions.[28]

In their decisions, judges invariably left the gender of the unwary purchaser vague. In most circumstances, consumption in terms of providing for the home was usually gendered female. Certainly many of the commodities that came up in these cases would affirm this reading: female apparel, food, and other household goods. While the gender politics of consumption likely contributed to the image of the purchaser as unwary, it cannot account for the doctrine as a whole. If anything, in the wording of these opinions the purchaser lacked a specific gender and judges interpreted consumption and advertising as drawing out the deceivable self in everyone. Below the surface of the autonomous, self-controlled, liberal self brewed passions and a potential for deception that the commercial world engaged.

This judicial understanding of the consuming subject had a specific function within these decisions. The ordinary purchaser was seen as lacking the common sense, lived experience, or accrued knowledge that would help navigate the deceptive terrain of the marketplace. That these characteristics were the qualities judges drew upon in crafting their decisions only highlighted the gulf between their perception of consumer behavior and their own self-understanding.[29] This move allowed judges to focus exclusively on deciding the visual and auditory similarities among commercial packaging, labeling, and branding. Lawmakers invoked the unwary purchaser in order to cut off empirical investigations involving differences in perception and psychological ability among consumers. The purchasers of these decisions were lacking in social distinctions such as gender, class, or race. Actual consumer behavior was presumed rather than introduced as a form of evidence. Although granted subjectivity within this legal network, consumers were not given agency. They were the law's implicated actors.[30]

Indeed, petitioners did not have to provide actual customers to testify that they had been deceived by the imitator. Rather, the application of the law centered on how the judge determined the probability that a customer may have been deceived; even when the testimony of deceived purchasers was available, the courts did not place great weight upon it. For example, in *McCann v. Anthony* (1886), a case where the plaintiff provided two customers who swore they had been deceived by the competitor's similar packaging, the court did not find their testimony to be evidence of critical importance. Instead, the presiding judge relied upon his own judgment of the competitors' labels to determine whether it was probable that the purchasing public would be misled. In such cases, the judges held that in all probability the competitor could provide alternative testimony. William Kneeland Townsend dismissed a different form of testimony in a case concerning two stores selling similar lamps. He ignored the evidence presented by designers, arguing that their expertise was irrelevant since it was in the production of the goods and not in how the ordinary purchaser was likely to perceive them.[31]

These potential witnesses posed another obstacle insofar as the law was crafted for the protection of a default reasonable person.[32] Instead, when it came to trademark cases, the courts were most likely to view purchasers as "extraordinarily foolish persons, devoid of common sense" and as outside the purview of this highly regarded legal standard.[33] This bracketing off of actual consumers fit with the primary aim of these legal decisions. Their goal was not really the protection of the consuming public from fraudulent merchandise or harmful commodities.[34] *Williams v. Brooks* (1882) made the intention of the court explicit:

> This proceeding is not primarily to protect the consumer but to secure to the plaintiffs the profit to be derived from sale of hair-pins of their manufacture to all who may desire and intend to purchase them. It is a matter of common knowledge that many persons are in a greater or less degree careless and unwary in the matter of purchasing articles for their own use.[35]

Because common knowledge dictated that the ordinary purchaser was careless, it was necessary for the state to police imitators to protect the profits of honest businessman.

These Gilded Age cases of trademark infringement made explicit the psychological aspects of consumption decades before such appeals became common within the advertising industry. For much of the nineteenth century, advertisements were expected to communicate to the purchaser in an informative manner and they constituted the typical consumer as a "rational man." In the first decade of the twentieth century, applied psychologist Walter Dill Scott offered a vision of humanity as creatures easily swayed by suggestion. Although the advertising trade made a definitive shift from an appeal to consumer rationality to emotional associations around 1910, the question of consumer psychology became a prominent legal concern in the late nineteenth century before it became a widespread marketing technique.[36]

Legal observers soon noted that judges failed to adhere to a standard measure when contemplating the psychological makeup of the ordinary purchaser. Where one judge may have seen the conditions necessary to induce deception among an average assortment of individuals, another may not.[37] In terms of labels and packaging, surely objectively determined points of visual similarity could be mapped out and infringement determined based on the number of points of resemblance? Judges were skeptical that they could ever establish any such mechanically applied standard for every case. In *McLean v. Fleming*, the case whose precedent did much to shape later decisions, the justices made the inability to arrive at a universally applicable standard explicit: "What degree of resemblance is necessary to constitute an infringement is incapable of exact definition, as applicable to all cases. All that courts of justice can do, in that regard, is to say that no trader can adopt a trade-mark, so resembling that of another trader, as that ordinary purchasers, buying with ordinary caution, are likely to be misled."[38] Commonsense knowledge about the behavior and cognitive abilities of the average consumer as determined by judges on a case-by-case basis was seen as the more desirable and feasible alternative to a systemically applicable standard.

These decisions stand in contrast to an influential narrative about the emergence of objectivity as a political ideal in modern society. Science studies scholars have stressed the social power of such tools as quantification, standardization, and precision in negotiating trust and certainty among heterogeneous actors.[39] Despite the attempts by certain lawyers and psychologists to produce a standard for deception informed by experimental psychology, a more precise definition of the ordinary purchaser's attention or memory was never achieved. Instead, with the expanding police powers of the state during the Progressive Era, the commonsense knowledge about consumption became institutionalized within a government bureaucracy.

The Police Powers to Prevent Fraud and Deception

The legal assemblage of the consumer as an unwary purchaser prone to deception first articulated in cases of trademark infringement soon became enrolled in the police powers of the state. These powers entail the state's authority to regulate individual behaviors within its territory in the name of the common good and a "well-regulated society." Lawmakers advanced this regulation in the name of protecting public safety and morality.[40] Initially, manufacturers took advantage of the space within tort law that permitted them to seek relief from imitators whose acts, although wrong, were not criminal. In the 1880s state and federal legislatures extended their police powers to include the prevention of fraud and deception against purchasers. The notion of an easily deceived purchasing public functioned well within a regulatory tradition that stressed the state's role as patriarch over its citizens.[41]

The animal-fat-derived butter substitute, oleomargarine, was the commercial object repeatedly used to define the police power to prevent the presence of deception in the mind of the purchasing public. A rather unremarkable food today, margarine provoked much controversy in the second half of the nineteenth century. In the 1860s the French government commissioned Hippolyte Mège-Mouriès to investigate how chemistry might contribute to the amelioration of the domestic economy. In an attempt to make the agricultural sector more lucrative, his particular task was to transform beef fat, a waste product of the abattoir, into something useful. His invention hinged on the observation that cows deprived of foods containing fats still produced milk that generated cream. He concluded that the only potential source of milk's fat was not some external nutrient, but rather was the product of the cow's own bodily processes. If the animal's organic chemistry could make such a conversion, then, Mège-Mouriès reasoned, he could reproduce it on an industrial scale. In 1869 oleomargarine first appeared in France, presented as a cheaper alternative for the laboring classes unable to afford the soaring prices of butter. With the addition of dyes, the new commodity closely resembled the texture, taste, and color of butter.[42]

From the beginning of its manufacture, margarine had a contentious history. Portraying it as a chemical forgery of true butter, members of the dairy sector and their political allies denied the product's origins in organic matter. They played on fears associated with margarine as a product of waste, that it was an adulterated good that harmed one's health. Its boosters insisted that the industrial synthesis simply mimicked the biochemistry of the cattle's body. They held that "oleo" ought to be considered a natural product. Margarine was introduced into the United States in 1875, but in 1881 Missouri became the first state to prohibit the manufacture, sale, and possession of the substance. Such legislation was quickly copied and Maine, Minnesota, Wisconsin, Ohio, Pennsylvania, and Michigan passed similar laws in 1885, with New York having adopted a statute the previous year. The geography of this legislation was notable. Although milk's perishable status ensured that the nation's five million cow owners were distributed throughout the country, commercial production concentrated in New York and the Midwest. Family farmers did not constitute the entirety of the butter industry as it also included factory owners who processed the milk into consumer goods and boards of trade that exchanged and promoted the product. With the invention of the centrifugal cream separator in 1878 and the industrialization of dairy production itself, this rather diffuse network of farms became increasingly integrated. This commercial network soon mobilized for political ends when the popularity of margarine risked usurping butter's place on the table.[43]

In spite of their successes at the state level, members of the dairy lobby found the existing legislation inadequate. They formulated the recommendations that formed the basis of the 1886 federal Oleomargarine Act. Invoking the state's police powers to protect its subjects from harmful products, the federal government passed a series of taxes specifically targeted to discourage the sale of margarine. While the policing of margarine as a dangerous substance may seem amusing in retrospect, contemporaries saw the debate as representing one of the most pressing governmental issues of the day. In 1887 one political scientist described the act in rather apocalyptic terms, as the opening salvo of the government's attempt to use its police powers to control the everyday, private choices of the citizenry.[44] Without embracing such pessimism, this legislation did successfully alter the scope of the state's police powers by opening a new avenue for governing consumption.

The government justified its opposition to the sale of margarine on two different grounds: first, that the good posed a health risk to the consumer and, second, that the commodity deceptively imitated another product, butter. States originally couched their actions in the claim that they should benevolently if paternalistically protect their citizens from the risks of a potentially hazardous consumable. Legislators emphasized their role in ensuring public health rather than the less-secure claim of preventing fraud and deception. The Supreme Court found this was a legitimate exercise of a state's police power and did not interfere with the liberties guaranteed under the Fourteenth Amendment in *Powell v. Pennsylvania* (1888).[45] Despite this legal victory, the health claims were always shaky and did not hold up to much scrutiny when dealing with properly processed margarine.

As it became increasingly difficult to justify the constraint of trade on strictly medical grounds, legal thinking came to stress the likelihood of deception in the minds of the purchasing public. Legislators argued that what needed protection was not the person's physical health, but the freedom of individuals to choose the product they actually desired. Margarine's deceptive appearance risked usurping this right. New laws stressed how ill-informed customers would likely fail to distinguish between genuine butter and the "oily menace." In the face of this perceptual confusion, the butter-wanting person risked accidentally acquiring the counterfeit. To preserve the freedom of purchasers, governments claimed they needed to restrict the options available by limiting access to a dubious if not dangerous good.

When the next major margarine case came before the Supreme Court, in *Plumley v. Massachusetts* (1894), the argument used by the Court to endorse the policing of margarine had shifted accordingly. The Massachusetts law specifically prohibited the coloring of margarine to give it a yellow, buttery appearance. The *Plumley* decision drew particular attention to the appearance of the substance and the ways in which it was labeled and marketed. Justice John Marshall Harlan explicitly invoked the "unwary purchaser," now familiar from trademark infringement cases:

> The real object of coloring oleomargarine so as to make it look like genuine butter is that it may appear to be what it is not, and thus induce unwary purchasers, who do not closely scrutinize the label upon the package in which it is contained, to buy it as and for butter produced from unadulterated milk or cream from such milk.[46]

The Court could not rely upon the easily deceived purchaser to carefully consider the information provided by the label. What made margarine an object amenable to governmental restriction was its propensity to induce confusion in the consumer through its deceptive appearance. As Harlan noted in his opinion, the novelty that the case presented to the Court was whether the state's police powers included the protection of the public against deception and fraud in the sale of commercial goods, specifically food products. *Plumley* enshrined a new kind of police power: to prevent deception and fraud. With two justices dissenting, the Court affirmed that the state possessed such a police power to protect persons from being deceived.

Deception in an Age of Federal Regulation

This state project of policing deception in the name of public health and consumers' freedom culminated in the passage of the Pure Food and Drugs Act (FDA) of 1906. A number of muckraking tracts provided the immediate impetus for the creation of an administration to monitor the purity of these products. Upton Sinclair's exposé of the meat-packing industry, *The Jungle* (1906), detailed the struggles of the Lithuanian immigrant Jurgis Rudkus and his family to survive the deceptive landscape of Chicago's packing yards, which promised plenty but failed to deliver. The novel included graphic descriptions of the unsanitary working conditions in the slaughterhouses. His novel brought attention to the squalor in which the nation's meat products were handled and resonated with Samuel Hopkins Adams's contemporaneous reportage on the patent medicine trade in the pages of *Collier's* magazine. Contemporaries credited these works for stimulating the greater regulation of consumer goods, but as the story of oleomargarine indicates, interest in policing "adulterated" goods had been already a concern among both state and federal governments for a number of decades. Identified as a Progressive move against corporate trusts, the margarine precedent illustrated how large commercial interests frequently played a role in shaping antiadulteration legislation in order to eliminate competition.[47]

Illinois factory inspector Florence Kelley was another proponent of the 1906 legislation. *Some Ethical Gains through Legislation* (1905) advanced her vision of a politically progressive citizenry. Kelley wanted to build a viable cross-class political alliance centered on the unifying experience of consumption. She insisted that "the older recognized right of the purchaser is to have his goods as they are represented." For Kelley, the ordinary purchaser was an unwary one not because of a cognitive failing, but because the complexities of the modern industrial system blinded people to the conditions of production. The immigrant laborers in the sweatshops she inspected possessed local knowledge about how they produced garments, but, due to a *willful ignorance* on the part of the purchaser, such information was not held in common. She did feel that typical purchasers were at the mercy of the producer and distributor of goods, but this was because they voluntarily chose to act with credulity. Purchasers abetted the adulteration of goods because they preferred the illusion that what they consumed came from distant lands. They lacked only the "enlightened imagination" to inquire into the production of the goods they purchased.[48]

For Kelley, the 1906 act was as much about refashioning the purchaser into a political subject capable of responsible decisions as it was about regulating and restraining large corporations. With better political machinery for identifying both the content of goods and the conditions of their production, a wary, politically mobilized citizenry could come into being. She bound together the right not to be deceived by manufacturers with the political duty to become more knowledgeable about the things one consumed. This entailed a very different understanding of what made the purchaser unwary than the courts offered. Kelley believed that greater clarity and transparency in labeling goods would unleash a canny purchaser truly exercising a freedom of choice. The creation of this freer and more prudent purchaser required governmental policing of industry. Kelley held that political progress and social justice required the democratization of information and the distribution of responsibility onto the individual consumer citizen.

Despite Kelley's hopes for this legislation, its applications after 1906 adhered more to the common law tradition's view of the cognitive abilities of consumers. The most visible attempt to enforce the act occurred in 1911, when government agents seized a major shipment of Coca-Cola in Chattanooga, Tennessee. Created by John Pemberton of Covington, Georgia, as a temperance-friendly substitute for alcohol in 1885, the drink became a nationally recognized brand soon after Asa Griggs Candler secured exclusive rights to its manufacture in 1888. Harvey Washington Wiley, the federal government's top chemist and an architect of the 1906 act, strongly objected to the marketing of the drink to children and struggled for years to curtail its production.[49] Although geared primarily toward questions of adulteration, Wiley offered a broad interpretation of the act's police power to encompass the long tradition of governing in the name of the public's health. He attempted to prove in court that the prohibition of Coca-Cola was justified since the drink constituted a dangerous and harmful substance due to its high levels of caffeine. Wiley believed that even with more accurate labeling the consumer remained largely ignorant about goods and open to manipulation. Where such claims were typically mobilized to bolster the position of established manufacturers and distributors in trademark cases, in Wiley's hands the charges had a distinctly anticorporate valence.

The company prepared a robust response, including the testimony of a young New York–based psychologist, Harry Hollingworth. He testified to the psychological and physiological benefits of caffeine such as heightened energy, attention, and alertness. To provide this testimony, Coca-Cola commissioned and funded Hollingworth's research into the effects of caffeine on mental and motor performance under laboratory conditions. To maintain his credibility as a witness, he designed his experiments so that he was personally unaware of which subjects were performing under the influence of caffeine and which were not. Wary of the tendency for observations to meet expectations, he did not trust his own ability to withstand the influence of his wealthy patron. He opted for a form of intentional ignorance to guarantee the validity of his results. The company easily won its case in a trial held in a sympathetic southern jurisdiction. The case was probably more important for Hollingworth, who benefited greatly from the commercial patronage and built upon this success to become a leading applied psychologist.[50] At Columbia University's Applied Psychology Laboratory, he developed a number of experiments on the psychology of advertising. It would not be the last time a Columbia psychologist designed experiments to meet Coca-Cola's legal needs.

The same year as Wiley's defeat, the federal Supreme Court decided, in a case concerning a purported cancer cure, that the 1906 act did not cover all forms of deceptive advertising. In *United States v. Johnson*, the Court determined that the act's phrasing limited its scope to the mislabeling of ingredients. Unlike in trademark infringement cases between competing commercial entities, the Court did not invoke the specter of the unwary purchaser. The majority's opinion defined "deception" more narrowly. They argued that the act covered only deceitful declarations in the form of false material claims about a product's chemical content and not the creation of deceptive impressions in the mind of the purchaser.[51] In the wake of the *Johnson* ruling, Adams penned a new exposé series for *Collier's* detailing "the tricks of the trade" and the connections among "the law, the label, and the liar."[52] President Taft demanded that Congress amend the original act. The resulting Sherley Amendment (1912) clarified that the act did indeed include the prohibition of false statements about "the curative or therapeutic effect" of drugs and not merely their substance. The amendment expanded upon the kinds of claims that might deceive, but it further entrenched the notion that manufacturers had to intentionally mislead their customers in order to face prosecution. Frustrated with these constraints, Wiley resigned his government office to pursue his campaign through a regular column in the pages of *Good Housekeeping*.[53]

Despite the setbacks of the Chattanooga trial and the *Johnson* decision, the federal government's use of police powers to prevent deception continued to both grow and take on an increasingly bureaucratic form. The federal government created the Federal Trade Commission (FTC) in 1914 to police the nation's commercial trusts. Instead of the federal government having to constantly pass legislation to prohibit specific practices or goods, the FTC existed as a standing body to monitor interstate commerce. The FTC was given the mandate to prevent "unfair methods of competition." The FTC commissioners soon interpreted this rather vague definition to include the prevention of the use of deception in marketing and advertising.

Although a number of legal challenges contested the FTC's authority, it soon received acknowledgment of having the proper power to police deception and not merely break up monopolistic trusts. The commission's cases most frequently centered on goods directed at the consumer rather than within industry.[54] Like their counterparts in trademark infringement cases, these individuals were not expected to have an expert's knowledge about the products they purchased, nor to have the time or attention to give the products close inspection. The federal court outlined the power and knowledge to be deployed by the commission in its policing deceptive or fraudulent commercial activities when the Sears, Roebuck Company challenged its authority in 1919. There the court held:

> The Commissioners are not required to aver and prove that any competitor has been damaged or that any purchaser has been deceived. The Commissioners, representing the Government as parens patriae, are to exercise their common sense, as informed by their knowledge of the general idea of unfair trade at common law, and stop all those trade practices that have a capacity or tendency to injure competitors directly or through deception of purchasers.[55]

Like its nineteenth-century court-based predecessors, the FTC's authority derived from the state's role as patriarch over its citizenry. In the absence of an exact standard for what constituted unfair trade, commission agents were to deploy their "common sense" in detecting and policing deceptive and fraudulent commerce. There were no fixed standards to determine whether a fraud had occurred, but rather its definition hinged upon the perceptions and beliefs of those involved in the transaction. The deception in misleading advertisements was based on the likely impression made in the mind of the consumer rather than by an external measure of physical similarity.[56]

When the commission explored the scope of its police powers in a test case, it grounded its authority in the existence of an unwary purchaser easily deceived. In 1927 it issued a cease-and-desist order against the Indiana Quarter Oak Company, prohibiting it from using terms such as "mahogany" or "Philippine mahogany" to describe nonimported wood. When the company challenged the order and a federal court heard the case, the deciding opinion drew upon the same logic as earlier tort cases. The judge argued that it was not necessary for the commission to prove intent. Rather, unfair competition occurred when "the natural and probable result of the use by the petitioner of such woods was deceptive to the ordinary purchaser and made him purchase that which he did not intend to buy."[57] Once again, the judge constituted the purchaser as an individual without proper knowledge forced to act against his own volition.

The creation of agencies like the FTC meant that a particular configuration of the deceivable self became institutionalized in the nascent regulatory state. This bureaucratization did not result in greater precision when it came to defining the unwary purchaser. Where previously judges determined for themselves whether the deception was a probability on a case-by-case basis, a similar kind of judgment was transferred onto the officials of the FTC. The language of the commission's act concerning what constituted unfair trade practices was vague and defined largely through a series of cases and court challenges. This was not the only possibility available at the time. Just as the notion of the unwary purchaser became institutionalized within government bureaucracy, a group of lawyers and psychologists challenged the utility and empirical basis of the concept from the perspective of the laboratory.

From "Haunting" Grocery Stores to the Quiet of the Laboratory

One legal professional in particular highlighted the overwhelmingly psychological character of the common-law understanding of the purchasing subject. Starting in 1909 Edward Rogers, one of the nation's leading copyright and trademark lawyers, wrote a series of articles discussing the law of infringement. As the series progressed, his approach shifted from an analysis of formal legal doctrine toward greater attention to the findings of perceptual psychology. Like many of his contemporaries of the Progressive Era, he saw in the sciences the potential remedy for ineffectual political institutions. He recognized that the legal arbitration of unfair business competition had taken on a distinctly psychological quality without the authorities responsible for making these decisions actually consulting the findings of this science.[58]

Rogers quickly emerged as the most prolific critic of the law concerning unfair competition. His criticisms focused on the figure of the unwary purchaser and revolved around issues of accuracy and standardization. On the one hand, he argued that these decisions did not take into consideration variations among the purchasers in terms of their intelligence, education, age, or station in life. At the same time, the application of the law lacked consistency, as various judges would attribute differing mental capacities to the purchaser. The courts simply assumed rather than investigated the mental processes at work during the act of buying a good. Rogers feared that certain judges had a tendency to read their own ability to detect imitations in the courtroom onto the purchaser, bestowing upon him a greater care and power to discriminate among goods. The result was an uneven application of the law. Where one court could argue that insufficient proof was offered that ordinary purchasers were actually deceived by the infringement, another might insist that the law was for the protection of the careless.[59]

Rogers argued that psychology, in contrast to the court system, had accrued considerable evidence about the actual mental performance of ordinary people. Rogers initially developed what he called a "practical psychology," which derived from his "haunting" grocery stores and carefully observing consumer behavior as shoppers purchased goods.[60] In an era when the majority of social surveyors still focused their attention on "problematic," minority communities (what later social scientists would call the "underclass"), Rogers anticipated a turn toward researching the propensities of the purchasing public as a population.[61] Responding to Rogers's frustration with a lack of scientific data on his specific topic, the editors of the *Illinois Law Review* urged him to write the famed psychologist Hugo Münsterberg.[62] When he did, Rogers complained that he was "forever running into that mythical person, the unwary purchaser, in lawsuits." Although the topic had yet to be directly addressed, the Chicago lawyer hoped "that the work in psychological laboratories had tested the capacity of enough normal people so that the unwary purchaser could safely be regarded as a real person and not a judicial abstraction."[63]

To address the lawyer's needs, Münsterberg designed a series of experiments to pinpoint the likelihood of an ordinary purchaser being deceived by the similarity between two commercial products. He published the results not in a psychology journal, but in the magazine *McClure's*.[64] He later included an account of these experiments in his highly regarded book, *Psychology and Industrial Efficiency* (1913). There he argued that the law concerning the deception of consumers lacked a precise measure for determining "the exact point at which the similarity becomes legally unallowable." In Münsterberg's formulation, the law's understanding of the self became mere "general conceptions" without basis in empirical reality. Furthermore, he interpreted the adversarial nature of the legal system, with its various contestations over the interpretation of the law, as merely "endless difficulties."[65] Münsterberg offered his experiments as an alternative source of authority to the common knowledge used by judges in cases involving questions of deceptive imitation. His goal was to assess which particular kinds of imitations would most likely confuse the eye or deceive the ear of the average person. These experiments on recognition sought to transport the mind of the purchaser from the chaotic world of the marketplace into the quiet space of the laboratory where mental processes could be isolated, repeated, and quantified to produce the aggregate behavior of the average human subject.

The experiments that Münsterberg carried out required that his laboratory subjects detect the subtle differences in various pictures and he opted to use mass-produced postcards to represent trademarks. He asked his participants whether they could perceive the variations in images viewed over a short period of time. Initially, he exposed the subject to two sets of six postcards, each viewed one at a time for five seconds through a shutter. The second set of postcards was identical to the first except for a variation in detail in one of the card's images. For example, a picture of a church may have been substituted for a similar building, or one photograph would picture a vase while its substitute would have a vase with flowers. Under this arrangement, the psychologist could manipulate a number of factors to probe the threshold for deception. He could reduce the exposure time to the image to two seconds, alter the interval between viewing the two sets of postcards, or augment the differences between the two images. Because of the control it offered, Münsterberg held that the psychology laboratory should serve as the site for establishing a standard for litigation based on the percentage of subjects deceived under these conditions. Psychologists could precisely measure the degrees of similarity beyond which the experimentally produced and statistically refined average observer would not be able to perceive the difference.[66]

Münsterberg also encouraged his graduate students Gustave Feingold and Harold Burtt to pursue and refine his experiments.[67] During these years, the Harvard psychology laboratory became a productive site for an array of psychotechnical surrogates for legal common sense. For example, William Moulton Marston, another Münsterberg student, served as an experimental subject in the laboratory's investigations into recognition while conducting his initial experiments that led to the lie detector.[68] Similarly Burtt contributed to research on deception tests and the psychological investigation of trademark infringement. These interconnections exemplify the cross-fertilization of psychological approaches toward criminology and industrial efficiency that occurred within the particular space of the Harvard laboratory. Such dual interests were a prominent feature of Münsterberg's own career, but this twin focus also shaped the applied psychology he fostered. For example, in studies on how changes in the environmental context of observation affect the capacity to recall and discriminate among already perceived pieces of information, Feingold suggested that the experiments designed to elucidate the problem of trademark infringement had further legal application for rationalizing the problem of eyewitness recognition of potential suspects in a police lineup. Such tests would serve as a counterpart to fingerprinting in assisting the detection of suspects without police records.[69] Psychological means of detecting and governing deceptive persons and deceptive things were coproduced in conversation with one another in the same space.

While applied psychologists privileged the laboratory as a manageable space, experimenters needed to produce data that would seem to recapture the experience of purchasing in the eyes of their anticipated legal patrons.[70] This required that the experimenters create situations that would be understood as analogous to the market. Feingold expressed worries about adequately simulating the conditions of the "crowded business street" or the "teeming department store" where "thousands of stimuli knock at the door of consciousness." By exposing his subjects to a large number of trade names in a short, fixed period of time and forcing them to recall the names and their pattern, he felt that he was "able to produce in our subjects the states of mind that they experienced on the city streets."[71] These psychologists sought to produce a carefully choreographed and precisely measured state of confusion.

Trademarks and the Psychology of Recognition

It does not appear that Münsterberg ever served as an expert witness in a trademark infringement case and Rogers later approached another scientist to offer testimony about the ordinary purchaser's mental processes. In 1915 Rogers acquired the services of Richard Paynter, a member of the Applied Psychology Laboratory at Columbia University, to assist in preparing a brief opposing the registration of the name "Chero-Cola" as a trademark. Rogers never said directly why he asked Paynter, but the choice of a Columbia graduate student does not seem accidental. Paynter's mentors, Hollingworth and Edward K. Strong, had done considerable work on industrial topics. Moreover, Hollingworth had served as an expert witness for Coca-Cola during the 1911 Chattanooga trial. With such precedents in place, Paynter's interest in the psychology of branding and brand recognition was not too surprising. Rogers requested that Paynter introduce no procedural innovations into these experiments, insisting on commonplace methodologies that were uncontroversial within the discipline. In doing so, Rogers hoped to use scientific standards and authority to bolster his client's claim that a competitor's trademark infringed upon its recognizable brand.[72]

Among American corporations in the early twentieth century few were as strident in protecting their trademark as Coca-Cola, a stance that frequently led to extended litigation. Born out of Asa Candler's evangelical zeal to reestablish the South on the national stage through entrepreneurialism, the beverage had become an omnipresent feature on the American commercial landscape by 1900. As company executive Harrison Jones noted in 1923, its marketing goal was "to make it impossible for the consumer to *escape* Coca-Cola."[73] The brand posed a particular challenge to jurists due to the tremendous success of this strategy. As became apparent in the Supreme Court case *The Coca-Cola Company v. The Koke Company of America* (1920), the company's saturation of the market had led to its trade name becoming a generic descriptor of all cola drinks. In that decision, Oliver Wendell Holmes Jr. defended the rights of the Coca-Cola Company, famously declaring: "It hardly would be too much to say that the drink characterizes the name as much as the name the drink."[74] Another Georgia-based pharmacist, Claude A. Hatcher, began marketing Chero-Cola in 1905 and it quickly grew to prominence as a regional alternative to Coke. The two corporations soon clashed and in 1914 Coca-Cola contested the registration of the Chero-Cola's trademark with the federal Patent Office.[75] Rogers, alongside Coca-Cola's in-house legal counsel Harold Hirsh and Francis Phelps, recruited Paynter as part of the resulting litigation.

In his deposition for the court, Paynter described his specialization as "the psychology of recognition, memory, and confusion and psychological matters relating to advertising."[76] He offered two classes of experiments to elucidate the memory, attention, and cognition of the average consumer. Both investigated the process of name recognition, one focusing on sight and the other on sound.[77] While in the past judges had dismissed direct testimony that any one customer had been deceived, here the experimental scientist could provide empirical evidence about the probability of how an average individual would react. Rather than remaining dependent on common sense, a judge could take into account a statistical portrait of how actual persons had responded in a controlled situation.

In the first set of experiments, Paynter sought to quantify the "visual recognitive confusion" generated by the two names. He showed his subjects twenty slips of paper for one second per slip; on each sheet a different trademark name was typewritten in the same unadorned, black script. Fifteen seconds after viewing the final slip of paper, the subject was given a stack of forty slips of paper with twenty additional trademark names written. The subject had to identify which names he had seen during the first phase of the experiment and his relative certainty in this decision. During the second phase of the experiment, the word "Chero-Cola" replaced "Coca-Cola." Of the forty people tested, eleven, or 28 percent, confused the two brand names. Furthermore, nine of those eleven said they were certain in their identification.[78]

In a second experiment, twenty-five new research subjects were given basically the same test, but now the name of the commodity was included on the slip of paper. Therefore, the slips read "Coco-Cola Soft Drink" or "Chero-Cola Soft Drink." Once again, the experimental subject was given a second to view the original slips of paper and was asked to identify those sheets that had just been seen from a pile of names. In this instance, the rate of confusion or deception was even greater, with seventeen out of twenty-five people incorrectly identifying "Chero-Cola Soft Drink" as "Coca-Cola Soft Drink." The greater rate of deception was attributed to a combination of a longer phrase to be read in the same amount of time and the fact that observers tended to associate certain trade names with specific goods. Despite the greater rate of confusion, the vast majority of observers remained confident about their incorrect identification.[79]

The final visual experiment placed the Coca-Cola/Chero-Cola pair into a set of nine other pairs of trade names that had been subjected to past infringement litigation. The goal was to rank the confusion generated in the cola case along a scale of similarly confusing names. Among the competing trade names used in the experiment were examples of both instances where infringement and noninfringement had been determined by the courts. This allowed Rogers to rank the deception in this instance in relationship to other cases known to the judge. Paynter found that the Coca-Cola/Chero-Cola pair ranked fourth in terms of the confusion generated among the experimental subjects. Furthermore, since many subjects found certain trade names that had not been determined to be infringing more confusing than those that had, this experiment presented a not-so-implicit critique of the legal decisionmaking process commonly used in these cases.[80]

In the final series of experiments, Paynter shifted his attention away from the purchasers' visual capacity and examined their auditory perceptions. A new set of subjects was given the same list of competing trademarks from the previous test, but this time only in auditory form. Paynter asked them to rank the pairs in terms of the difficulty they experienced in terms of distinguishing one name from the other. This final test served a number of functions. By testing another sense, one could get a clearer understanding of the confusion involved in the act of purchasing. Was it merely the eye that was deceived by the resemblance or would the ear prove to be as prone to error? If confusion was generated simply by the sound of the trade names, the auditory test would serve as an independent confirmation of the results in the earlier portion of the investigation. Second, since the product in question was a soft drink, the customer would frequently order it at a soda fountain without ever seeing or referencing the packaging. In these instances, the consumer was entirely reliant on the auditory sense when it came in detecting deception. Finally, the mental processes involved in this experiment closely paralleled what was asked of judges in arbitrating these trade disputes. The key difference was that instead of relying upon the perspective of a single or a few judges, this experiment, through the repetition of the process, offered to eliminate the bias of individual judgment with the creation of a statistical aggregate.[81]

C. L. Parker, the legal counsel for Chero-Cola, vehemently objected to the psychologist's testimony, dismissing it as simultaneously "illegal, incompetent, irrelevant and immaterial." Parker protested that the experimental data did not meet the legal standard for truth-telling: Paynter's subjects had not been under oath during the experiments nor was opposing counsel present in the laboratory to cross-examine them about their responses. Moreover, Parker attacked the laboratory as an inappropriate site for addressing the question of deceptive trademarks: "The conditions of the experiments were not at all similar to the conditions which prevail in transactions leading to the purchase of the goods of the parties." In his testimony, the psychologist had claimed to render "the probability of confusion" through the statistical analysis of the responses to his experiments.[82] Rather than embracing the aggregate as a representation of the probable response, Parker argued that all Paynter offered was a conclusion derived from the testimony of witnesses not present in court.

Although the court decided in Coca-Cola's favor and acknowledged the novelty of the psychological evidence, it did not give great weight to Paynter's testimony. As Constantine Smyth noted in the court's decision, "we are invited to listen to the teaching of psychology on the subject. None the less the question in dispute is a simple one, and the principles by which its solution may be reached have been often declared and applied by this court."[83] Smyth, Rogers, and Paynter all agreed that the typical purchaser of Coca-Cola was easily deceived. In both the scientific and legal sphere, this deceivable self was the product of frugal mental mechanisms confronted with an abundance of novel sensory experiences. Despite this consensus, the judge favored the body of knowledge provided by the common law in making his ruling over the innovation provided by psychology.

In many respects, this lukewarm reception of scientific expertise was unsurprising. Most jurists looked skeptically at attempts to usurp their individual judgment by a laboratory-generated test. This skepticism received its most famous articulation two years later when the very same court delivered its influential *Frye* decision concerning the admissibility of the lie detector. Long before Rogers sought to tighten the legal standard for trademark infringement, expert testimony had a contentious history. The British case *Folkers v. Chadd* (1782) helped establish new mechanisms for the admission of scientific expertise into court proceedings, but in a very specifically circumscribed form. The scientist as witness testified on behalf of a particular litigant in the adversarial legal process rather than as a neutral friend of the court. Many came to question the sincerity of expert witnesses as they seemed to serve the needs of two potentially conflicting masters, both the Book of Nature and their particular clients. When Hollingworth testified on behalf of Cola-Cola, he felt the need to keep himself in a state of ignorance during the experiments in the hope of preserving his credibility. Beyond this general distrust of the scientist as consultant, psychology posed even greater difficulties. The psychologist seemed to specialize in assessing a person's state of mind and this role traditionally belonged to members of the jury. Not merely offering evidence about physical nature, the psychologist was viewed by many judges as usurping the crucial operation of judicial deliberation.[84]

Rebuked by their potential legal audience, experiments on trademarks did expand the scope of psychological research. Compared with other faculties like memory and attention, recognition had received scant interest within turn-of-the-century experimental psychology. Certainly nineteenth-century psychophysics involved the process of discriminating differences in sensation, but it focused on the judgment of immediate sensations. In contrast, these psychologists saw recognition as a species of memory, the capacity to identify past impressions after an extended period of time.[85] Branding as a commercial practice was about the consumer's capacity to reliably recognize a specific product based on a previously seen image in a catalogue or the recommendation of a friend. As a result, wealthy patrons had much invested in learning humanity's talents and weaknesses in this area.

Consumer culture provided more than the impetus for these experiments to occur. It also furnished the experimental tools to be used, whether the mass-produced postcards, full-page advertisements, or trade names so central in the early science of recognition. Aiming for applied results, these psychologists argued that they had demarcated an important mental process from closely related ones. The commercial context of policing trademark infringement and the potential patronage from courts led to the delineation of recognition as something distinct from attention and recall. Certainly people spoke of recognition earlier, but the topic had failed to sustain the experimental research that properties like attention, memory, or feeling had. Furthermore, it was understood as an affective process, independent of conscious cognition.[86]

The appeal of recognition as an object of investigation had much to do with an expanding, increasingly branded consumer market of the early twentieth century. In the 1910s psychologists invariably linked their bringing the subject's capacity for recognition under experimental purview back to these commercial and legal applications and patronage. In this regard, the other major site for experiments on recognition consisted of psychologists helping their corporate audiences make their branded commodities more recognizable to the same purchasing public at the center of trademark infringement cases. Even the most seemingly esoteric study of recognition invariably made some reference to its utility for improving the design of trademarks and brands.[87] Among the experimental psychologists there were some telling slippages as recognition was closely linked to consumer choice. For example, Feingold argued that discrimination was analogous to selecting a tie among many in a store.[88] The bringing of recognition into the American laboratory was in large part an effort to make sense of how consumption functioned in everyday life.

The identification of the psychology of recognition with market research became even more pronounced in the following decades.[89] At the University of Iowa, future pollster George Gallup developed another "recognition test" intended to gauge the mind of the consumer. Gallup wanted an accurate measure of reader interest in various newspaper sections in order to improve marketing. He felt that direct questioning was ineffective insofar as it relied upon an individual's imperfect recall of what he or she had read. More important, he suspected that people might be tempted to misrepresent their reading interests to the interviewer so as to present a more serious-minded appearance. To circumvent these difficulties, Gallup designed a study in which he went through an unmarked, older newspaper with his subjects and asked them to identify those items whose content they immediately recognized. Gallup's recognition surveys became an industry standard after he completed a heavily publicized study for *Liberty* magazine in 1931. He deployed a team of interviewers to visit 15,000 homes in six cities to assess their recognition of the content of the country's leading weeklies such as *Saturday Evening Post, Collier's*, and *Literary Digest*.[90] In the spring of 1932 the Psychology Corporation coordinated the efforts of sixty psychologists in different cities across the country in a "cooperative" experiment to test the effectiveness of advertising, especially among housewives. Instead of asking which advertisements the subject could best recall based upon her unaided memory, the psychologist inquired whether she recognized the specific products associated with particular catch-phrases.[91]

In all of these experiments, the mental process of recognizing an object was intimately related to one's capacity to recognize particular brands, advertisements, and trademarks. Such developments complicated the boundary between applied and experimental psychology, for immediate commercial concerns were constitutive of research into the structure and function of the mind. The need to manage and rationalize mass consumption both served as the inspiration for these experiments and provided psychologists with the tools they used. In this regard, the identification of the laws of nature governing recognition resulted from psychologists' attempts to alter the laws governing commerce.

The Legal Measurement of Deception

As the psychologists involved made clear, the aim of the psychotechnical reforms was not to replace the law with science, but rather to introduce an "accurate" scale into legal proceedings. Münsterberg suggested that determining what degree of similarity constituted a commercial deception remained what he perceived as a social convention and, therefore, within the realm of legal decisionmaking. What psychology could provide was what he deemed an objective scale grounded in experimental evidence of the varying degrees of similarity that the ordinary observer could recognize. Consisting of lists of previously litigated trade names whose confusion had been tested under laboratory conditions, this scale would serve as a map of the human mind's ability to notice differences in visual or auditory stimuli. With this scale in place, jurists and patent officials could then decide what point on the scale represented the moment when deception occurred, legally speaking. Paynter emphasized the technocratic potential of adopting a standardized scale. By deferring to psychological authority, decisions could be accurate and efficient with both private enterprise and the state able to "economize in time, money, and labor" by avoiding litigation.[92]

When American lawmakers eventually turned toward measurements for determining when a commercial product was deceptive, they did not choose psychological ones. The Wheeler-Lea Act (1938), which expanded the FTC's powers to discipline commercial deception, especially advertising, in the name of consumer protection largely abandoned the purchaser's mind as a legitimate site of governance. The 1938 act replaced the psychological with an increased emphasis on defining commercial deception in terms of falsifiable claims made in labeling or advertisements. At issue were the merchants' material assertions about their products, not the consumer's impressions. The commission would deal with outright deceit rather than deceptive appearances. Such a shift in focus was seen as easier to administer, as governance based upon the psychology of the supposedly ordinary purchaser had been controversial from the outset.[93] Despite such moves in redefining commercial deception within the purview of the FTC, the unwary purchaser and the likelihood of generating confusion within its mind remained the legal standard in cases of infringement.

For his part, Rogers became increasingly dedicated to the cause of tighter government regulation of trademarks. Between 1920 and 1924, he chaired a committee on behalf of the patent division of the American Bar Association to draft a new trademark law. After some failed attempts in the late 1920s, Texas congressman Fritz Lanham introduced a version of the American Bar Association's model legislation in 1938. The bill made it unlawful to either introduce or deliver a copy, counterfeit, or imitation of any federally registered trademark. The bill renewed the tradition of defining infringement in terms of the deception or confusion induced in the mind of the purchaser. The centerpiece of this legislation was the creation of a national registry for trademarks. Registration of a mark was no longer contingent upon its prior use in market. In this regard, it went further than the standing 1905 Trademark Act, whose strictures had largely gone unheeded by the nation's jurists.

These changes to the federal trademark law occurred against the backdrop of a renewed call for greater consumer protection. The movement's leading intellectuals, engineers Stuart Chase and Frederick J. Schlink, saw advertising as a corrupting influence on American civic life. Due to the prevalence of dishonest puffery, they contended that the consumer could no long trust advertising and branding as guarantors of a product's quality. Like Florence Kelley before them, they viewed the consumer as vulnerable and in need of protection due to the inaccessibility of accurate information. Books like Chase and Schlink's *Your Money's Worth* (1927) had a pronounced influence on leading New Deal policymakers, especially Rexford Tugwell. Certainly attractive to prominent Washington intellectuals, the consumer movement had considerable grassroots support on a national scale, including women-led meat boycotts and the New Negro Alliance's "Don't Buy Where You Can't Work" campaigns.[94]

At their most radical, these consumer advocates proposed replacing the existing system of using corporate branding to identify commercial goods with a government-enforced labeling system based on grades of quality. Canned foods, with their content hidden from visual inspection by the consumer, became a focal point in this war over labeling.[95] To promote the passage of a vigorous Food and Drug Act, Tugwell initiated a traveling exhibition, which became known as the Chamber of Horrors, which detailed the life-threatening health risks faced by a public deluded about the content of food and pharmaceuticals. Champions of the existing advertising system opposed this call to universal standardization, claiming that a brand's reputation in the mind of the purchasing public still constituted the best assurance of quality. Furthermore, critics like journalists Charles Carpenter and Christine Frederick argued that the engineering perspective was misplaced because consumers did not behave in a rational fashion. The typical purchaser did not select an item based on some calculus of price and quality, but rather consumption occurred in response to subjective wants. To curtail this freedom of desire was to solicit Russian-style collectivism where housewives would turn to a catalogue and order "a can of B-64."[96]

Although a variation of this proposed legislation, known as the Lanham Act, passed both houses in 1946, the changes Rogers introduced met with considerable opposition. Representing the Boston Patent Law Association, Harrison Lyman objected to the securing of a trademark through registration with the federal government from the outset rather than "by natural use and growth" in a specific market. Lyman insisted that trademarks needed to be established through the gradual cultivation of goodwill toward a product in the mind of the purchasing public and not by de jure pronouncement. Unlike patents, trademarks acquired value "because a man uses them and because the public mind associates a mark with him." The existence of a national depository risked having "sharpsters" registering trademarks already in use in one locality, thereby preventing the existing business from moving into new territories.[97] Milton Handler, a Columbia University law professor and the legal counsel for Pepsi Cola, opposed the bill on similar grounds. He expressed concerns about how the new act might constrain the public's use of trademarks that had become generic descriptors of products, like cellophane and shredded wheat. Handler insisted that a trademark should not be made the property of a single company because the success of words referring to brands hinged upon their acceptance among the public and not in the ingenuity of their design. The most successful trademarks came to refer to a category of products rather than their particular proprietary source and so were held in common in the collective minds of the purchasing public. He concluded that "it is the essence of the competitive system that one may share in the good will of the product unless we are going to permit the law of trade-marks to be used as an instrument for creating monopolies."[98]

Another set of critics emerged from the Department of Justice. During the 1942 hearings concerning the bill, Elliott Moyer, a special assistant to the attorney general, argued that the proposed legislation would interfere with a demand stemming from the Food and Drug Act that a product include on its packaging the description by which the public was best acquainted with it, such as aspirin for acetylsalicylic acid. Moyer worried that "an incontestable trade-mark, if it has become the generic or common name used by the public generally to designate an article, originally unpatentable, might result in monopoly."[99] Where a patent might expire, allowing other companies to sell similar products, the new trademark law would constrain their circulation by not permitting the honest and transparent labeling of the competing products. All of these critics accused Rogers of using the power of the federal government to create undesirable and "artificial" monopolies, which would inhibit business competition.

Rogers countered that that securing national rights for trademarks was the only means of guaranteeing free competition in a modern economy. When Rogers testified before Congress, he argued that trademarks functioned as "valuable instrumentalities of commerce." In an era defined by an anonymous market on a national scale, trademarks served as technologies for protecting competition by ensuring consumers made the right choice in the face of a confusing variety of products. They "enable anyone easily to pick out the goods he wants to buy and to avoid the goods he does not want to buy."[100] Later in his testimony, Rogers further disputed the argument that the greater protection of trademarks would engender more monopolies:

> trade-marks are what move goods from the manufacturer to the consumer. It is the confidence that the people have in articles marked in a certain way that makes the public buy them. Consequently, it is just as much a means of getting goods from the manufacturer to the consumer as the railroads and trucks are which provide the physical transportation.[101]

Just as Congress possessed the means of regulating the objective means of transporting goods, the federal government had the responsibility to manage "the subjective means of transportation."[102]

Trademark reform was necessary for the protection of companies, not consumers. Indeed, the proposed changes went against the more transparent labeling practices advocated by New Dealers. Rogers's goal was to secure those routes of cognition and feeling along which commodities flowed in an economy of distance. As this congressional testimony made evident, he advanced a psychological conception of the purchaser in order to protect the access that established businesses had to their customer base. Throughout his career, Rogers emphasized the deceptive nature of advertising and brands alongside the consumer's cognitive vulnerability to enshrine trademarks as a kind of property. By the 1930s, he concluded that the standardization and regulation of trademarks through a national repository was the best means of furthering this goal. Ultimately, grading did not come to replace branding, although the final wording of the Lanham Act tempered Rogers's vision by placing greater emphasis on the need for trademarks to be in use prior to their registration.

Conclusion

Whereas the legislatures moved to emphasize material misrepresentation and deceitful labeling in the realm of drugs, judges became increasingly forthright about the commonsense psychology that animated their decisions concerning trademark infringement. Supreme Court Justice Oliver Wendell Holmes Jr., a major architect of the reasonable man as a key legal fiction, was particularly articulate in this regard. In a 1924 decision he continued to emphasize how trademark infringement was constituted not by measurable similarities between names, but entirely by the mental effects such similarities were likely to exercise in the minds of the purchasing public. After all, "when the mark is used in a way that does not deceive the public, we see no sanctity in the word as to prevent its being used to tell the truth."[103] Another Supreme Court justice, Felix Frankfurter, advanced a closely related commonsense psychology of the purchaser. In a 1942 decision he argued, "The protection of trade-marks is the law's recognition of the psychological function of symbols. If it is true that we live by symbols, it is no less true that we purchase goods by them."[104] In this view, during the act of purchasing the consumer recalls previously recorded symbols and compares them to the present stimuli in order to determine if the product is recognizable. For Frankfurter, the mind of the consumer was a semiotic machine that processed symbols and calculated their similarity in a fashion quite analogous to the laboratory-designed scale proposed by psychologists but rejected by his judicial predecessors. In these examples, neither Holmes nor Frankfurter felt the need to take recourse to the authority of the psychologist.

Focusing on the failure of Münsterberg, Feingold, and Paynter to get their expertise recognized by the courts underestimates the extent of the shadow cast by psychology over the legal regulation of the commercial sphere. Both before and after their intervention, the law concerning trademarks and deceptive advertising was saturated by talk about the nature of the consuming self and its mental traits. By invoking the figure of the unwary purchaser, judges and lawyers did not seek to replace the reasonable man as the default legal standard for human nature. Instead, they identified consumption as an activity that compromised this default status and rendered the individual deceivable. While rejecting the discipline's methodology, they saw consumption as an arena where psychology's vision of the individual as a deceivable self was particularly salient. Ironically, this equation of consumption with deception in the legal sphere was made most forcefully by the defenders of large national corporations, like Edward Rogers, rather than these institutions' critics.

As the discipline gained increasing prominence within the wider culture, the language of psychology, if not its personnel or methods, did come to shape how judges articulated their commonsense understanding of the individual's mental constitution. A 1949 dispute between the California Fruit Growers Exchange and a baking company over the word "Sunkist" well illustrates this. In this case, the judge invoked a certain aptitude on an intelligence test as the standardized measure for determining an individual's inclusion within the category of ordinary purchaser. "We cannot believe that anyone whose I.Q. is high enough to be regarded by the law would ever be confused or would be likely to be confused in the purchase of a loaf of bread branded as 'Sunkist' because someone else sold fruits and vegetables under that name."[105]

The aggressive marketing and careful branding of consumer goods on the part of leading corporations came to undercut Rogers's hope for a more stringent standard for trademark protection. Here the presumed psychology of the consumer again played a crucial role. In the 1920s judges found it increasingly more difficult to presume that certain trademarks were unfamiliar to consumers. They contended that the power of habituated mental associations actually served the needs of these most familiar brands. As Justice Andrew Cochran noted with reference to Coca-Cola in a 1929 case, "Its trade-mark has been burnt into the consciousness of people generally. Instinctively one recalls in memory its appearance and sound."[106] Such a decision was in marked contrast to the Chero-Cola case. These commonsense judgments about how the consuming public was likely to behave in certain situations played a critical role in the legal constitution of the parameters of the sanctioned, legitimate economy. The unwary purchaser may have lacked empirical support, but it proved an irresistible tool for judges to deploy.

CHAPTER FIVE

Diagnosing Deception: Pathological Lying, Lie Detectors, and the Normality of the Deceitful Self

Casting himself in the role of a modern-day Diogenes, A. A. Lewis penned a 1939 survey of methods for identifying an "honest man." He argued that in recent years scientists had developed three techniques "that will actually pierce the surface of outward demeanor, however pharisaical, and show a man up for what he is." Lewis argued that psychology had provided the means for assessing an individual that undercut deceptive self-presentations. The deceits of human hypocrisy would now dissolve under the scientist's harsh gaze, promising new forms of sincerity in daily life. His survey opened with an assessment of the highly publicized technological fix known as the lie detector. Lewis argued that his Diogenes now possessed two other potential lanterns alongside this proprietary technology. The first was the "biographical" lantern, or the case history, which he associated with work conducted in juvenile courts. The other was the "statistical" lantern, namely the analysis of aggregated character traits found in large populations using experiments and mental tests. Lewis contended, "The short-cut of the laboratory magician with his lie-detecting set-up (his polygraph which registers its several findings simultaneously) is at the other pole from the procedure which characterizes the biographical techniques."[1] This chapter investigates precisely the historical relationship between technological and biographical techniques for parsing the honest individual from the deceitful self.[2]

Pairing the histories of these techniques and the understandings of deception they materialized offers a novel perspective on the development of the forensic sciences. A number of scholars have emphasized the continuity between nineteenth- and twentieth-century attempts to make criminality more visible through the visual inspection of bodily traces. Certainly early twentieth-century criminologists were well aware of such precedents as Thomas Byrnes's famed photo-array in *Professional Criminals in America* (1886), the search for atavistic stigmata in the criminal anthropology of Cesare Lombroso, Alphonse Bertillon's anthropometric classification system, and Francis Galton's composite photography.[3] Yet an exclusive focus on continuity is misleading. The traces generated by bodily reactions during a lie-detector examination work to identify the criminal, but the test also represented a rupture with the photographic approach to criminal detection. The visual techniques crafted in the Victorian era focused upon the identification of fixed, bodily markings such as the shape of an ear or the ridges on fingers. In contrast, the power of the lie detector resided in the claim that it traced dynamic emotional states that changed from moment to moment.[4] Deceit was an activity that regularly occurred during an individual's life course; it did not define a person's character.

Both the technological and biographical approaches assumed the existence of a deceitful mind interconnected with a fundamentally "honest" body. Much like psychoanalysis, which also garnered considerable interest at the time, the lie detector promoted a vision of the self as governed by petty deceits and self-deceptions. The mind's deceits generate the bodily responses studied by the expert, be they changes in physiology measured by the lie detectors or neuroses unpacked by the psychotherapist. The polygraph provided a technologically enhanced, "short-cut" version of the therapeutic promise offered by the crafting of a case history. Certainly the nature of deception examined differed in crucial respects. The case study method aimed to reveal unconscious deceits rooted in forgotten past experiences submerged deep within the individual's psyche.[5] In contrast, the lie detector aimed at revealing conscious deceits deliberately concealed from others. Despite these differences, the techniques Lewis discussed underscored a common shift in the science of criminal detection toward individuals who *happened* to commit crimes rather than focusing exclusively on fixed criminal *types*.[6]

This shift in forensic science owed much to the mental hygiene movement that turned psychiatry's focus away from the confinement of the insane toward guiding "maladjusted" persons in their daily lives. Historians who have studied this transformation focused on changes within the asylum and hospital, but the place of psychosomatic medicine in urban policing served as an equally important site.⁷ The success of the polygraph was in part due to the space William Healy created for dynamic psychological explanations and therapeutic interventions within American criminology. Working as a psychiatrist for the juvenile court in Chicago, he introduced the biographical approach into the debates over the causes of youthful delinquency.⁸ Although many advocates of the lie detector largely eschewed Healy's therapeutic impulse and he remained skeptical of the technology, the two projects remained connected by place and common concern with the scientific study of human testimony.

Ironically, Healy's own work on deception illustrated the unevenness in this transition away from fixed types toward a more dynamic understanding of human deception. In a 1915 monograph he coauthored with his first wife, Mary, he championed the existence of a specific psychiatric type characterized by deceitful behavior, the pathological liar. While rooting the condition in the patient's lived experience, the Healys still suggested that there existed a specifically deceptive kind of person. Yet, in contrast to their European counterparts, the Healys strongly argued for the need to distinguish these individuals from the epileptic, the feeble-minded, and the insane. In its Americanized form, pathological lying constituted a form of mental maladjustment, a condition amenable to the therapist's intervention.

Between 1905 and 1930 there emerged two distinct forensic approaches to detecting and explaining the deceitful self. Both ultimately stressed the normality and unavoidability of deception within ordinary mental life. There was also an important irony in the reception of these traditions. Grounded in their respective mechanized tools that sought to discipline the judgment of the individual psychologist, each technique remained mired in charges of fraudulence and commercial charlatanism. Because these techniques often required deception on the scientist's part, commentators frequently questioned the sincerity of the scientists involved. This suggested to some commentators that forensic psychology risked becoming a kind of patent medicine.⁹

Deception Tests and Patent Medicines

Both the pathological liar diagnosis and the lie detector as a technology had their origins in the psychological study of legal testimony. In Germany William Stern led the way as he translated the interest among early laboratory psychologists in memory and perception into a rudimentary applied field.[10] Swiss psychiatrist Carl Jung established his reputation by championing the measurement of reaction times during word-association tests as a means of accessing the patient's emotional landscape. The legal possibilities of Jung's method soon became apparent. Reaction-time tests could make visible the suspect's consciousness of guilt in the form of a noticeable delay in the time it took to arrive at a nonincriminating response to the select words meant to invoke the crime. In 1909 both Stern and Jung were given the opportunity to present their ideas before an American audience when G. Stanley Hall invited them, along with Sigmund Freud, to help commemorate the twentieth anniversary of Clark University.[11] In short, the reception of psychotherapy and a legal psychology of testimony in the United States shared, from the beginning, an intertwined history. While advocates of each would work to disentangle one from the other, they continually jostled in the public sphere as means of explaining a simultaneously deceivable and deceitful self.

Hugo Münsterberg emerged as the leading advocate of the psychology of testimony in the United States. With his familiarity with the German scene, he commenced his investigations in this arena prior to the Clark lectures. In a series of popular essays for mass-circulation magazines, later collected as *On the Witness Stand* (1908), he sought to replace the evaluation of a person based on his moral character and physiognomy with the measurement of individual reactions. Ultimately, he hoped to design what the press called a "machine to detect liars."[12] His ventures into the realm of legal testimony were similar to his contemporaneous exposures of the paranormal. They constituted another avenue for publicly performing his expertise by extracting the truth from persons marked as deviant. His attempt to craft such a space within the legal system was controversial from the start. Judges suspected psychologists not only of trying to introduce new forms of evidence, but of replacing the deliberative process of the jury system in its entirety.[13]

Münsterberg's advocacy of the inherent unreliability of humans placed him in opposition to the legal view of the self as a reasonable and responsible individual. In a chapter on "Illusions," he set out his position by asking his reader to picture a case involving an automobile accident. Fortunately, there were two independent witnesses to give first-hand testimony; unfortunately, they could not agree upon what they saw. One claimed that the road was wet, while the other insisted it was dry. One remembered that the offending driver had sufficient time to brake, while the other heartily disagreed. Münsterberg stressed that their differences of opinion were not due to deceit: "Both witnesses were highly respectable gentlemen, neither of whom had the slightest interest in changing the facts as he remembered them." The problem was that the individual's unaided observation was not a reliable kind of evidence regardless of the person's moral character; nor could one arrive at the truth merely through a commonsense evaluation of their respective testimonies.[14] Beyond questioning the reliability of eyewitnesses, Münsterberg argued that his psychotechnics could elicit from suspects the truth behind their deceptions. Ironically, when Münsterberg sought to introduce this approach into an actual legal setting, his self-aggrandizing methods led to his own dismissal as a commercial charlatan.

In 1905 an assassin's bomb killed Frank Steunenberg, the former governor of Idaho, and the police soon arrested a man known as Harry Orchard as their suspect. In exchange for a lesser sentence, Orchard confessed to the crime and provided details about how he killed Steunenberg on the orders of the Western Federation of Miners (WFM). Such a claim was a political windfall. The WFM was part of the International Workers of the World, the most radical union in the country. Soon its leader, "Big Bill" Haywood, was placed on trial for soliciting the former governor's murder. Despite the prosecutor's enthusiasm to implicate Haywood in the assassination, events from Orchard's past threatened to invalidate his testimony. Haywood's defenders claimed the murder was not a political crime, but rather resulted from a private grievance. It appeared that Orchard had invested in the area's mines and that actions taken by Steunenberg had cost him a considerable amount of money.[15] The question at the heart of Haywood's trial was whether Orchard could be trusted to tell the truth.

With a commission from *McClure's Magazine* to carry out an examination and publish his observations, Münsterberg left for Boise in June 1907. He hoped to subject Orchard to a series of experiments to determine whether he was truthful about implicating Haywood. Reeling from the departure of his leading muckrakers following revelations about his own extramarital affair, Samuel McClure saw the Haywood trial as a means of reestablishing his magazine's reputation as a cutting-edge investigative organ. He had previously published a heavily edited version of Orchard's autobiographical "confession" and hoped that Münsterberg's unique psychological approach would garner further attention.[16] Receiving the state's permission, Münsterberg claimed to have subjected the assassin to a barrage of psychological tests over the course of seven hours. Without revealing his precise methodology or results, Münsterberg contended that the tests confirmed Orchard's story and the state's case against Haywood.[17] Yet his own secrecy placed Münsterberg in at best a paradoxical, if not hypocritical, position.

This was how Haywood's attorney felt when the content of Münsterberg's psychological profile was leaked to the press. Clarence Darrow accused Münsterberg of simply being McClure's mouthpiece, implying that the psychologist was a man whose opinion could be easily bought. He insinuated, "It may have been simply that you were used by him, but you were called by him to my certain knowledge, and you went at the matter in the same way he did, in a one-sided and partial manner." Speaking to the press, Darrow suggested that Münsterberg's account was "merely a paid testimonial."[18] These were significant words. At the time that Darrow made these charges, testimonial advertising was held in particularly low regard. With its roots in P. T. Barnum's spectacles and campaigns of competing endorsements, testimonials involved the promotion of consumer goods—such as soaps, cosmetics, and, most important, patent medicines—through the recognizable personality of an individual celebrity. Rather than advancing the minimalist realism of "truth in advertising," the testimonial harkened back to the industry's roots in puffery.[19] Darrow's turn of phrase identified Münsterberg and his science with all that the scientist claimed he detested in American culture: the crassly commercial, the feminine, and the fraudulent.[20]

By transforming the psychology of deception into a public spectacle, Münsterberg jeopardized his image as impartial scientific observer. His reputation for securing patronage made it easier for his opponents to question the sincerity of his position. That the jury found Haywood innocent of the charges further undercut the psychologist's authority. Despite his battery of tests, Münsterberg felt sufficiently unnerved by the verdict that he did not want his scientific expertise to be seen as contradicting it. He wrote a quick note to the *McClure's* editorial office asking that the article be temporarily scrapped and then quickly sent another letter to Darrow insisting that he wanted to end their public quarrel over the case.[21]

The dispute with Haywood's lawyer set the stage for much of the response to Münsterberg's legal psychology. The legal community depicted him not as a sincere student of nature, but as a scandal-mongering journalist. The *American Law Review* labeled Münsterberg's brand of exposé "yellow psychology," the scientific equivalent of sensationalistic yellow journalism. John Wigmore, dean of Northwestern University's law school and the leading authority on the nature of legal evidence, exemplified this reaction when he penned a hostile review of *On the Witness Stand* that took the form of a parody of a libel trial. Wigmore warned Münsterberg privately, "I expect in a month or two to poke some fun publicly at your indictment of us, which doubtless you will not grudge me, in view of the fun you had of publicly putting our profession in the pillory." The Harvard psychologist did not take the ribbing well. When Wigmore wrote to Münsterberg in 1913 to request permission to reprint some of his writing, he took the precaution of reassuring the psychologist that he would not "attempt to take advantage of the occasion to continue the sarcastic controversy of three years ago." Attempting to convert the legal profession to his psychotechnics, Münsterberg instead cultivated ridicule from even its most sympathetic members. A 1909 newspaper editorial condemning the idiosyncratic character of legal decisions declared that "every magistrate is his own Münsterberg."[22]

It was not only other professions that attacked his public immodesty, but also former allies within his own discipline. Not for the first or last time in his career, insincerity became a problem for Münsterberg's advocacy of truthfulness in the public sphere. Lightner Witmer, professor of psychology at the University of Pennsylvania, was highly critical of what he perceived as Münsterberg's enthusiasm for commercial gain. He condemned his Harvard colleague for "crying his psychological wares in the marketplace."[23] According to Witmer, Münsterberg's decision to publicize his findings through sensational pronouncements in the popular press without backing the claims with detailed scientific articles was irreparably damaging the discipline's reputation. Witmer singled out Münsterberg's behavior during the Orchard affair as particularly egregious, probably retarding the legal acceptance of psychological testimony for decades. Others concurred. As Boston physician Morton Prince confided about Münsterberg after his death, "there was a less admirable trait and which I think was at the bottom of the distrust of him by his colleagues insofar as he was a popularizer of psychology. In this respect it was thought with considerable reason that he was not intellectually honest."[24] Because of his means of communicating and his aggrandizing persona, the expert in detecting deception was seen as unreliable by his peers. Much like in the case of spiritualism, Münsterberg demanded full transparency from those he investigated while his actions courted suspicions about his own hidden motives.

Tellingly, Witmer's condemnation of Münsterberg appeared as part of a polemic against the Emmanuel Movement, a form of psychotherapy then popular in Boston. Founded by Episcopalian minister Elwood Worcester in 1906, the movement initially garnered support from leading members of the Harvard faculty, including Prince, William James, Richard Cabot, and James Jackson Putnam. The majority of these men came to withdraw their support over concerns about the role of ministers in offering medical treatment, a move that risked blurring the boundaries with faith-healing movements. Despite such setbacks, the Emmanuel Movement was the first widely practiced form of psychotherapy in the United States and did much to seed the ground for the reception of psychoanalysis a decade later.[25]

One of the ways in which the Emmanuel Movement framed the reception of psychotherapy was this association with the potentially fraudulent use of suggestion as a therapeutic technique. The physician's use of suggestion in the form of the placebo came under particular scrutiny during this period because of Samuel Hopkins Adams's journalistic exposé of the dangers posed by patent medicines.[26] Medical ethicists responded by railing against the pernicious influence of any form of dishonesty in the doctor-patient relationship. For example, Cabot argued that the use of deception ran counter to a therapeutics targeting specific disease entities, which the physician could accurately identify with the assistance of the increased diagnostic precision provided by new medical technologies.[27] He condemned the liberal use of patent medicines by the public and extended the criticism to include family physicians, whose deployment of effective placebos constituted an unethical deceit. The connection between deception and unethical commercial behavior was important for his understanding of the placebo. Cabot made the telling analogy between this kind of medical practice and a fictional financial advisor who recommended certain stocks without explaining to the investor why they might be profitable.[28] Cabot demanded that transparency govern a healthy doctor-patient relationship. Without this virtue in place, the physician risked becoming the moral equivalent of the speculator, secretly exploiting the trust invested in his person. Cabot's attempt to demarcate the respectable physician from the commercial predator led him to expunge the long-standing therapeutic practice of bedside deception from the canon of acceptable medical practice.

This equivalence between psychotherapy and patent medicines featured prominently in the initial publicity surrounding psychoanalysis in the United States.[29] Journalist Max Eastman expressed the difficulty he faced as he tried "to avoid the language of the patent-medicine advertisement" as he described his enthusiasm for the talking cure's marvelous effects. Another early advocate, writing for *Good Housekeeping* in 1915, seemingly answered Eastman's worries. He insisted that because psychotherapy was a slow process "there are no financial rewards to tempt the charlatan or the merely avaricious." Despite these reassurances, metaphors of financial exploitation saturated the earliest representations of psychotherapy in the United States.[30]

Analogies to financial speculation also featured prominently when in 1911 Henry Goddard, one of the chief architects of America's intelligence testing program, presented a test to detect a criminal's guilt to the public. In addition to measuring reaction times, Goddard used a galvanometer to measure changes in the body's electrical current. The major failing of his test was that it targeted elevated emotional states in general, without really demarcating deceit as a separate mental event. As Goddard candidly told a *New York Times* reporter, he was "unable to tell from the record whether the emotion aroused is pleasure or pain, shame or triumph, guilt or mere nervousness." He confessed, "If I were to examine a stock gambler with the psychometer and mention stocks he would record a strong emotion, but I should be unable to determine whether he had won or lost."[31] Goddard's illustration—so distant from the clinical population he typically faced—demonstrated how imagery of the marketplace was never far from view when scientists spoke about deception. Despite his excitement about the possibility of his psychometer, Goddard still felt that the project remained a speculative gamble. From the outset, then, both psychotherapy and lie detection emerged as techniques for exposing the individual's deceits and self-deceptions, but were themselves mired in charges of fakery and charlatanism.

Psychiatry and the Socialization of Justice

The notion that certain forms of lying ought to be interpreted as pathological originated in the asylum-based psychiatry of the German-speaking world. In 1891 the Swiss psychiatrist Anton Delbrück reported on a number of cases under the rubric of *pseudologia phantastica*. In such instances, the patients had such a great propensity for fabricating false life stories that they convinced both themselves and others of their authenticity. A former maidservant who claimed at various times to be a Romanian princess, an impoverished medical student, and a descendent of the Spanish crown served as his representative example. Delusions and intentional lies were inseparable for this patient, as she could not distinguish between what she had made up and what she perceived. Delbrück associated pathological lying closely with a distorted faculty of imagination, as the sufferer fabricated wonderful and exciting self-narratives.[32]

European psychiatrists linked pseudologia phantastica to those forms of mental degeneracy that were plaguing fin-de-siècle Europe. Delbrück described another patient who would ingratiate herself with strangers by claiming to be a relative from a distant city, only to rob them of goods once their trust was secured. Although the legal system was uncertain of the extent of her responsibility, he determined that she was both a hysteric and epileptic. The majority of European cases similarly involved the crossing of class lines. They depicted laborers successfully passing for their aristocratic betters, deluding themselves and others with imaginings of lives of wealth and luxury. Such was the case of "J. M.," a supposed pathological liar and hysteric admitted to the Royal Edinburgh Asylum in the winter of 1916. He claimed to be the son of the director of Lloyd's Bank and regaled the asylum staff with his youthful adventures, including an odyssey to Canada and experience in the war trenches in France. It soon became apparent that the young man was the illegitimate child of a local maidservant.[33] In his entry on pseudologia phantastica for his textbook, Swiss psychiatrist Eugen Bleuler identified the mental pathology particularly with panhandlers and swindlers.[34]

These European concerns about class passing were more muted in the case studies collected by William and Mary Healy. Certainly, anxieties about the relationship between their clinical population and the marketplace remained, but dramatic tales of peasants living as nobles were absent. The Chicagoans claimed to eschew reporting the more "picturesque" adventures of the long-time sufferers so prominent in the European literature. Unlike European doctors, who derived their cases largely from the long-term inmates of asylums, the American investigators gained their insights from consulting with patients in a clinic. The Healys focused on the manifestation of pathological lying in juveniles, a group whose shorter life histories they hoped would facilitate identification of the root causes of the behavior. This shift in population led to a transformation from a focus on class to an emphasis on gender and sexuality. Reflecting the influence of Freud's psychoanalytic theories, the Americans associated manifestations of pathological lying with sexual knowledge and experience, but denied that sex was an exclusive cause. Furthermore, pathological lying in the United States seemed to afflict a disproportionate number of females compared to males.³⁵ Despite this emphasis on gender, the Healys insisted that pathological lying was a distinct mental disturbance from that other prominent female malady, hysteria.

This Americanization of pathological lying derived from research conducted by William Healy and his associates at the Juvenile Psychopathic Institute. In 1899 members of Chicago's Hull House mobilized to establish a distinct Juvenile Court for Cook County. In their crusade to regulate the city, they privileged the young, usually from immigrant households, as the proper locus for reform activity. Reformers like Florence Kelley and Jane Addams protested the prevalence of child labor in the nation's urban centers, arguing that these employment practices threatened the moral fabric of the nation. They were further concerned with the commodification of young women, both as industrial workers and in terms of the commercialization of their leisure time. Even those children not employed in industry were threatened by the morally corrupting influences of the city and its entertainments. These stimuli, combined with poverty and ignorance, led to the moral depravity of urban youth as they engaged in conspicuous consumption, truancy, and petty theft. In 1904 Hull House member Ethel Sturges Dummer raised the funds for the Juvenile Psychopathic Institute in the hope of discovery the underlying causes of delinquency.³⁶

Healy's investigative procedure shared certain features with the methodology championed by these social reformers. The women of Hull House sought to combat urban ills through the collection of sociological data on working and living conditions in Chicago's tenements and factories. Through face-to-face contact with residents and the filling out of questionnaires, Hull House sociology privileged first-hand experience of city life over abstract thinking. By collecting and aggregating information on the individual household, these women generated new ways of perceiving the city and its problems. They translated the immediacy of their encounters into standardized graphical representations of living conditions.³⁷ Like Hull House investigators, Healy sought to map the subterranean origins of social problems. He and his staff questioned and catalogued the individuals brought to their attention through contact with the juvenile court. While his surveying techniques were tied to the political projects in Hull House sociology, Healy's place as a court-appointed arbiter, outside the traditional adversarial structures of legal witnessing, led him to embrace an ideal of disinterestedness, rather than empathetic observation, toward his objects of study.

Appointed as the original director of the Juvenile Psychopathic Institute in 1909, Healy wrote prolifically during the next decade, sharply criticizing nineteenth-century understandings of crime and delinquency.[38] The majority of earlier criminology had been grounded either in ideas about inherited criminal types or in general environmental explanations about the depravity of the urban poor. In contrast, Healy's primary targets for understanding criminal behavior were dynamic psychological experiences within the life history of the subject. The title of Healy's major work, *The Individual Delinquent* (1915), indicated his original tack.[39] Eight hundred pages in length and combining case histories with statistical analysis, this eclectic text established Healy as the leading psychological authority on crime in America. Healy received praise for the concreteness of his approach, which carefully integrated examples from his clinical experience into his theoretical models. He stressed the dynamic nature of delinquency: criminal behavior simply could not be reduced to a single cause, either hereditary or environmental. He shifted the criminologist away from the study of social classes or inherited natural kinds toward the study of individual personalities.[40] For Healy, deviance remained largely a question of the functioning of mind and individualized psychiatric pathologies. Families could have numerous children while only having a single offspring who engaged in delinquency. Healy's biographical approach emphasized habit formation as the key to understanding behavior. He saw the repetition of certain acts as molding the individual psyche and engraining behaviors into the will.

His expositional style involved presenting the case histories of various delinquents, grouped as representing particular conditions. The publication of edited case histories served an important pedagogical function for Healy; he hoped that future specialists reading his work would acquire the ability to collect subtle forensic clues from the cases, a skill to be applied in their own investigations. Reading case histories was a means of training one's observational skills to be able to detect similar cases in a clinical situation. Late in life, Healy insisted that inspiration for such a methodology did not derive from medical practice, but from the case-history pedagogy then being developed in business and law schools.[41]

He integrated into these observations and interviews a series of recently developed psychological tests, the most famous being the Stanford-Binet intelligence scale. Healy did not believe that mental conditions could be made manifest simply through the application of a mechanical procedure; his was a clinical art of interpretation. Certainly he played upon the cultural authority of mechanized tests that produce regular, standardized measures, but he was adamant that these tools could not be used indiscriminately by just anyone. Critical to their success was the insertion of a trained eye as a crucial aspect of the assemblage that made these technologies work.[42] More broadly, Healy's interpretive art began from a carefully cultivated relationship with the subject. Of uppermost importance was "a temperament or an attitude of mind calculated to develop friendly cooperation with the offender and his relatives." Healy felt that women, in the tradition of Hull House, could play a role in such investigations and were capable of administering the tests. He would collaborate with both of his wives, Mary Healy and Augusta Bonner, on major psychological monographs. Yet the importance of an interpersonal relationship of trust was such that Healy did not feel that women could serve as the primary investigators since they lacked the moral authority of men and Healy was skeptical that young men would reveal their inner troubles to women. Although embracing mechanical tools for visualizing deception, Healy belonged to the ranks of psychological experts seeking to secure a place for their own carefully crafted judgments.[43]

Among the hundreds of cases he examined for the juvenile court, Healy argued he had uncovered a select number of delinquents with a particular maladjustment: pathological lying.[44] He dedicated his second monograph, coauthored with Mary Healy, to the explication of this disorder. They defined the condition as "falsification entirely disproportionate to any discernible end in view, engaged in by a person who, at the time of observation, cannot definitely be declared insane, feebleminded, or epileptic."[45] In his role as a court psychiatrist, Healy regularly encountered young people who engaged in petty theft and delinquency, and often these individuals would attempt to manipulate or deceive an authority figure like Healy. He argued that pathological lying, however, was different in kind. Unlike the actions of other youths, whose lies could be understood on a rational basis, there was no reason for the pathological liar's falsehoods. What marked this condition as a distinct form of conduct was that the lies told did not benefit the falsifier, nor did the liars seem to derive pleasure from their deceptions. In certain instances, the lies involved self-incrimination in crimes that the individual had not committed or that had not occurred. Contrary to similar conditions described by European psychiatrists, the designation of pathological lying, argued Healy, should be reserved exclusively for otherwise *mentally normal* individuals, demarcating these kinds of individuals from the usual targets of eugenics: the "insane," the "feeble-minded," the "epileptic." In other words, he wanted to distinguish pathological lies from the delusions of the insane, as well as from more everyday self-interested lies.

The narratives that the Healys published in 1915 feature tales of misconstrued sexual knowledge, female mobility, and financial vulnerability. A woman they called Inez M. provided the most dramatic of their cases.[46] Inez, claiming to be seventeen, arrived at a Chicago boarding house for young women with few possessions and little money. She first came to the Healys' attention when she falsely confessed to being a lost heiress when questioned about the possibility by the police. Inez did not possess any of the telltale physical signs of deviancy or degeneracy in accordance with the logic of the time. In their case history, the Healys specifically remarked on her "markedly strong, regular, pleasant features, including a set of teeth well cared for."[47] Her outward appearance was supplemented by her verbal skills and intelligence, both of which were deemed to be well above the norm.

Despite this initially positive assessment, as the full details of Inez's life history were brought to light, the Healys determined that she was actually the perfect manifestation of the pathological liar. When first interviewed, she recounted how she had lived in a series of abusive homes in the American South before migrating to Chicago. Inez also claimed to suffer from a series of ailments and serious illnesses: persistent pain from an appendicitis scar, incurable diabetes, Bright's disease. She even produced a blood-soaked handkerchief as evidence that she was suffering from tuberculosis. During an operation, the doctors removed an encysted hairpin that had been straightened and inserted through an old appendicitis wound. Inez was willing not only to simulate ailments but also to inflict bodily harm upon herself to maintain her deceptions. When confronted with skepticism, Inez fled, but wrote to her court-assigned guardian a few months later asking for money to pay for further medical work. She would eventually return to Chicago and the institute a number of times more. Through correspondence with other state institutions, the Healys pieced together her life history. She was not in fact an adolescent of seventeen but a woman of twenty-seven, although she consistently refused to admit to it. For nearly a decade, she had crisscrossed the nation, coming into contact with the embryonic social services in these regions and ultimately alienating herself due to her inconsistent, but perpetual lies. During an eight-year span, she had been hospitalized eighteen times. Her illnesses would often cease upon her release from hospital as a malingerer. Even after her birth mother had been located, Inez refused to change her story. The case study emphasized how she maintained her lady-like demeanor, her well-modulated voice, and her exquisite verbal expression throughout these events.

Not all of the cases were this dramatic, but the Healys catalogued a long list of deceitful behavior that did not seem to benefit the individual.

One young woman approached a professor at her university to confess to a murder that had not even taken place. Another teenager appeared at a social service center claiming that her enlisted brother had just died in a local hospital. When the social worker accompanied the young woman to the hospital, it was soon found out that no such individual had ever been admitted.[48] What was striking about these deceptions was how easily they were uncovered by the parties involved. The liars themselves were eager to bring their confidants into situations that soon revealed the artifice of their tales.

Such was certainly the case with Janet B., whose example best illuminated the connections among commerce, profitability, and sexuality in the definition of the pathological liar. At first glance, Janet appeared to be a self-sufficient, autonomous young woman. Running away from home at nineteen, she had successfully managed to find gainful employment in New York City. Within a few weeks, she approached her department's manager informing him that she had to support not only herself but an elderly man who had saved her life years before after an accident. According to her story, her parents had paid a pension to the gentleman, but they had passed away the year before and Janet could not afford to support them both. The story was fabricated and the man did not exist, but this was no ordinary swindle, for Janet refused to accept money. Janet also became known for telling elaborate tales of her family's former wealth and influence within the community. In a similarly boastful vein, Janet wrote to her parents, whom she had claimed to the department manager were dead, claiming to be earning fourteen dollars a week at her job, despite making only a little more than half that amount.

Janet's fabrications around money management first indicated the possibility of pathological lying, but Healy soon linked her distortions to a more deeply rooted problem, one of a sexual nature. Her parents claimed that her deceptions had become extensive prior to her leaving their home, culminating in her claims that a local boy had proposed marriage when he had not. Such fabrications had proved to be both embarrassing for the family and detrimental to Janet's reputation. In New York she had become similarly infatuated with a young man who was staying at the same boarding house. Despite her parents' claims about her moral rectitude, Healy soon uncovered repressed episodes of sexual knowledge through his interviews with her. He determined that Janet's propensity to deceive, beginning at age twelve, coincided with her secret friendship "with a certain group of girls on a hillside and she indulged in many conversations about sex matters."⁴⁹ From these older girls, Janet had learned the means of "self-gratification" and her tendency to lie was born out her attempts to conceal her practice of masturbation. From her initial attempts to cloak her new sexuality, her deceptions proliferated so that they came to dominate many aspects of her life, to the extent that the young girl even reported difficulty sleeping from being so consumed by sexual thoughts.

Healy felt that pathological liars warranted particular attention not because of their large numbers, but because they posed particular problems for the justice system. The nature of their behavior brought them into much greater contact with the police and the courts than was typical of the general population. Due to their habit of deceiving, they made highly unreliable witnesses and lawyers needed to be able to detect them in order to avoid miscarriages of justice. An even more pressing concern was their propensity toward false accusation and self-incrimination. The false charges leveled by them cost the legal system financially while also threatening the proper administration of justice. For these reasons, forensic knowledge of the otherwise normal-seeming pathological liar was a necessity.⁵⁰

It is significant that a theory of pathological deception emerged out of the particularities of the juvenile court system. As an institution, the court was designed to govern over an unruly and disruptive population that it also helped constitute: juvenile offenders. The court aimed to regulate juvenile behavior through a therapeutic regime designed to cure them of their deviancy rather than punish them as adults. In practice, the juvenile court expanded the police powers of the state so that individuals and actions that were previously not deemed criminal came under the court's jurisdiction. In invoking its *parens patriae*, the municipality quite literally became a surrogate parent for a host of new subjects. As Healy later recalled, the first individual who proved to be a pathological liar was brought to his attention by a social worker.⁵¹ Because of this more interventionist policy, certain individuals came into greater contact with scientific experts.

Pathological Lying in the Culture of Appearances

The parsing of deceptive appearances was central to clinical services Healy offered to the court. In his earliest report on juvenile thieves, he drew attention to the potential contrast between the delinquent's physical appearance, especially an apparently "innocent" face, and his criminal behavior.[52] The pathological liar took this disconnect to new extremes. False visual and auditory cues defined the pathological liar. As one commentator noted, "Their countenances mirror whatever the lie is trying to convey, whether it is horror, repulsion, dramatic interest or a profound conviction of truth."[53] They were often well dressed, had seemingly healthy bodies, and passed as ordinary members of respectable society. Unlike the usual targets of juvenile psychiatry at the time, pathological liars were not deemed feeble-minded, but rather drew notice for their superior intelligence. A closely related characteristic of pathological liars was their superior verbal skills, which enabled them to generate such grand fabrications.[54] It was only with extended contact and conversation that the illusionary nature of this visual façade became apparent and the enormity of the deceptions evident.

Like his contemporaries working in hospitals, Healy turned to diagnostic instruments to probe beneath the surfaces of the body.[55] In particular, he highlighted the pathological liar's propensity for invention when responding to the visual clues in the "Aussage," or testimony, test. Healy had been working alongside members of the institute "to discover the power of the subject to report faithfully what he has seen." By 1914 Healy boasted that, when it came to the study of testimony, his "own series of studied cases is much larger and more carefully correlated with other information about the individual than any that has heretofore appeared." The aim of the test was the establishment of "correlations between laboratory work and the individual's reactions in social intercourse." The Healys claimed that the nature of the pathological liar had first emerged through these investigations.[56]

In this version of the testimony test, a clinician presented the subject with a familiar scene featuring a variety of observable objects and actions relevant to the setting. At the Juvenile Psychopathic Institute, the illustration was of a butcher shop, an image chosen for its familiarity to a wide array of observers. The subject viewed the scene for ten seconds and was then called upon to describe the picture in as much detail as possible without prompting. Once the description was complete, the examiner started to question the subject about the content of the image.[57] The purpose of this was twofold: on the one hand, the psychiatrist could see what other details could be elicited with a little aid, while on the other, the subject was asked about seven objects that were not present in the image but could have been. This aspect of the investigation aimed to observe the subject's openness to suggestibility and her unreliability as a witness. The Healys contended that the pathological liar would take these prompts as an opportunity to begin her fabrications.[58]

The etiology of the pathological liar was complex and followed Healy's general principle for analyzing delinquency: that there was no unique cause. Heredity certainly contributed to this disorder but was not a definitive cause. Instead, individuals may have possessed "inherited instabilities" that socioenvironmental factors could agitate. Much like the female reformers funding his research, Healy stressed the importance of evaluating the moral quality of the child's household. One such moral influence was the exposure to unacknowledged familial secrets. Also important was the potential presence of repressed mental conflicts in the minds of the young delinquents. They suggested that "anything deeply upsetting, such as the discovery of the facts of sex life or questions about family relationships, are the incidents which cause the trouble." They obliquely drew upon Freudian theories to suggest that such repressions were the result of early sexual experiences or contact with sexual knowledge. After all, "for students of modern psychology nothing more need be said on this point."[59]

Where European psychiatrists tended to stress the pathological liar's degenerate, incurable status, the Healys argued that a therapeutic regime could ameliorate the person's condition and restore her to an honest life. Although none of the cases presented in *Pathological Lying* offered sufficient long-term observation to definitively prove a lack of recidivism, they believed that a number of the subjects had dramatically altered their habituated behaviors. Effecting such a change required that the psychiatrist confront the sufferer with her deceits in a therapeutic setting as well as identifying the underlying mental conflicts and repressions that initially generated the deviance. Former pathological liars needed to abandon associates of bad moral character and locate a proper, moral outlet for their superior imaginative sensibility. The capacity to invent and create could be transformed from a failing into a virtue if channeled into a productive end, such as journalism.[60]

Not everyone agreed that the qualities associated with the pathological liar were entirely undesirable. For example, a 1916 article boldly proclaimed that "the pathological liar is superior to the truthful man." The author noted how this individual was bestowed with particular gifts far superior to the ordinary person, namely a "power of rhetorical presentation." It was not every day that one came across a supposed delinquent who excelled in certain realms of human endeavor. Verbal acuity, the capacity to convince and persuade others through the use of mere words, was a highly prized ability. Moreover, the pathological liar was a man bestowed with a special creative capacity; he was unlikely to steal ideas from another, because his type was so prone to invention. This popular account reread pathological lying as a masculine mental talent rather than a feminine mental disease. The author reinterpreted the pathological liar as an identity fit for the times, one attuned to the dynamics of money making and a society that placed great emphasis on the cultivation of surface appearances. A vibrant, magnetic personality and ability to influence others were deemed gifts, not defects.[61]

Among forensic specialists, the concept of pathological lying met with a mixed reception. Two elements were especially controversial: the almost exclusive reliance on case histories as a mode of presentation, and the Healys' insistence that pathological lying should be reserved as a designation for those who did not suffer from insanity. Richard Cabot succinctly captured the second criticism, arguing that "pathological story inventions, like the delirium of fever, are to be treated, not believed or disbelieved. They are neither honest nor dishonest, but symptomatic."[62] He believed that the Healys mistook a symptom of an underlying mental illness for a unique form of maladjustment. Herman Adler, who succeeded William Healy as director of the institute, complained that the Healys' narrative style of exposition lacked the precision that the replication of the quantitative data from the various mental tests would have provided. As a result "the cases are literary rather than scientific," the analytic methods simply the age-old ones of the "laity."[63] Adler also noted how the focus on the individuality of the pathological liar put the Healys at odds with a prominent and growing trend in psychology. Psychologists at this time negotiated a highly visible niche for themselves through their increased focus on the statistical aggregate.[64]

The Healys' notion of pathological lying, as transmitted to the legal community, had an even more controversial reception in the courtroom. Legal experts found that the insights from their psychological studies of testimony and deceit could be used to advance quite different interpretations than those originally intended. Wigmore cited the Healys' examples of pathological liars falsifying cases of sexual abuse as evidence that the testimony of women was inherently unreliable in these matters. Where the Healys had emphasized that pathological lying as a type of mental behavior was a relatively rare phenomenon, Wigmore's uses of the concept erased the notion of a specific pathology. Citing their work on pathological liars as the scientific basis for his argument, Wigmore suggested that "no judge should ever let a sex offense charge go to the jury unless the female complainant's social history and mental makeup have been examined and testified to by a qualified physician."[65] He incorporated the qualities identified with pathological individuals into the rules that framed how courts ought to treat the testimony of women in general. An early example of this logic can be found in a 1927 case in Chicago where a principal stood accused of having contributed to the delinquency of two pupils, aged twelve and thirteen. The jury based its acquittal of the accused on the defense's arguments, grounded in medical evidence, that the girls were "pathological liars" and hence their testimony could not be trusted.[66] Some within the legal community rejected this expansion of pathological lying to encompass all female testimony. These commentators insisted the Healys' study proved that only in very rare circumstances did female "sexual immorality" lead to consistent mendacity. This interpretation stressed that the investigations of pathological mental states should not be read back into the legal understanding of "normal" individuals. When it came to the legal question of whether one's bad character could be used to impeach one's testimony, C. W. Hall, writing for the *North Carolina Law Review*, found that the Healys' study could not be used to sustain such a view since their object of study was precisely the extraordinary.[67]

Bernard Glueck was among the American psychiatrists most responsive to the importance of pathological liars. Like Healy, Glueck's interest developed from his immersion in the legal system, in his case as founding director of the psychiatric clinic at Sing Sing prison. According to Glueck, pathological lying closely paralleled another stereotypically female mental maladjustment, kleptomania, or "pathological stealing." Both diagnoses pathologized the lack of self-control among women operating in increasingly fluid social situations.[68] In both instances, Glueck considered that the exhibited behavior was indicative of a mental disease because it did not benefit the sufferer. As Glueck wrote, in language closely mirroring descriptions of the pathological liar, "the misconduct is disproportionate to any end in view." Women did not profit from their actions in terms of monetary gain. Pathological lying and kleptomania inscribed economic rationality in the definition of one's mental health. Pathological lying resonated with Glueck because he believed that "the transition from absolute mental health to distinct mental disease is never delimited by distinct landmarks, but shows any number of intermediary gradations. Nowhere is this better illustrated than in the pathological liar."[69] The psychological mechanisms of lying were the same for the healthy and for the diseased person, so the study of the pathological liar could help explain the ordinary person's recourse to fantasy, deceit, and lying. The significance of the pathological liar lay in the ways in which it illuminated the continuum from mental health to abnormality to insanity and helped explain the prominence of deceit in the lives of supposedly normal persons.

Inspired by both progressive jurisprudence and psychoanalysis, William Healy tried to shift the focus of forensic science away from the criminal anthropologist's study of types and kinds toward dynamic processes. His clinical practice with the courts brought the question of testimony to the forefront of his thinking. This interest made visible a class of subjects who seemed to deceive without regard to benefits they might accrue from such behavior. The Americanized version of the pathological liar diagnosis that he and his wife presented ultimately stayed true to his vision of denying the existence of fixed types in criminology by presenting these individuals as abnormal and maladjusted, but within the purview of therapeutic reform.

William Moulton Marston's Detective Stories

By the 1920s this psychotherapeutic account of the self as ruled by numerous deceits was garnering considerable interest in American newspapers, plays, and popular songs.[70] At this time, a number of scientists revived Münsterberg's project of generating a legally recognized deception test. In contrast to the analyst's use of ambiguous symbols and complex life narratives, the technique known as lie detection was a deliberately straightforward and unabashedly physiological approach for locating sincerity. The advocates of the lie detector insisted that the "deception complex" was not a marker of pathology, but rather a temporary phenomenon expressed by every normal individual. Their approach presumed that the changing rates of a person's basic bodily processes constituted an index of emotional exertion and could identify deceptive statements. Building on this proposition, a number of inventors assembled a hybrid technology that combined psychological theories, medical technology, and the commonsense epistemology of the police officer. The result was a technique whose scientific and legal status was questioned from the start, but that nonetheless shaped policing in the United States.

These inventors typically denied or downplayed the groundwork for their success laid by Healy and other psychiatrists operating in judicial settings. In a 1924 article William Moulton Marston complained of a lack of psychological investigations of "normal" adult testimony. Similarly, when John Larson examined the intellectual antecedents to his own approach to the detection of deception in criminal cases, he quickly dismissed the applicability of the Healys' studies of pathological lying.[71] To a certain degree these criticisms had merit. Although the identification of the disorder's typology might help the courts deal with a particularly problematic class of witnesses, it did not address the more general concern about the overall reliability of witnesses, nor how to scientifically extract true testimony from more run-of-the-mill suspects.

While advocates of the deception test downplayed the relevance of psychotherapy, the era's crime fiction prominently featured the affinities between these fields. In 1910 Arthur Reeve introduced his "scientific detective," Craig Kennedy, whose adventures sold over two million copies over the next two decades.[72] In contrast to Sherlock Holmes's use of deductive reasoning to solve mysteries, Kennedy relied entirely upon a host of wondrous technologies from his laboratory to reveal the culprit. Reeve suggested that this brand of truly scientific detective story "began when several writers tried to apply psychology, as developed by Prof. Hugo Muensterberg of Harvard and Prof. Walter Dill Scott of Northwestern University, to either actual or hypothetical cases of crime." The resolution of Kennedy's cases frequently involved "psychometers" designed to identify suspects by their altered emotional states during questioning. Perhaps most prophetically, the very first Kennedy story hinged upon a scientific interrogation of a female college student, using reaction times and a plethysmograph to monitor her changing blood pressure. Kennedy did not avail himself solely of physiological psychology. In "The Dream Doctor" (1913), he demonstrated his familiarity with "the latest untranslated treatises on the new psychology from the pen of the eminent scientist, Dr. Freud of Vienna." Reeve's detective stories played an important role in framing public expectations about the psychological approach to crime, whether physiological or psychoanalytic. He portrayed these diagnostic tools as unambiguous truth machines requiring no interpretive judgment on the scientist's part. Reeve presented both forms of detection as thoroughly mechanized procedures, untouched by the vagaries of human judgment.[73]

The interplay between science and fiction also played a crucial role in the development of the actual deception test. From the outset, Marston's investigations into what he called the "lying complex" fused together experimental physiology with the drama of stagecraft. While a student in Münsterberg's laboratory, Marston conducted a series of experiments that aimed to demonstrate the correlation between acts of deception and changes in blood pressure. He contended that monitoring a person's blood pressure provided the most reliable index of deceit compared to the calculation of reaction times or changes in the skin's resistance to electrical current. Marston built upon the work of the Italian physiologist Angelo Mosso, who linked fluctuations in blood pressure to the expression of intense emotions and the experience of fatigue. The research of Harvard physiologist Walter B. Cannon served as another important influence. Cannon argued that emotions originated from impulses within the subcortical regions of the brain. What Marston found interesting in Cannon's theories was his contention that these affective states had a dramatic effect on the workings of an organism's glands and muscles. Cannon had correlated six basic emotions with differing branches of the automatic nervous system. According to this schema, a rise in blood pressure represented the bodily expression of certain primal emotions, especially fear. Marston interpreted the expression of this emotion as the subject's not wholly conscious worry about the exposure of his consciously crafted deceptions. To bolster such claims, Marston had his initial test subjects record the introspective events they experienced during the test. While many respondents noted that they derived a sense of pleasure in pulling off the deceit, Marston found that the majority of subjects described a feeling of anxiety over the risk of being uncovered.[74]

Marston drew not only on physiological theory to explain deception, but also on its apparatus. The measurements recorded by a sphygmomanometer served as the centerpiece of his deception test. Following a 1901 European tour, Harvey Cushing introduced the blood-pressure cuff to the United States as a simple diagnostic device requiring little skill or interpretation. While the cuff was initially met with suspicion by doctors, its use had become a routine aspect of care in Boston hospitals by 1910, albeit supplemented by the physician's judgment utilizing the stethoscope.[75] Rather than develop a novel apparatus, Marston relied on the "Tycos" brand cuff, an already standardized and widely marketed medical instrument. During the course of the experiment, he would take a reading of the subject's blood pressure after each question with this unremarkable device. Marston enrolled the instrument's clinical reliability with the aim of convincing skeptics of the validity of his own findings. In this regard, it was significant that he emphasized that his approach constituted no revolution on the level of scientific practice. Marston explained that in developing his deception test he had invented no new devices, but simply used already standardized medical equipment.[76]

Grounded in experimental physiology, storytelling and role-playing also featured prominently in Marston's laboratory practice. Trying to establish legal applications in his very first publication, he cast his experiment as a fictional court case, complete with witness, prosecutor, and jury. He asked his subjects to play the role of a witness before a prosecuting attorney, "resolved to save a friend who was accused of a crime." Marston requested that these university students simulate acts of deception—with the feigned emotional investment of saving a friend—while he monitored changes in their blood pressure. In other words, he did not ask his subjects to lie or tell the truth, but rather lie or tell the truth within the narrative he provided for them. Fiction was not the antithesis of experimental realism. He argued that this dramatic aspect actually heightened the validity of his experiments as "all the subjects took the task of deceiving the jury very seriously, doing their utmost to outwit both jury and experimenter."[77] Building on precedents established by his mentor, Marston employed an investigative style that rendered laboratory experimentation into a species of showmanship from which the detection of deception would rarely veer.

The United States' entry into the Great War furnished Marston with an opportunity to take his test into the world. Having secured a commission from the National Research Council's Psychology Committee to determine the applicability of his test to the war effort, Lieutenant Marston took his lie-detection technique into a number of judicial settings. He first used his test in a Boston courtroom, where he selected cases from the docket in which the testimony given during the experiment could be verified later that day through a medical examination. In one instance, the police had arrested a former female drug user since they found a syringe in her place of residence. Marston's test indicated that she was truthful about no longer using drugs. A medical examination later that day indicated that no needle marks were found on her body and she had gained weight, two physical indications she was no longer using narcotics.[78] His most ambitious project in this regard was his attempt to introduce his test into the court-martial trials of allegedly treasonous soldiers. While the military brass recognized the appeal of Marston's approach, in 1919 they remained convinced that it required further refinement.[79] In 1918 he obtained a law degree and three years later Harvard granted him a PhD for his work on the deception test. Upon receiving his doctorate, he joined the faculty of American University in Washington.

In his initial 1917 report, Marston made clear—as Goddard had a few years earlier—that his test could not provide a direct measure of a subject's truthfulness. Rather, he claimed to have developed "a practically infallible test of the *consciousness of an attitude of deception*."[80] His test targeted only intentional deceits, not errors of judgment. Following Cannon's physiology of the emotions, his was a measure of the embodied fear generated by the anxiety of being caught in a lie. In other words, the acuteness of a subject's conviction that his or her detection was imminent was central to the test's reliability. Subsequent researchers recognized that heightening a subject's sensitivity and unease in the testing situation rendered them more legible. Reflecting on this dynamic, Marston raised the possibility that his test might be a kind of patent medicine, those proprietary remedies whose value lay "in the suggestion exercised upon the mind of the patient by the mystery of the contents."[81] In this respect, Marston's version of the deception test offered little improvement upon the word-association test despite the veneer of greater accuracy supplied by its increased technological complexity.

Dormitory Sisters and Other Case Histories

Across the continent in Berkeley, California, Marston's initial publication inspired further experiments, conducted not in a university laboratory, but under the auspices of the city's police department. August Vollmer, the city's chief of police since 1907, was nationally renowned as the leading reformer of police services. He argued that traditional police practices inefficiently managed resources, lacked accuracy, and led to the corruption of officers. Possessing no university degrees himself, Vollmer turned to scientific researchers at local universities to provide knowledge about chemistry, biology, and medicine to assist with difficult cases. Vollmer felt that relying on such experts as resources external to the police department was inadequate. Instead, he desired the transfer of knowledge from the scientific community to under the patrolman's cap, and sought an overhaul in police training in order to merge the everyday, commonsense knowledge of the beat cop with the latest advances in forensics. He raised the minimum educational requirement for the Berkeley force to at least a high school education, although most police officers had taken some university courses. As part of this project, Vollmer introduced a program to recruit men and women with advanced degrees in medicine, social work, and psychiatry as police officers. Under his leadership, the Berkeley department emerged as a paragon of scientific policing. This movement aimed to demarcate the bureaucratic-administrative functions of government from the vagaries of machine politics and party loyalties. It entailed the reform and expansion of government services while trying to limit the influence of electoral politics.[82]

As historian Ken Alder has argued, two competing visions for the deception test emerged out of the Berkeley police department. John Larson served as the prototype for Vollmer's "college cop" and he labored to establish the test's scientific validity. He possessed a doctorate in physiology from the University of California, later acquired a medical degree from Rush, and then studied psychiatry at Johns Hopkins University under Adolf Meyer. Larson's protégé, Leonarde Keeler, possessed a very different sensibility. He had barely completed his bachelor's degree in psychology at Stanford and undertook no medical training despite Vollmer's persistent urging. Keeler believed that the future of the test lay in the commercial realm, as a patented commodity marketed to police departments, corporations, and the federal government. Larson's preoccupation with the careful calibration and validation of the test did not appeal to the entrepreneurial Keeler.[83] By and large these two men adhered to these dichotomous roles, although Keeler insisted that he pursued commercialization to fund future research and Larson could be persuaded to engage in commercial ventures, especially when what Vollmer called his "paranoidal personality" led him into personal financial difficulties.[84]

These two visions were not diametrically opposed in the beginning. Larson originally sought to validate his test through headline-grabbing cases that mirrored Craig Kennedy mysteries and became central to polygraph lore. In 1921 Larson reported that he had tested actual police suspects. The residents of one of UC-Berkeley's female dormitories, College Hall, seemed besieged by a thief living among them who had stolen approximately $600 worth of property over the course of several months. The Berkeley police could not narrow the list beyond three suspects. Without a resolution in sight, the officer in charge requested that Larson attempt to solve the case with his novel test. In his written report, Larson emphasized how he remained ignorant of the short list of suspects and began by testing twenty-five residents whose rooms were in the area of the thefts. By the time this first group had been screened, Larson was convinced he had his culprit. The measures of all of the girls but one demonstrated a "marked uniformity." The record of the one exception exhibited an involuntary holding of breath alongside noticeable fluctuations in her blood pressure. When confronted in a follow-up examination, the suspect protested the questions, kept eyeing her graphical record, and stormed out of the interview. A few days later, she confessed, confirming Larson's suspicions.[85]

Where Marston had developed his technique through simulated situations in a laboratory, Larson would subject all suspects in select criminal cases to the test to see if it could detect deception under less controlled conditions. A confession from a suspect convinced of the machine's efficacy was the only confirmation that Larson's tests could receive. Contemporaries recognized the difference between these two methods of validating the deception test. Larson's work was grounded upon the belief "that deception tests can be applied much more effectively to persons to whom the examination has a very real interest, than to such persons who lie just for the sake of experiment."[86] To elicit a truly deceitful response something tangible had to be at stake. The subject needed to be fully invested in the test's outcome, a condition that Marston could hardly achieve through the dramaturgy of his early Harvard experiments. In these earliest practical applications, Larson used his cardio-pneumo-psychogram in cases where there were a large number of suspects for a single crime. The large number of individuals tested served as a type of experimental control since the guilty party's responses would stand out in stark contrast when asked the same questions as others.[87]

Vollmer hoped that the technology could deliver both more accurate police work dedicated to precise findings and a more sincere police force free of corruption. As suggested by Münsterberg two decades earlier, psychological tests offered an alternative to the thuggish violence of the "third degree." It promised a less violent method of obtaining confessions from hardened criminal suspects. Instead of officers resorting to physical intimidation, Vollmer hoped to defer such interrogations to a scientific expert. The suspects would not have to confess in words—their own bodies would betray them.

Putting the Deception Test on Trial

The designers of the deception test managed to develop an identification technology that had considerably more reach than the Healys' diagnosis of the pathological liar, but the technique failed in obtaining the endorsement of legal, as opposed to policing, authorities. As early as 1923, the ruling in *Frye v. United States* seriously curtailed the admissibility of the lie detector into court proceedings. Marston did not anticipate that the presiding judge would accept the test; rather, he hoped its use during the trial would function as a spectacular provocation to get a Supreme Court decision of about the utility and legality of the device.[88] The trial judge, Irvin McCoy, initially excluded the test from being administered on the witness stand. The Court of Appeals of the District of Columbia heard the challenge to the test's admissibility. This was the same court that had deemed irrelevant Richard Paynter's psychotechnical expertise in *Coca-Cola v. Chero-Cola* two years earlier. In the Frye case, Josiah Van Orsdel ruled that scientific evidence was only admissible if there was an overwhelming consensus within a given professional community that it was valid. The judge argued that the deception test had not attained such universal assent among either physiologists or psychologists.[89]

Best remembered as the case that set the legal standard for admitting novel forms of scientific evidence, *Frye* unraveled in a very specific time and place. Marston sought to authenticate his technique by proving the innocence of an African American youth accused of murder in the southern, racially segregated city of Washington, DC. When the American legal system touched the question of "race" during this period, the courts tended to curtail the rights and freedoms of African Americans. This role for the courts had been reasserted following the failure of Reconstruction in *Plessy v. Ferguson* (1896), where the Supreme Court held that the concept of separate-but-equal access to facilities was constitutional, a decision that sanctified formal segregation and legitimated institutionalized racism following the abolition of slavery.

The case began on Thanksgiving weekend in 1920, when an unknown assailant shot the prominent African American physician Robert Brown in the consulting office he maintained at his home. Although people witnessed a man enter the office and heard an altercation, none could identify him. Brown was a prominent community leader and president of the Mutual Benefit Insurance Company, and his murder drew considerable attention within the capital's African American community, whose press put the city's police under considerable pressure to apprehend a suspect. His family also posted a reward of $1,000 to solicit information leading to the conviction of the killer. Significantly, individual police officers, and not just members of the general public, were eligible to receive the reward.

In August 1921 the police arrested James Alphonso Frye as a suspect in an unrelated robbery and, while in custody, he confessed to the Brown murder. He would soon rescind his confession, proclaiming that he had been persuaded by two police officers to confess to the Brown murder in order to have the robbery charge dropped. Knowing he had an excellent alibi and would later be exonerated, Frye claimed that he and the officer would then have shared the reward money. Frye's story strained credulity. His first lawyer wanted his client to plead guilty and seek leniency from the court, but Frye refused, insisting on his innocence of the murder charge.[90]

His next attorney, Richard Mattingly, a novice court-appointed representative, was equally confused by the tenacity of Frye's claims. Mattingly did not contact Marston in order to exonerate his client. Rather, he hoped that this psychological expert could demonstrate to Frye how untruthful his story was, leading to a plea. Both Marston and the lawyer were startled to find that the test indicated that Frye was not deceitful. Mattingly tried unsuccessfully to get the results of Marston's deception test admitted into the proceedings a number of times. Perhaps most dramatically, he wanted Frye to undergo a deception test on the witness stand, demonstrating his innocence in a spectacular display before the jury. McCoy refused this request and ultimately excluded the lie-detector test as a legitimate form of evidence.[91]

Through the *Frye* case, Marston tried to craft a public persona for the psychotechnician as a humane and liberal administrator of justice. Rather than cooperate with the police in trapping suspects, he offered his services to the defense pro bono. Underneath this persona resided a showman bent on demonstrating the emotionality of humans. His earlier research had emphasized "the extreme emotionality of negroes," but in this case he portrayed the deception test as a race-neutral technology. His restraint was short lived, and Marston quickly returned the deception test to its roots in showmanship. In 1928 he orchestrated a spectacular demonstration of his apparatus to test the relative emotionality of blondes versus brunettes, using Ziegfeld showgirls as his experimental subjects.[92]

The judicial rejection of Marston's lie-detection technique and the growing skepticism expressed by academic psychologists did not curtail the test's impact.[93] In 1929 Keeler accepted a position at Illinois State Penitentiary, where he captivated Wigmore with a demonstration of the test. Deeply suspicious of Münsterberg's legal psychology twenty years beforehand and conspicuously silent during *Frye*, the senior statesman of legal scholarship enthused, "I write to thank for your very lucid and convincing demonstration of your apparatus for testing the truth of a witness's story. My former skeptical attitude has been very much altered since I listened to your statement and demonstration." In a letter written to Vollmer later that year Keeler described his excitement about a new plan to fund for research on the lie detector. Hearing rumors that department stores like Marshall Fields "credit a loss of about fifty thousand dollars a year to their 'sticky-fingered' employees," Keeler proposed that he place a machine operator in every store. At regular intervals, the operator would test each employee for knowledge about any dishonest behavior. The key to the plan involved having the operator serve as a permanent and highly visible fixture to "weed out the 'lifters' and gangs that I know to be working the store, eliminate these, and put the 'fear of the Lord' in the others." In short, Keeler hoped to promote employee honesty by casting an ever-present shadow of surveillance. The fees raised by such a service would fund a scientific institute to refine the test.[94]

The fortunes of lie detection benefited greatly from the Depression-era "war on crime." The fallout from the Saint Valentine's Day Massacre (1929), a brutal reminder of Chicago's gang warfare, served as the impetus for founding a Scientific Crime Detection Laboratory. Chicago lacked a competent ballistics expert to testify at the coroner's jury, and an examiner, Calvin Goddard, had to be found in New York City. Greatly impressed by the performance, jury foreman Burt A. Massee became convinced that Chicago needed a standing scientific unit to investigate similar crimes.[95] The laboratory emerged out a mixture of civic boosterism—its advocates emphasizing American backwardness in crime detection—and concerns for crime control at the tail end of Prohibition. Part of the laboratory's mission was the production of a new periodical, the *Journal of Police Science*, an organ intended to publicize relevant scientific advances to the national policing community. The journal reflected the crime control ideology of the laboratory's champions, who eschewed "the psychiatrical and reform articles" they saw as prevalent in other criminology publications.[96] The sciences represented among the laboratory's specialties included ballistics, toxicology, microscopy, handwriting analysis, and the use of photography. Ultimately a "Psychological Department" was added, but it would be a far cry from the psychiatric therapeutics offered by Healy. Instead, "psychology" almost exclusively referred to Keeler's use of the lie detector.

The Scientific Crime Detection Laboratory was formally affiliated with Northwestern University from 1929 to 1938. Such an association was not surprising considering that Wigmore was dean of the law school. Although severely hampered by the Great Depression, the acquisition of the nation's first crime laboratory promised to bolster the university's reputation. The affiliation with a university was emblematic of the political vision of the laboratory's founders. Central to the institution's success was convincing the public of the inherent reliability of its agents and their findings. Its supporters were adamant that it could not fulfill its service to city and country if it was administered by the municipal government or run as a freestanding corporation. The former would make its findings partial to the influence of graft while the latter lacked the requisite openness. To guarantee both "moral integrity" and "expert capacity," the university affiliation was "the feature that will avail most to assure the public confidence."[97] To this end, the laboratory was incorporated as a nonprofit body with a staff paid exclusively by fixed salary. Indeed, the imminent threat of the university selling the laboratory in 1937 prompted Wigmore to recall this founding vision. After reassuring the university president, pioneering applied psychologist Walter Dill Scott, that members of the laboratory were "real scientists, and not merely high class technical men," Wigmore explained why the laboratory could only function within the framework of the university, insulated from the political system. He insisted that

> *no* laboratory attached to a police department will *ever* have civic full trust and confidence. The powers of such a laboratory, as an engine of jealousy and spite and graft, to divert justice from truth and to injure the innocent, are enormous. Their possibilities of abuse are terrifying; because these scientific methods of proof are irresistible.[98]

Wigmore insisted that such a venture remain independent for the sake of the scientific credentials needed to best serve the public interest.

With the creation of the crime laboratory, Chicago became a magnet for those interested in applying science to the detection of crime. Calvin Goddard was persuaded to relinquish his private practice to become the laboratory's first director, while Clarence Muelberger joined the staff as toxicologist. While psychology and the detection of deception had not been part of the original vision for the laboratory, Keeler received an appointment as a full-time, salaried staff member in August 1930. He concentrated on coordinating the various apparatus and experiments assembled around deception into a standardized and patentable machine.

Although university affiliation supposedly served as the guarantor of disinterested status, the laboratory was nonetheless a commercial venture, with its advocates eager to secure publicity. Just as Joseph Jastrow had worked to legitimate the new experimental psychology among the consuming public through his anthropometric exhibit at the Columbian Exposition in 1893, the polygraph formed the centerpiece of the Scientific Crime Detection Laboratory's contribution to the policing exhibit within the 1933 Century of Progress Fair's Hall of Science. The promotional materials for the exhibit played upon the potential voyeuristic tendencies within the audience. *World's Fair Weekly* noted, "Here you get all the thrills, with none of the disadvantages, of having a burglar in your house." Prominent among these titillating displays was "the 'lie detector,' that tell tale machine which registers the jump of your nerves when you tell even a white lie."[99] Where Jastrow had attempted to quantify fairgoers' mental abilities in 1890s, in the 1930s the most notable psychological apparatus on display was intended to expose hidden emotions and personal secrets.

Under the auspices of the Crime Laboratory, Keeler advanced a coercive deployment of the deception test, one braided with the era's crime-control campaigns. For example, newspapers reported how African American suspects claimed to be have been swindled by police into taking the test that did not exonerate them.[100] This movement involved the repudiation of the progressive justice embodied by Healy and replaced it by "applying business methods to combat organized crime." Members of the movement dismissed the therapeutic sensibility of the Progressives, which they characterized as "providing for criminals flowers, libraries, athletics, and hot and cold running water, social visiting organizations, probation, paroles, pardons, and a lot of other things, until what was previously intended as punishment is no longer punishment." With its application in the courtroom curtailed by judicial skepticism, the lie detector found a more receptive audience among police and corporations. Keeler's persona centered on his role as a self-aggrandizing, masculine crime fighter. Larson complained that Keeler and his associates "almost routinely now complicate the test procedures by adding such questions as 'Are you a homosexual?' Have you had intercourse with more than two girls on campus?" If Marston's actions during the *Frye* case represented an attempt to align lie-detection technology to Progressive political culture, Keeler embraced the ideology of crime control.[101]

Keeler was not the only person linking secrecy, coercion, and crime fighting. In February 1930 Chicago's Association of Commerce announced the formation of the "Secret Six" as the response of "the business interests" to gangster violence. Avoiding partisan politics, the group promised to "supplement the work of the constituted authorities." Because of the threat to their personal safety, their spokesman, Robert Isham Randolph, declared that the identity of the force's members was a necessary secret. Keeler's approach closely resonated with the stance advocated by these contemporaries. Both championed private, commercial solutions to crime control where progressive, state approaches failed. Moreover, they all agreed that truth and justice required deceit and secrecy on the crime fighter's part. Indeed, the Secret Six's career ended in disrepute when their own members were charged with graft and corruption in 1932.[102]

The "commercialization" of lie detection as both a patented technology and as a fee-based service enraged Larson. He held the leadership of the Northwestern laboratory in low regard and viewed the entire operation as a business secretly run by Massee.[103] His self-image as *the* defender of science when it came to the deception test led Larson to end his friendship with Keeler in 1931. In a letter from that year, he praised Keeler's ingenuity when it came to his machine's design, but chastised his former ally for his promotional techniques. Larson insisted that the deception test was "not a technique which should be turned loose and used by laymen or even by men who have only been trained superficially for a short time." He felt that the test's long-term success hinged upon constraining access to the device in the near future. Larson insisted that he "had this question fairly well controlled until you started putting out machines, because until then this sort of work could only be accessible to university people trained in laboratories. Now anyone whom you or your firm release a machine to can go ahead and mess things up." For years to come, Larson complained about how he constantly had to "debunk and pull Keeler's chestnuts out of the fire after he has commercialized and charged for what should be a scientific investigation."[104]

The Deception Test and the Therapeutic Society

A number of the deception test's advocates argued that it possessed the added advantage of offering some therapeutic relief to the suspect. They found that a criminal's demeanor and attitude changed dramatically once his lies had been detected. Most suspects would provide a full confession when confronted with the revelation of a single lie. Once the confession was made, the physiological strain that had generated the higher blood pressure would dissipate. The result was a calmer and mentally healthier subject.[105] This led Larson and Marston to explore the boundaries between lie detection and psychotherapy.

Such a program began in earnest in 1923 when Larson became assistant criminologist for the Illinois Department of Public Welfare. Despite a cool personal relationship, both Larson and his immediate superior, Herman Adler, championed a role for the deception test in therapeutic penology. In a 1925 address before the state's Probation Officers Association, they discussed the dilemma facing psychiatrists wanting to work with prisoners. Psychoanalysis hinged upon securing rapport with the patient, but "with criminals who have become perfected in deception it is almost impossible for the psychiatrist to feel certain he has secured this rapport." The psychiatrist cannot rely upon the sincere commitment of an involuntary patient because of the "very faculty of being able to appear honest which plays such an important role in the lives of most criminals." They suggested that the deception test could, at the outset, determine whether the prisoner was genuinely devoted to personal change. The test would clear the way for a more efficient therapeutic encounter.[106]

Using their access to wards of the state, Larson and Adler also explored the relationship between the deception of others and self-deception. They wanted to determine whether they could detect the deception complex among institutionalized schizophrenics when discussing their hallucinations. These experiments suggested a new application for the test: to find the extent to which individuals would feign the symptoms of mental illness in order to minimize punishment for their crimes. While Adler and Larson did identify a class of psychotics whose responses indicated that they had some sense that their sensations were false, the tests also indicated that schizophrenics were genuinely convinced of their delusions.[107] Walter Summers, a Jesuit psychologist, reported on the similar response of a former asylum inmate who claimed he heard the voice of God. Summer was convinced that the "subject really believed, and the record substantiated his belief, that he was receiving direct communication from God."[108] These experiments probed the limits of the lie detector's applicability by reaffirming that the test was only capable of detecting conscious deceit. If individuals were sincere, the measurements of their bodies would not indicate a deceptive response. Larson deemed lying detectable in a healthy person because he understood it as a momentary lapse in an otherwise properly functioning body. In contrast, hallucinations, delusions, and self-deception became the very essence of the insane person's condition and, therefore, could not be isolated and measured. Each mode of deception was different in kind, thereby requiring a distinct disciplinary approach.

Marston went further, drawing the parallel between a criminal interrogation and a psychoanalytic session. He believed that his test not only could be used to detect the truth in particular instances but also, if properly understood, could potentially eliminate crime. Marston argued that at the heart of criminal nature was "the power to deceive and the consequent habit of deception." In his popular 1938 book on the lie detector, Marston asserted that deception could not simply be reduced to the physiological emotion of fear. Central to the young person's experience of deception was the thrill of fooling others. Due to the pleasure derived from this activity, the embryonic delinquent repeated deceitful activity until it had become an unstoppable habit. At this point, the criminal required a psychotechnical intervention to return to an acceptable kind of behavior. By undergoing analysis, the patient was forced to confront the deep, repressed emotional complexes that lay at the heart of dysfunctional behavior. This was a painful but necessary process to reformulate a more healthy sense of self. Similarly, when an examiner confronted hardened criminals with the fact that they could no longer successfully deceive, their power was broken. Once the criminal was fully convinced of his inability to deceive, "it becomes easy to break down all his habits of lying and build up instead mental habits of telling the truth. Then the person is no longer a criminal."[109]

Marston advocated the wide diffusion of knowledge about these aspects of the polygraph examination. He argued that dishonesty was at the heart of all American life from business to politics. He stressed the moral cost of "building civilization on the quicksands of deception."[110] With the wide diffusion of the lie-detection methods, such deceptions could no longer form the foundation of social life. Faced with constant exposure, the lies of the businessman, the politician, the parent would dissipate. The end result of a society endlessly subjected to the deception test would be a new moral order peopled by perfectly readable citizens. Marston's vision of a fully sincere, honest society revealed the perennial moral project that underwrote the scientific study of deception.

Marston's utopia had rather pecuniary roots. In 1936 he wrote Larson trying to entice him into forming a joint venture in New York City tentatively called "the Truth Foundation." Marston proposed that they pair together to combat the "the key to evil in the world—deception." At the time, Larson was somewhat receptive, having been recently fired from his position as assistant state criminologist for Illinois. Marston promised that "emotion-measurement can be used—and paid for highly—commercially." The project would encompass Keeler's program of employee screening, but also "further research on psycho-physiological procedures, personality research, writing, lecturing, getting out a psychological code of ethics and morality, training courses for children and adults, a magazine, sponsored publications, and so on almost ad infinitum." Marston also insisted that "all this enterprise on my part is your brain-child." Ironically, it was Fred Inbau, a close associate of the entrepreneurial Keeler, who condemned the grandiosity of Marston's claims. Inbau argued that Marston had forgotten that in the drama of the deception test he was "supposed to play the role of a scientist and not that of a popular magazine writer, or newspaper reporter, or special guest on some advertiser's radio program." In other words, despite constant protestations throughout the 1930s, each of these scientists advanced the commercialization of the deception test.[111]

Its limitations notwithstanding, Marston's imagined utopia revealed much about the shared reception of his deception test and psychotherapy in the United States. For example, his Harvard contemporary A. A. Roback concluded a 1917 survey of "the psychology of confession" in the legal sphere with its counterpart in psychoanalysis. Münsterberg, who served as a mentor to both Roback and Marston, had suggested this parallel between scientific alternatives to "the third degree" and the emerging "Vienna School" as early as 1907. The lie detector aroused modest curiosity within the country's nascent psychoanalytic community. For example, Viennese émigré Paul Schilder provided a sympathetic preface to Larson's 1932 exposition of the technique. He argued that the Berkeley experiments validated fellow analyst Theodor Reik's theory that there existed "a universal tendency in human souls to confess." This led to the interesting paradox "that human beings lie with their consciousness, but are truthful with their unconscious, and when they do not confess with their mouths, then they confess with their bodies." Other analysts remained more reticent. William Healy remained skeptical of Marston's claim that the measurement of blood pressure might reveal "repressed ideas and emotions" more efficiently than psychotherapy. Yet central to both psychotherapy and lie detection was a vision of the self with a mind mired in petty deceits, which were constantly betrayed by the honesty built into the body. Despite their differing opinions on the lie detector, these psychiatrists defined psychotherapy itself as a kind of truth machine, albeit one that demanded a lengthy and laborious procedure.[112]

Conclusion

In 1924 Doris Blake, champion of the "lovelorn," anticipated Marston's fantasy of universal sincerity in one of her nationally syndicated advice columns. Speculating on the latest fad in California, she queried whether in the near future her readers might enjoy "a little lie detector in your home." Blake answered with a definitive no, proclaiming that she would "rather have a washing machine." She recoiled at the thought of the "hideous world we'd live in if the truth were delivered daily, hourly, either by or to us, particularly by those with whom we have most intimate daily contact."[113] Her vision of the lie detector as a household appliance, analogous to the then-novel washing machine, indicates how some Americans understood the deception test from the outset as a commodity designed to make daily life more rational.

Blake's daily columns further illuminate a number of notable features about the reception of the lie detector and related forensic techniques. A self-professed reactionary on sexual matters such as casual dance-hall petting, Blake nevertheless challenged central tenets of the existing advice literature genre. Starting in 1913, she regularly reminded her readers of the normality and necessity of lying in certain social situations, especially within the bounds of matrimony. She reaffirmed female superiority in the realm of deceit, a talent for calculated insincerity necessary for preserving domestic harmony. Blake openly acknowledged that middle-class "appearances" were frequently fraudulent, but they nevertheless provided the foundations for both blissful private relations and a stable social order grounded in marital happiness. "Fibs" and "white lies" maintained the ease with which people interacted with one another.[114] While clearly rejecting Marston's utopia of a transparent society, Blake's columns marked a triumph for the normalized deceitful self that lay at the heart of both the deception test and psychotherapy.[115]

The introduction of psychology into the practice of urban policing was an important episode in the emergence of a therapeutic ethos in the United States. This therapeutic imperative originated in the political culture of the Progressive Era with its emphasis on the shared, social responsibility for crime. Even the lie-detector advocates occasionally partook of this vision. Champions of the polygraph shared with William Healy's clinical therapeutics a common commitment to identifying individuals who happened to commit crimes rather than delineating fixed criminal types. For contemporaries like A. A. Lewis, both psychological approaches offered the opportunity to pierce what he saw as the deceptive shell of outward appearances and bring into public view an authentic inner self. They accomplished this goal through the introduction of a host of newly standardized diagnostic technologies, whether the Aussage test, the case history, the galvanometer, or the blood-pressure cuff. In this regard, the practice of forensic psychology also reflected the increasingly technological character of American medicine during these years. Certainly, newspaper stories and detective fictions emphasized this technological dimension, but these scientists, much like physicians in the hospital, also insisted that their technologies required the interpretive judgment of the trained expert. Moreover, a constant refrain that these two techniques closely resembled the suggestive patent medicines of old shaped their reception. From their very beginnings, both the clinical and the technological deception tests were shrouded in charges of fraud. At the same time, their circulation ultimately worked to normalize the prevalence of a deceitful self.

CHAPTER SIX

Studies in Deceit: Personality Testing and the Character of Experiments

In 1956 William H. Whyte Jr. offered his readers some practical advice on "how to cheat on personality tests." It appeared as an appendix to his bestselling *The Organization Man*, a scathing analysis of corporate America's managerial bureaucracy. According to the *Fortune* magazine editor, modern enterprise abhorred individuality and creativity and fostered instead persons whose identity consisted primarily of belonging to an organization. He laid much of the blame on the science of human relations—especially psychology—for this sorry state of affairs. The personality assessments that managers administered privileged "the most conventional, run-of-the-mill, pedestrian answer possible." To enable his readers to beat this system, he offered some tips on how to answer the questions in order to "mediate yourself a score as near the norm as possible without departing too far from your own true self."[1] According to Whyte, people possessed an authentic inner nature, but the public presentation of this personhood required vigilant self-management due to the demands of modern, commercial civilization. He deemed such advice necessary because he presumed that his readers approached psychological apparatus with an unquestioning sincerity and that this credulity blurred into a dangerous conformity. The preservation of individuality and true democracy required some carefully circumscribed deceit.

Whyte's warning to his readers seems to reflect what anthropologist Catherine Lutz has called "the epistemology of the bunker." By this she refers to the ways in which the suspiciousness endemic to the political culture of the Cold War informed how people understood the self. She points out how Americans during the 1950s were preoccupied with conformity, obedience, and, most troubling, brainwashing.[2] Put slightly differently, the decade was a period in which what I have been calling the deceivable self posed an acute threat to the liberal political order, an order that depended upon an individualism that was not wholly trusted. This concluding chapter offers an alternative genealogy of these concerns by locating the problem signaled by Whyte in the cultural response to mass consumption and the corporation during the 1920s rather than the political culture of the 1950s.

Deception operated in two fashions in the emergence of the widespread personality testing: as the object under study and as an aspect of methodology. An important quality that early personality tests sought to measure was an individual's deceitfulness. Scientists designed tests for this trait because of a broader concern about the future of the nation's citizens. At the time, there were numerous jeremiads against the pernicious influence of new forms of mass culture that threatened the moral character of the nation's youth. Moreover, corporate leadership expressed worries about what it saw as the proliferation of embezzlement among seemingly trustworthy employees. These concerns led to renewed attention to "character education" in schools, churches, and civic groups. Psychologists attempted to measure the efficacy of these institutions in reducing dishonesty. They wanted to test not so much the youth's knowledge of morality, but rather actual behavior in situations that allowed for lying, cheating, and stealing.

Attention to the historical development of these studies and how they became central to the discipline of psychology places Whyte's concerns about vocational testing in a new light. Whyte complained of overly credulous test-takers, but the scientists who crafted those scales constantly expressed frustration about the opposite problem: uncooperative subjects whose potential knowledge risked undermining the test's validity. In 1928 psychometrician Louis Thurstone raised serious concerns about designing an objective scale for measuring a person's attitudes. Put bluntly, he worried, "The man may be a liar." Eliminating intentional misrepresentations did not preclude the possibility that the people would modify their publicly expressed opinions "for reasons of courtesy, especially in those situations in which frank expression of attitude may not be well received."[3] The validity of personality tests depended on the subject's willingness to disclose socially undesirable traits. Prominent solutions to this problem included the staging of dramatic testing situations, lying about the test's purpose, and observing subjects from hidden locales. Through their studies of character and personality, these psychologists came to the conclusion that deceitfulness formed a normal and even necessary element of a person's mental constitution. They also argued that deception formed a normal and even necessary element of psychological experiments and mental testing.

This chapter tracks the changing meanings of the character of experiments in these two registers. It maps how scientists and educators understood character as a mental quality, especially in its relationship to deceitful behavior. It also elucidates how they grappled with the moral valences of a discipline that required deceit to study character and personality objectively. The tension between these two forms of deception—that of the subject and that of the scientist—and their ultimate integration into psychology received its clearest articulation in Hugh Hartshorne and Mark May's *Studies in Deceit* (1928).[4] The book confronted directly the debates over whether deceitful behavior originated with the child's inner moral development or in the demands of the social situation, strongly advocating the latter stance. Perhaps most significantly, their book pushed the use of deception by psychologists to the center of the discipline. The introduction of deception into the design of American experimental psychology is often attributed to the innovations of Kurt Lewin and his students.[5] By the time he emigrated from Germany in 1933, the practice of deception was well integrated into American psychology. Since at least the 1890s, psychologists had used deceptive apparatus in the study of illusions, but outright deceit had been reserved for particular kinds of subject, namely the deceitful spirit medium and the criminal suspect. In early personality testing, psychologists assumed that due to the sensitive nature of the objects under study—one's vocational inclinations, sexuality, and, yes, honesty—that the ordinary subject was likely a deceitful one. Responding to the specter of these deceitful selves, psychologists designed tests with deceptive and often deliberately misleading appearances. Seen as a loss of liberty by Whyte, this materialization of the deceivable self within the test itself constituted one of the greatest victories for an earlier generation of psychologists. Deception served as the means through which their science could achieve success via widespread application while avoiding the damaging effects of exposing its methodologies to suspicious subjects.

Cultivating Character

In the early twentieth century, a number of civic organizations promised to train the child's moral character to impede what were seen as the corrosive effects of "the acids of modernity."[6] Two British imports—the YMCA and the Boy Scouts—were particularly prominent. The first Young Men's Christian Association began in London in 1844 and spread to the United States a decade later. Backed by local businessmen, the organization originally emphasized prayer meetings and Bible readings for men ranging from their late teens to their thirties. It was only during the 1890s that the YMCA began to reach out to younger boys with its physical activity programs. In 1910 a number of men involved with the YMCA founded the Boy Scouts of America (BSA). On the surface, the new organization resembled earlier youth groups like Ernest Thompson-Seton's failed Woodcraft Indians, but such appearances were misleading. The naturalist's earlier movement had involved an antimodern withdrawal through the fantasy of playing the role of a primitive Indian. What made the BSA different was its combination of homosocial camping with an embrace of civics education. Aimed at preadolescent boys, the BSA hoped to instill the moral character that modern industrial life demanded, yet eroded.[7]

In the 1920s the leaders of both organizations came to target urban boys due to fears about the corrupting influence of the city's commercial amusements. Following the First World War, many critics blamed novel forms of mass consumption (namely the automobile, the radio, and the cinema) for ushering in a new era of immorality. Numerous religious, civic, and business leaders advanced the claim that an American civilization increasingly oriented around consumption, commodities, and moneymaking was creating a moral vacuum and a socially irresponsible generation. In turn, spokespersons for the youth claimed to reject the "Victorian" morality of their parents. They claimed to adhere instead to a more authentic and honest attitude toward life by abstaining from the fraudulent artificiality of decorum.[8]

Many organizations founded in the 1920s specifically targeted deceit as the greatest threat. In 1922 the chairman of the National Surety Company, William B. Joyce, founded a national Honesty Bureau. Its stated aim was the cultivation of ethical behavior in the young to raise a trustworthy generation of corporate employees. Joyce complained that a lack of traditional values would ultimately unmake American commerce. Similarly, E. A. St. John, another officer of the National Surety Company, noted that the recent increase in illegal activities was in the realm of "financial crimes and not in crimes of violence." Young boys had to be taught "that a code of strict integrity is demanded in modern business, that upon this ideal of honesty financial credit and positions of responsibility are founded."[9] Dishonesty in the corporate sphere had important ramifications during an era dominated by the political ideal of "the associative state." According to its main architect, Republican Secretary of Commerce and future president Herbert Hoover, such a strategy should replace the Progressive Era's regulatory policies. Hoover was convinced that the proper organization of the nation's economy could be achieved without coercive state intervention. In this vision, trade associations served as a model for the integration and rationalization of industry. With individualism preserved, voluntary cooperation could arrange matters of common concern such as standards, prices, and wages.[10] Such a system required ethical self-management. Writing a year before the stock market crash of 1929, Hoover noted, "Our whole business system would break down in a day if there was not a high sense of moral responsibility in our business world."[11]

The organizers of the Honesty Bureau understood modern commercial culture as the cause of moral decay as well as the very thing in need of protecting. In a 1923 interview Honesty Bureau director William Byron Forbush laid the blame for the new morality on consumer desire. He told the reporter that "good times on unearned money is one of the dangers of the age."[12] Likewise, St. John blamed dishonest corporate behavior on a lack of self-control in the day's youth due to abundance of consumer pleasures available in the postwar economy. "We are the most extravagant and luxury-loving people in the world today and an extraordinary amount of energy has been devoted to the invention of expensive methods of pleasure seeking."[13] In other words, they understood the abundance of their particular historical epoch as seriously compromising individual honesty. The bureau's leadership feared that America as a civilization would fall victim to its own financial success. Forbush warned, "no nation has ever been able to stand prosperity and overcome its own internal corruption." To preserve the nation, the organization engaged in what it called "counter-education."[14] To this end, its officers develop a course in "business honor" to be run through local branches of the YMCA and prepared a freely distributed text, *The Honesty Book*, to be used by parents and teachers. Eschewing abstract moral lessons, the book presented its readers with concrete situations, either drawn from everyday life or relevant to the corporate world, that would permit reflection on the conduct such situations would require. Its aim was to inculcate the proper tendencies in future employees by training them from an early age to act as responsible corporate agents. To assist them, the Honesty Bureau turned to a large number of professionals, including "ministers, educators, theater-managers, playwrights, magazines, newspapers, moving picture scenario-writers."[15]

Organizations like the Honesty Bureau located deceitful behavior in the development of the child. Inspired by G. Stanley Hall's "genetic psychology," Forbush had been studying what he termed "the boy problem" since the beginning of the century. Deceit was a question of inner psychological constitution rather than circumstances. Individuals embezzled funds from the corporations that employed them because of defects in their moral upbringing, not because the momentary situation motivated them to do so. Certainly a person's character was not fixed at birth. Indeed, education was the favored solution because it could channel development along the right course. In the 1920s these organizations boasted of their ability to shape character in order to save America's commercial civilization from itself.[16]

The Era of Mental Measurement

Concerns about the absence of traditional moral moorings for character, at work in American culture for several decades, cannot adequately explain the shape that character education took. The movement came to embrace scientific modernism in the name of shoring up moral values. The Character Education Inquiry and related investigations sold themselves as a means of appraising the claims of character educators like the Honesty Bureau. They justified their science in the language of economic rationality. For example, psychologist Mark May noted, "Hundreds of millions of dollars are probably spent annually by churches, Sunday schools, and other organizations for children and youth with almost no check on the product—a negligence of which no modern industry would be guilty."[17] He defined character as a product, a commodity that had a commercial value with the psychologist providing quality assurance for various pedagogical strategies claiming to nurture moral development. The problem with existing institutions was not that they were improper, but that they might not be the most cost-efficient means of achieving their worthy ends. These psychologists represented the liberal wing of American Protestantism, one that embraced scientific methods for achieving religious ends.

The model of intelligence testing was paramount. In the first two decades of the twentieth century, psychology gained considerable public visibility through the prevalence of intelligence tests. In France Alfred Binet designed a series of tests for assessing intelligence levels that, in a modified form, became integrated into the everyday routines of American society to a greater extent than in any other country. In a geographically expansive and social heterogeneous polity, standardized intelligence testing promised an impartial means of ordering individuals based on their inherent merits. The most important figures with regards to propagating and domesticating intelligence testing in the United States were Henry Goddard and Lewis Terman. In 1916 Terman published an Americanized version of Binet's scale while Goddard institutionalized such measures within the medical community. During the First World War, enterprising psychologists led by Robert Yerkes offered their scientific knowledge to the military bureaucracy as a means of rationalizing recruitment and promotion. With their assistance, the soldier's capabilities need not be proved through experience on the battlefield; rather, persons could be sorted in terms of mental abilities based on their performance on a common standardized test. To meet the needs of the army brass, Yerkes redesigned Binet's intelligence test so that it could be administered to a multitude of individuals simultaneously and, consequently, be graded easily.[18]

The army psychologists' apparent success in creating a universal measure for intelligence and its distribution encouraged the application of mental tests in various realms. For example, Thomas Edison made national headlines in 1921 when he announced that his future employees would have to achieve a passing grade on a 163-item questionnaire the inventor had devised. Dubbed the "brainmeter" by the press, the vocational test was condemned as unreliable by academic psychometricians.[19] Regardless, Edison's use of the questionnaire spoke to the growing authority of mental testing. Not everyone was as sanguine about the relevance of the army tests to the demands of industrial life. In 1920 the *New York Times* cautioned its readers about the extension of testing into other realms. The paper warned,

> Psychology is a portentous word to the layman, implying many weird notions from mental telepathy to the ouija board. Therefore, when there is talk of applying the principles of psychology to industry or business, it is only natural for the average graduate of the school of experience with no fancy looking letters after his name to grow a bit suspicious.[20]

The article went on to claim that the science's most visible achievement was likely to invoke "a smile, a downright guffaw" from those who, as "a small and humble unit in the late army of millions," were subjected to its inaccurate assessment of ability.[21] The nation's leading newspaper depicted the academic psychologist as just another commercial predator preying upon the goodwill of the unwary businessman. From the paper's perspective, the discipline remained indistinguishable from the charlatans its members labored to expose. In 1922 Walter Lippmann further skewered the validity of the army tests in the pages of the *New Republic*. He condemned the psychologist's attempt to reduce humanity to a single metric. Two years later, the humorist Stephen Leacock mocked the trend of using a generalized mental test to determine for people "what particular business they are specially unfitted."[22] Despite the tests' appeal in a number of organizational settings, popular criticisms of the limits of what they could measure became a recurring theme in the national press by the mid-1920s.

For their part, the nation's leading educators announced that their role was not simply to instruct the intellect, but also to guide the child's moral development. This was a task whose progress intelligence tests seemed ill equipped to evaluate. As one journalist noted, "many a child with a soaring I. Q. may be quite lacking in human and moral virtues."[23] Intelligence testing, then, provided a dual legacy for the psychological study of character. On the one hand, it offered a model of inquiry for future researchers to emulate. The measurement of intelligence in both military and vocational settings provided the crucial precedent for the scientific assessment of morality. As a sympathetic psychologist remarked, "Elated with their success with the measurement of intelligence during the World War, psychologists turned their attention to the measurement of personality and character."[24] On the other, the new field promised to address a dimension of the person that people frequently accused psychologists of neglecting. Character education promised to measure what the intelligence test could not.

It was not only that the measurement of moral character had a certain resemblance to intelligence testing; also, the communities of practitioners overlapped extensively. In 1919 Terman declared, "There is nothing in one's equipment, with the exception of character, which rivals the IQ in importance." Furthermore, Terman argued that he could study both these traits using similar methodologies. In 1923 he wrote, "it would not seem absurd to predict that within a few decades we shall be measuring delinquency tendencies almost as accurately as we can now measure intellectual ability."[25] Not surprisingly, E. L. Thorndike also became interested in the possibility of measuring character. A member of the psychological committee organized to meet the army's needs, Thorndike was among the more vocal critics of Terman's understanding of intelligence as a unitary mental process. He did not object to the scientific investigation of intelligence, but he disagreed that intelligence consisted of a single entity. Thorndike was dissatisfied with Terman's Stanford-Binet Scale and crafted tests not only for what he called abstract intelligence but also mechanical intelligence, the capacity to visualize the proper relationships among objects, and social intelligence, the ability to operate acceptably in interpersonal situations.[26]

At Columbia University's Teachers College, Thorndike supervised the doctoral research of Paul Voelker. His 1921 thesis assessed the claim made by various civic organizations that they instilled upright moral behavior in adolescent males. Working in collaboration with the Boy Scouts, he focused on whether the ideal of trustworthiness, central to that organization's credo, was transmitted to those who participated in its activities. Voelker privileged "trustworthiness" as an object of investigation because it seemed an unqualified good in American society. He argued that regardless of whether one was an individualistic businessman or a collectively minded socialist, one would concur that the ability to rely upon others was a necessary element of social cohesion. The interdependency of modern living made honesty a necessary virtue.[27] A 1924 survey of "four hundred of the leading men and women of the United States whose names appeared in the latest edition of *Who's Who in America*" reported that the majority of respondents selected honesty as the most desirable trait to be inculcated in the young as a guarantee of good citizenship.[28] Citing a speech by steel executive Charles Schwab, Voelker insisted that integrity, or honesty defined as transparency, was the key to achieving personal success. In examining trustworthiness, he hoped to calculate "the irreducible minimum" for good citizenship.[29] Well into his forties when he undertook his doctorate, Voelker did not spend his professional life within the discipline of psychology. Education and school administration were his primary vocational interests. Upon completion of his dissertation, he was named president of Olivet College, in the Michigan town of the same name. He continued along this path and was promoted to school superintendent for the state.

His dissertation considered both practical and epistemological questions about the use of human beings in experiments. Already in Voelker's account, the very act of observing in such a manner as to guarantee valid results was at the forefront, as he was concerned with actual behavior and not simply with the knowledge of right and wrong. He insisted that his "tests must confront the individual with real, not imaginary, situations."[30] It was insufficient to question the boys about their moral knowledge, since earlier research had already demonstrated that the capacity for moral discrimination was distinct from the exercise of moral conduct.[31] What mattered was not the individual's rote memory of ethical principles, but rather the habits and behaviors they exhibited. In this regard, examinations of moral character ultimately demanded a different approach than those concerns with measuring intelligence.

To overcome such concerns, Voelker contemplated conducting a covert ethnography of the Boy Scouts. He mused whether "a boy should be followed for a period of days, weeks, or months, and carefully observed under all conditions and all situations, to see how he would respond." Much like the earlier exposure of psychics, the psychology of character involved the fantasy of becoming a private detective. Tempted, Voelker concluded that such an approach was not practical: "If a trained observer could be found who would 'shadow' a boy for a period of time it is doubtful whether he would not soon be detected by the boy. Moreover, how could this trained observer, or anyone else, guarantee that the boy during this period of time would confront every situation that we were able to present to him in the series of tests?"[32] The psychologist as private investigator was an attractive identity, but Voelker was unsure whether his collaborators possessed sufficient skills to remain undercover in their scientific espionage. He quickly dismissed it as an alternative, but the very suggestion indicated the continued appeal of the ideal of unobserved observation.

While he dismissed outright espionage as a methodology, Voelker's experiments required that the investigator approach the subjects using some degree of subterfuge. Ingenious experimental design could serve as a surrogate for actual detective work. As would become standard procedure when testing for moral character, the psychologist did not reveal the precise aims of the experiments to the research subjects. Voelker made no mention that he was particularly concerned with trustworthiness, even though he was convinced that some of the "brightest" may have guessed his true aims. He felt it imperative that he misdirect the vast majority of boys for fear that their knowledge could interfere with the manifestation of their typical behavior. Voelker's own deception went further than simply failing to disclose his true intentions to his research subjects. To complete his experiments, he enlisted the assistance of persons he called "confederates" who would present the experimental subject with false social situations. During the "purchasing errand test," he sent the young subjects to a local stationery store with a quarter to procure an item valued at nine cents. Unbeknownst to them, the woman operating the cash register was collaborating with the experimenter and she "accidentally" returned two dimes, a nickel, and a penny; after the transaction the subject possessed more money than he initially had. The test for trustworthiness was whether the subject acknowledged his windfall and returned the extra dime either to the cashier or to the experimenter upon his return. If, after being questioned, the subject still claimed that the money was his, then it would be determined that he was deceitful in this situation.[33] Such performances allowed Voelker to place to his study in a more naturalistic field, even though it was not a full-blown ethnography of juvenile behavior.

In important respects, prevalent attitudes about adolescence in the wider culture shaped Voelker's experimental design. Historians of the family have documented a novel sense of "bewilderment" that confronted the parents of the interwar generation. Mass consumption in the form of the cinema and the automobile led to new patterns of fairly autonomous youthful sociability involving both sexes. The resulting generational conflict rendered a wide swath of parents uncomfortable in their familial role. These parents came to regard even nondelinquent children with a certain level of distrust and suspiciousness.[34] Similarly, Voelker did not view children as vulnerable innocents in need of his coddling or protection. His experiments largely reconfigured parental distrust about the unobserved behaviors of their children into a testing situation. His understanding of the child as prone to dishonesty facilitated and excused his own use of deceit.

Following Voelker's experiments, a number of psychologists attempted to measure the deceitfulness of children using deceptive tests. In 1927 the *New York Times* reported that over two hundred psychological studies of children's moral behavior had been published between 1920 and 1925.[35] Professional journals regularly included massive bibliographies of the most recent experiments on character traits in an attempt to keep researchers abreast.[36] In addition to the work at Columbia, Terman's students at Stanford also produced a number of innovative studies. Where Voelker sought to compare the typical child with the particularly trustworthy one, Vernon M. Cady's study went in the opposite direction. He replaced Boy Scouts with the inmates of Whittier State School, an institution for delinquents. The two studies complemented one another; together they examined a distribution of moral attributes from the trustworthy to the delinquent.[37]

Character testing played an important role in Terman's longitudinal study of the mental development of a thousand gifted children. These studies took place soon after Terman argued that genius was not comorbid with forms of moral deviancy and that the "gifted" ought to serve as America's natural leaders.[38] In a 1922 grant proposal he outlined his plan to use these tests to challenge the common view that "bright children are more than ordinarily likely to develop eccentricities along emotional, volitional, and moral lines." He concluded that the superior results of his "Termites" on these tests demonstrated that "the danger of over-intellectualization at the expense of normal development along social and moral lines is probably much less than some have believed it to be." Despite their superior intellectual prowess, the gifted exhibited greater morality than the general population. In spite of this confident conclusion, in drafts of his report Terman expressed uncertainty whether character tests measured the temptation to cheat or simply a willingness to disobey instructions.[39]

Terman's favored test was Cady's refinement of the circle or "peeping" test first proposed by Voelker. The test presented the child with a twelve-square-inch chart consisting of a large circle with five smaller circles of varying sizes drawn along its circumference. Instructed to close his eyes, the child moved his hand around the large circle making an X where he estimated one of the smaller circles to be. The psychologist determined the subject's dishonesty by his accuracy in correctly marking the circles. The odds of the boy successfully marking all five circles with his eyes closed were astronomical. To succeed implied that he peeped during the exercise, thereby demonstrating a dishonest response to the situation. In Cady's version, the use of deceit on the part of the experimenter became an even more integral aspect of securing valid results. The psychologist deliberately lied to the subject, presenting "the tests as scientific experiments in measuring distances with the eyes closed."[40] Where Voelker had urged testers to remain vague when questioned about the test's ultimate goal, Cady held that conscious misdirection was more profitable. The reason for misleading subjects was, ironically, to secure an honest performance from them. Another Terman student most clearly articulated the need to distinguish between the methods used to analyze character and those for calculating intelligence. In 1925 Albert Raubenheimer observed that where it was permissible for intelligence tests to measure a subject's optimal mental achievement, studies of character had to elucidate a person's habitual behavior. For a subject to be at his most upright would defeat the stated aims of the test.[41] As early as 1927 Harvard psychologist A. A. Roback noted how "in the Voelker-Cady-Knight-Raubenheimer tests considerable ingenuity is displayed on the part of the devisers to ward off suspicion of the purpose of the procedure."[42]

Other psychologists deployed honesty tests to explain the nation's shifting racial formations. Katharine Murdoch, an alumna of the Teachers College, introduced the tests to the schoolchildren of Hawaii. Writing in 1925, Murdoch declared that she was an agnostic on the question of the relative import of heredity versus culture in the formation of mental traits and was noncommittal about her study's applicability to the day's debates over immigration and citizenship. Situating her work explicitly in a specific time and place, she sought to map racial difference in terms of both intellectual and moral traits. Working as a school psychologist with select twelve-year-old children from Honolulu and the surrounding plantation towns, Murdoch used both the Stanford-Binet intelligence test and the "peeping" test. She found that the Anglo-Saxon population was among the least trustworthy in the region, while the Chinese and Japanese "Oriental" population more successfully balanced ambition and honesty.[43] Others used the tests to ascertain that morally upstanding rural white children gradually bestowed upon Native Americans a culture of honesty.[44] Regardless of whether or not individual scientists upheld whiteness as the marker of honesty, the use of these tests worked to materialize an economy of character organized by racial types.

Building upon the optimism engendered by the wartime intelligence testing program, psychologists set out to measure character as a means of depicting the nature of youth under the threat posed by the acids of modernity. By and large, their results challenged certain truisms held by character educators. Perhaps even more troubling for those in the honesty business was the sense that character was not something that could be taught. This was because it was not consistent across differing situations. While earlier studies suggested this possibility, it became the central finding of the massive Character Education Inquiry.

"A Colossal Venture"

In a 1924 issue of *Religious Education*, Hugh Hartshorne announced that he had embarked on "a research extraordinary." With a three-year grant and the institutional support of Columbia's Teachers College, he and Mark May had begun a major inquiry into the efficacy of current strategies for character building.[45] The Institute for Social and Religious Research (ISRR) served as their patron. This ecumenical Protestant organization sought to facilitate international cooperation on humanitarian issues in the wake of the First World War. Its short-lived predecessor, the Interchurch World Movement (IWM), had sponsored the Indiana Survey of Religious Education, which originally secured Voelker's services. Both the IWM and the ISRR received considerable financial support from John D. Rockefeller Jr., the heir to the Standard Oil fortune. In the wake of the war, he felt that the unification of the various Protestant denominations under an ecumenical banner offered the greatest hope for international peace, the assimilation of the immigrant working class, and the maintenance of industrial stability. The younger Rockefeller held that a renewed, efficient Protestantism could resolve much of the industrial strife that was one of his father's major legacies.[46]

Its leader described the ISRR's purpose as unifying "the scientific method with the religious motive." This distinguished it from the IWM, whose primary aim was financing international missionary work. In the summer of 1922 the ISRR's chair, Raymond Fosdick, wrote Rockefeller to secure support for the new organization. Beyond his patronage, Fosdick felt that Rockefeller offered an important model for religious institutions, stressing "how anxious American Protestantism is for guidance along sane and business-like lines of development." A few months later, Rockefeller pledged $200,000 per annum, ensuring that the ISRR could undertake new projects and not simply see through the commitments of its failed predecessor. Originally the new group called itself the Committee on Social and Religious Surveys. They chose the name because they felt that "surveys" connoted "an objective Study and not the appraisal of intangible qualitative factors."[47]

Robert S. Lynd was undoubtedly the most famous beneficiary of ISRR funding. He had first come to national attention in 1922 as a vocal critic of the conditions he observed as a preacher in a Rockefeller work camp in Elk Basin, Wyoming. The ISRR commissioned Lynd to carry out a field study of religious practices in a "small industrial city." With his wife, Helen, he conducted extensive fieldwork in Muncie, Indiana, and the study's scope expanded dramatically. *Middletown* (1929), the resulting book, was a best-selling portrait of daily life as transformed by mass consumption.[48] The study was largely received as a tribute to the virtues of Midwestern normality. This had not been the authors' intentions; they wanted to capture how commercial values not only corrupted the character of swindling robber barons, but also infected small-town America.[49] The approach taken in *Middletown* reflected a change in direction for the kind of research conducted under the auspices of the ISRR. In 1923 the directors of the ISRR selected Lynd because they felt that in this survey "the psychological viewpoint is to be stressed." At the same meeting, they chose to change the organization's name to reflect how it was "increasingly concerned with psychological factors."[50]

The Character Education Inquiry (CEI) was the major outcome of this new interest in the psychological. In 1924 the ISRR reached an agreement with Teachers College at Columbia University to undertake an inquiry "to advance exact knowledge of those elements of character generally recognized as possessing moral and religious significance and of the ways in which they are modified." Under the agreement, the ISRR gave the CEI an annual budget of between $20,000 and $30,000 to cover the salary of the lead investigators, supplies, and materials.[51] This made the CEI one of the best-funded social scientific project of its era. For example, this amount represented almost twice what Lynd received for his study.[52] In 1926 the initial funding for the CEI was set to expire and Thorndike met with the executive to express that he was "confident of its solid value." His intervention meant that the ISRR extended funding to the inquiry for another two years.[53]

The personnel assigned to administer the CEI reflected its dual role as religious and scientific enterprise. Hartshorne, an ordained Congregational minister, received his PhD jointly from the Teachers College and Union Theological Seminary in 1913. As both a student and young professor there, he came under the influence of George Albert Coe and his psychology of religion. For both men, a psychological perspective on Christianity enabled the believer to make the practice of his faith more effective and the transmission of the gospel more efficient.[54] May, then a psychology professor at Syracuse University, was primarily a mental tester. Terman described him as "one of my 'sidekicks' in Washington during the war." During this phase of his career, May had acquired some experience testing traits that bordered moral questions when he was recruited to examine the psychology of conscientious objectors. Despite differing backgrounds and interests, Hartshorne and May maintained an amicable relationship. Each agreed that they should share the same title at Teachers College, receive identical salaries, and that they should divide responsibilities equally. Hartshorne explained that he did not "wish to be solely responsible for all the field work" and he was willing to share "the drudgery that may be connected with the statistical end of the work or the laboratory end of the work."[55]

Charged with arriving at a better understanding of how religious education functioned, May advocated that they take an indirect route. He reminded his colleague that "intelligence testing started by studying the absence of intelligence, rather than the presence of it. Why not begin the study of character in the same way?" Hartshorne concurred with the proposal that they should focus their work on "the study of failures in normal children." They adopted the intelligence tester's method of ascertaining the nature of character by studying its deviations.[56] They worked closely with educators in recruiting their experimental subjects. Selecting children enrolled in grades five through eight, they conducted their tests in three communities: a middle-class residential section of New Haven (group X); what the researchers called an "unfavored community," consisting mainly of the children of foreign-born parents also residing in New Haven (group Z); and the children from a small town in eastern New York whose socioeconomic condition was deemed to be between that of groups X and Z (group Y). The deception tests required a total of thirty classroom hours to perform.[57]

In selecting their tests, Hartshorne and May privileged those that offered "a natural situation." They contrasted the natural with the "experimental," situations deemed so artificial that they would never occur, save for the intervention of an investigator. Seeking out the natural did not mean opting for the unstructured fieldwork of the ethnographer. If a manufactured situation possessed low visibility on the part of the subject completing it and had a high correlation with what were deemed typical life situations, it could serve as substitute for uncontrolled life.[58]

If May's exposure to intelligence testing furnished the inquiry's methodology, Hartshorne's psychology of religious informed the interpretation of their results. In his 1913 doctoral dissertation, Hartshorne turned to the sociology and anthropology of religious rituals to distinguish primitive beliefs from the more rational and critical stance of the Christian. While never denying the superiority of the Christian faith, he felt something had been lost in its modernism. Avoiding "a reversion to the primitive 'crowd' type of emotional control," Hartshorne still felt that the psychologist's understanding of the emotions could enrich the form of Protestant worship while avoiding the excesses of evangelicalism. The aim of religious education was to achieve social control through the installation of "*habits* that correspond to the demands of prevailing social mores."[59] In a 1924 letter to May, Hartshorne explained that

> human nature tends to develop sectionally instead of as a unit, as the physiology of the self indicates it should develop. This sectional development is largely due to the acquisition of culture. Customs and conventions set up strains between

the autonomic requirements of the original nature operating in a primitive environment and the habits acquired through experience in regulated groups.[60]

Hartshorne argued that religion functioned as the primary mode for resolving this tension between what he called the autonomic and the habitual aspects of the self. He understood the task of the CEI as assessing "just what the present situation is among different groups of children in different places, with different religious and cultural background, at different age levels."[61] He envisioned the CEI as a study of acculturation that would consider the various situations in which the child faced moral challenges.

In interpreting the resulting data, Hartshorne and May argued that deceit as a discrete and unified character trait did not exist: "Honesty appears to be a congeries of specialized acts which are closely tied up with particular features of the situation in which deception is a possibility, and is apparently not greatly dependent on any general ideal or trait of honesty." The typical child was likely to deceive while operating in certain situations, but act honestly by avoiding lying, cheating, or stealing in others. There was no distinct dishonest personality type, for "whatever honesty a man possesses resides not in a secret reservoir of honest virtue, nor in the ideal of honesty which he may hold before himself as worthy of his best effort, but in the quality of the particular acts he performs."[62] Deceitful behavior was a passing experience within the course of the person's life. The dishonest person was not a morally or mentally aberrant, but rather was indistinguishable from the ordinary individual. The CEI placed the origins of deception in those stimuli that coaxed the individual to behave in a certain manner. Commissioned to make the cultivation of moral behavior more efficient, Hartshorne and May concluded that the prerequisite fixed character for such a project did not exist. They rendered morality pragmatic and situational. Born out of the younger Rockefeller's hopes of salvaging morality in an industrial world, the psychologists ultimately reaffirmed and made normal what many saw as the worst facet of his father's business practices: their amorality. As the psychologist A. A. Roback wryly observed, "that religious organizations have financed quantitative investigations of personality conducted from a behavioristic angle, as the stimulus-response terminology often shows, is a sign of the times."[63] Rather than an enduring trait, these scientists defined honesty as a situation-specific behavior.

They were able to make this claim because the psychologists presented their subjects with a wide array of deceptive situations. Their own deceits, however, generated tensions within their dual personae as disinterested, amoral scientists and morally upstanding religious educators. These roles required a balance between the need for reliable, valid results with an imperative to operate honestly and ethically as researchers. They outlined ten criteria for selecting tests. One of the conditions for a proper test was "The test should not put the subject and the examiner in false social relations to one another. The examiner should guard against being deceptive himself in order to test the subject." This imperative, however, was in conflict with the very next criterion, namely that their tests ought to have

> low visibility; that is, it should be of such a nature as not to arouse the suspicions of the subject. This is one of the fundamental difficulties in all such testing since the entire purpose of the test cannot be announced in advance. This criterion is all the more difficult to meet when coupled with criterion number four, for the examiner must keep secret one aspect of his purpose and at the same time be honest with his subjects.[64]

In other words, their test subjects should not be aware that they are participating, making these situations rife with fakery and deceptive conduct in order to observe genuine behavior. Trying to instill honesty as transparency in the nation's future citizenry, Hartshorne and May found that their professional roles offered no such luxury.

The case of the "Mystery Man" test best illustrated the tension between the deceit exhibited by experimenter and that of the subject. The psychologists held a party for young children where they could play certain games under close supervision. The children were seated in a circle with their hands behind their backs when the "Mystery Man" entered the room and placed in the hands of each an unknown item whose identity one had to guess. The child then went into an adjoining room and reported to a helper, who looked at the object still gripped in his or her hand. The helper instructed the child to go into the next room to deposit the object in a box where they were all being collected. Every other child had been given a ten-cent piece rather than a toy. The point of the test was to determine if the child would return or pocket the money, as the children did not believe that the observer in the final room could see whether or not they were depositing their object in the box. Yet Hartshorne and May used their own subterfuge: "the box is open on the side toward the helper so he can see what is dropped in and make a cryptic note of it on his paper opposite the child's name or number."[65]

The investigators worked on the assumption that the typical test subject had the potential to resist the intentions of the experimenter through deceit. Consciously surveying "average" children, these subjects needed to be approached in a fashion formerly reserved for the exposure of the spiritualist and the interrogation of criminal suspects. The experimenter's intentions had to remain hidden, the research subject distracted and co-ordinated through a stage-managed procedure. The psychologist could no longer rely upon the ordinary subject's cooperation. Every experimental situation risked being usurped by resistant or unreceptive, or even overly receptive, subjects because they did not share the same goals as the scientist.

To guarantee valid results, therefore, deceit needed integration into experimental design. Experiments on illusions during the 1890s had involved deceptive stimuli, but subjects knew that the experiment concerned perception in some respect. In this regard, tests of character differed from earlier forms of psychological measurement. In a 1931 survey of recent advances in the study of personality and conduct, Teachers College psychologist Percival Symonds argued for the importance of "keeping the pupil unaware that he is being measured. In order to disarm the person being measured, he must be thrown off the scent, as it were, by telling him to do one thing while at the same time giving him the opportunity to do something else."[66] Drawing inspiration from Hartshorne and May, Symonds advocated that the psychologist adopt an outlook similar to that of the tricky detective secretly trailing a difficult suspect.

The Honesty Index and Its Critics

Hartshorne and May began publicizing their findings in late 1926 through educational conferences and newspaper coverage. In this regard, their academic peers and a more general reading public received the information simultaneously. Like the studies of less-ambitious scope conducted earlier in the decade, the results of the inquiry both confirmed and confounded expectations about morality. For some, the study raised the possibility of a definitive "honesty index," understood as a direct parallel to the objective measure of intelligence.[67]

Newspaper reporters paid close attention to what the inquiry might say about differences in income and ethnicity. On this issue, the inquiry's results were quite ambiguous. For example, one of the earliest newspaper stories focused on the comparative honesty of "Jewish orphans" in the subject pool compared to their wealthier and Protestant counterparts. In contrast, when the *New York Times* covered a preliminary presentation by Hartshorne, the reporter noted "that children of day laborers practice deception more than those of the professional class." In discussing the influence of kinship, they noted, "Deception runs in families to about the same extent as eye color, length of forearm, and other inherited structures. This does not prove that it is inherited. But the general drift of the evidence inclines one to believe that, if all children received identical nurture, they would still vary in deception."[68] Compared with fellow ISRR grantee Robert Lynd, Hartshorne and May ascribed greater explanatory power to ethnicity in *Studies in Deceit*. They claimed that children born to Northern European and American parents were less deceptive than the descendants of Southern Europeans, who were more honest, in turn, than "colored children." Much of this deceit, however, was due to the "cultural and social limitations in the home background."[69]

When asked to account for leading factors that produced dishonest behavior, Hartshorne and May invariably emphasized the situational over the inherited. They favored explanations rooted in personal background and especially the proximate environment over grander explanations grounded in socioeconomic conditions and structures. They cited the importance of the individual's classroom associates as well as the relationship with the parent and teacher.[70] At the 1928 Character Education Research Conference, Hartshorne made an analogy between a tendency to deceive and a physical force. Under ordinary conditions, one could not measure "the 'pressure' of deceit because children are barred from deception and have little opportunity to do these things if they want to do them." His and May's experiments permitted children to behave without the stifling influences of adult supervision. By removing such barriers, the child's natural inclinations would appear. In his extended metaphor, these experiments "let the current flow out."[71] Despite this metaphor, Hartshorne did not want to turn deceit into a fixed, natural trait. Rather, the comparison was meant to suggest that such behavior was measurable, much like the phenomena of the physical sciences. At the conference, a number of delegates claimed that because deceit lacked consistency and reliability across situations, it did not exist as an actual trait. Deceit was less about the development of morality in particular bodies marked by race, sex, religion, or class and more about the particular conditions where the behavior occurred.

This understanding of deception reflected Hartshorne and May's controversial definition of personality. In a 1929 survey of the individual psychology of adults, May suggested that psychologists should consider "the type of adjectives that are popularly used to describe personality" in making their own definitions. Grounded in this common knowledge, he determined that one's personality had little to do with one's inner constitution. Rather, personality was simply "an individual's *social stimulus value*. It is the way in which he impresses others." Three years later, he described personality research and his work on deception in particular in terms of understanding "the virtues and vices which underly American business and American life."[72] In 1938 Henry A. Murray, director of the Harvard Psychological Clinic, equated this attention to "the fabrication of a 'pleasing personality'" with the values of corporate America. He identified a focus on "conduct rather than inner feeling or intention" with a culture that prized "Listerine and deodorants, the contact man, friendliness without friendship, the prestige of movie stars and Big Business, quantity as an index of worth."[73] Indeed, businessman Dale Carnegie emerged as the most eloquent and popular spokesperson for this vision of the self in May's own era. Carnegie's best-selling book *How to Win Friends and Influence People* did not appear until 1936, but he was a fixture on the corporate lecture circuit a decade earlier. Eschewing the heroic figure of the self-made man, Carnegie etched out a brand of advice literature from the daily practices of corporate intermediaries: lawyers, salesmen, and managers.[74] Much like Carnegie, May normalized the small-time deceptive hustles of the impression manager as a model of human nature.

May and Hartshorne's emphasis on the demands of particular situations put their interpretation of deceitfulness in direct conflict with character-building organizations. In November 1927 May announced some of preliminary findings about the situational aspects of character at a meeting of the Connecticut Council of Religious Education. The newspapers quoted him as saying, "Boy Scouts, who are presumably trustworthy and honest about the campfire and in the clubroom, would cheat in the classroom as much, and in some cases more, than boys who had not had this specific training."[75] May argued that developmental training could not instill a consistently honest inclination in the child. Rather, particular situations brought forth particular behaviors. Such claims were a significant attack on the social utility of the Boy Scouts as their mission was the creation of civic-minded young men. The BSA did not accept such criticisms and funded the New York University sociologist Henry Pratt Fairchild to conduct a follow-up study.⁷⁶ The conflict signaled the split between the experimental psychologists and the character-building institutions that served as their initial patrons over how to interpret human nature.

The study also received considerable attention in neighboring disciplines, namely in anthropology. In her *Patterns of Culture* (1934), Ruth Benedict stressed the affinities between the CEI and her brand of cultural anthropology. She enthusiastically reported, "Recent important experiments dealing with personality traits have shown that social determinants are crucial even in traits of honesty and leadership. Honesty in one experimental situation gave almost no indication whether the child would cheat in another. There turned out to be not honest-dishonest persons, but honest-dishonest situations."⁷⁷ Benedict had made similar arguments when she analyzed the habits and rituals of numerous cultures. Just as sex was a role played within a specific social situation, so was the supposedly immortal virtue of honesty.

Benedict's sympathy for the interpretation of behavior articulated in *Studies of Deceit* was not shared by many psychologists. A. A. Roback condemned the CEI study for reflecting the era's commercial excesses. He had expressed his concern with "the American character testers," observing that "while they move in a mechanistic and moderately behavioristic atmosphere, they yet are too content to busy themselves with virtues and vices (honesty, dishonesty, trustworthiness) instead of attempting to pick out the psychological warp and woof of these traits." He pointed out these psychologists' tacit orientation toward commercial life. He complained about how they confused the actual personality trait of dishonesty with a single form of its manifestation "in money matters." In a 1933 review of the status of personality research, he was even more scathing. He argued that the CEI, "which was carried out with all the technique and efficiency which makes American machinery sought after in every nook and corner of the civilized world, is reminiscent to a certain extent of the organization in the production of a major cinema." "Dazed" by its numerous tables, figures, scattergrams, and histograms, Roback saw more spectacle than reliable science in the endeavor.⁷⁸

Among psychologists none was more vehement in his criticism of the approach taken in the CEI than Roback's friend, the influential personality theorist Gordon Allport.⁷⁹ His opposition centered on two issues: the language of character and the argument that behavior was largely situational rather than governed by personality traits consistent within the individual. Since his graduate school days in the early 1920s, Allport had been the most vocal champion of a distinct psychology of personality. Raised in an evangelical Methodist household, he opposed the use of the language of character within psychology, an idiom he felt belonged in the realm of morality. In his autobiographical writings, Allport emphasized how he broke with his mother's faith, pursuing science as a vocation in its stead. In contrast to character, *personality* was a technical word, linked to the process of measurement and graphical representation with a psychogram. Allport was even more dismissive of these "religious education people" in his private correspondence, mocking them as "men who by nature ought to be clergymen or reformers who are sublimating their zeal in the direction of mean square errors."[80]

Yet, as historian of psychology Ian Nicholson has demonstrated, such pronouncements masked the ways in which his religious upbringing and a romantic commitment to an "undivided" individuality continued to inform Allport's theories.[81] His rejection of Hartshorne and May's situation-habit definition of character in favor of the constancy of personality was much closer in spirit to a Protestant definition of a person's unified inner essence as his or her eternal soul. The CEI's habit-oriented self, seen by Allport as deriving from behaviorism, was merely the amalgam of situation-constituted responses. The individual had no essence, but was simply the product of responses to immediate circumstances. Allport disagreed with the notion that personality traits were not continuous within an individual, that all behavior was merely learned as the response to the stimulus of a particular situation. He held that individuals had a propensity to respond consistently across particular situations and that such stability was emblematic of established personality traits. Personality had a cohesiveness that enabled a response to new circumstances without learning a new set of habits wholesale.[82]

Institutional factors also undoubtedly contributed to Allport's harsh assessment of *Studies in Deceit*. When he published his criticisms in the mid-1930s, he was engaged in a protracted struggle for control over Harvard's psychology department. Karl Lashley and Edwin Boring, though neither a true behaviorist, rejected Allport's more holistic personality psychology, arguing that the department should be devoted solely to experimental work. Allport felt that their stance, which depicted psychology as a natural science, provided far too narrow a framework for understanding the human psyche. The conflict came to a crisis point in 1935, when Lashley opposed granting tenure to Allport's closest ally in the department, Henry A. Murray, who championed a psychoanalytically inspired theory of personality.[83] In such a context, not only in terms of its methodology, but especially its Thorndikian version of personality, *Studies in Deceit* must have appeared truly misguided to Allport. Lacking in the CEI was any sense that the individual might possess regular qualities that the psychologist could and ought to assess. Allport felt that Hartshorne and May's behaviorist orientation was incapable of adequately explaining deceptive behavior. If their habit theory was correct, then the execution of every behavior would involve training the necessary neuromuscular processes anew.

The model of the self presented in *Studies in Deceit* received a mixed reception in both academic and popular press. Overall, Hartshorne and May placed their emphasis on the specific situations that elicit deceitful behavior, but they did make room for other explanations. For example, they argued that factors such as a child's socioeconomic background and nationality correlated with deception, while religious affiliation did not. These explanations received greater attention in the newspaper coverage of the story than in academic venues. What most preoccupied their fellow professionals was the claim about the nonunitary nature of the self. An influential 1937 social psychology textbook formulated the early history of personality in terms of this debate between those who held that there existed a "generalized honesty" trait and those who insisted that an individual simply possessed a large number of honest or dishonest habits tailored to specific situations.[84] There was an irony at the center of *Studies of Deceit*'s reception within psychology. Its authors' intention had been to develop a series of applicable tests for the detection of deceit, but instead their study was a watershed in the use of deception as an investigative method. Their closest scientific peers remained skeptical of their interpretation of behavior, but were much more receptive to their methodological innovations.

The Afterlives of *Deceit*

Despite these criticisms of their theoretical model, there were striking continuities between Hartshorne and May's approach and the ultimate shape of psychology in terms of scientific practices and techniques. As early as 1927, Goodwin Watson had signaled this future for personality assessment. After completing his 1925 dissertation on measuring "fairmindedness" at Teachers College, Watson carried out a study of the effects of YMCA summer camp participation on the conduct and attitudes of young boys. In surveying his field, he raised concerns about the "'fakability' of most of the measures." Watson felt that the problem facing personality testers was their imminent success. "If ever the great body of workers with school-children, delinquents, and industrial applicants were to start using tests of the present sort as extensively as intelligence tests are now used, the results would soon not be worth a whistle."[85] The more people took the tests, the greater the likelihood that the next subjects would come to them aware of their ultimate purpose. Whether they wanted to live up to the test's expectations or defy them, knowledge of its aims would compromise the results. For Watson, the world described by Whyte would spell the end of successful personality testing with numerous "industrial applicants" divining the purpose of widely distributed tests and altering their responses accordingly. The only solution in an age of mass exposure was subterfuge on the tester's part. By the early 1930s, numerous psychologists conducted experiments to investigate the ability of subjects to deceive the administrators of such tests. These scientists came to see the psychological experiment itself as a psychological problem. When it came to testing vocational inclination, sex differences, and, of course, deceitful behavior, deception seemed like the most viable solution.[86]

In a language closely echoing Roback, Allport conceded that the CEI "is justly famous for its ingenious methods, its extensiveness (being probably the largest experimental project ever carried out in the field of personality), for its accurate and painstaking treatment of results." It was these "ingenious methods" that were to have the most lasting impact on the discipline. Allport was all too aware that a major methodological limitation of existing procedures was that they depended upon "honest and sincere cooperation" on the test-taker's part.[87] The use of trickery, guile, and deceit to elicit a supposedly natural response from experimental subjects became a major component of twentieth-century psychology. Indeed, the very standardized language of mid-century social psychology betrayed an ironic inheritance that inspired little reflection from participants: the term for the assistant who played a fictional role during the course of an experiment was a "confederate," a label long given to the criminal's covert accomplice.

To criticize the CEI's particular findings, Allport turned to the work of Donald MacKinnon, who was then completing his doctorate as part of the Harvard Psychological Clinic and was working on a different kind of cheating test. MacKinnon conducted an experiment to see if individuals repressed memories that they associated with feelings of guilt. Although aimed at adults rather than children, his experimental design possessed striking similarities to that of Hartshorne and May. MacKinnon presented college-aged males with a series of twenty puzzles that they were to solve while left alone in a room. Also available were two pamphlets containing solutions to the various problems on clearly marked pages. The experimenter instructed those completing the test that they were permitted to consult the solutions to certain problems, but they were expressly prohibited from looking at the answers to others. MacKinnon's study targeted the "violators" of this simple prohibition. He designed the experiment to examine how such a transgression would later affect their memory of the experience. Would guilt over breaking the rule generate the phenomena of repression?

MacKinnon's project required the discrete, unobserved observation of subjects while they performed certain tasks. He stressed how it was "important that the subjects should not know that they had been observed."[88] Central to the experiment's procedure was the use of a one-way screen that permitted the experimenter to observe the behavior of the unknowing and misdirected subjects. Among his ninety-three men, MacKinnon found that 46 percent violated the instructions given to them about not consulting the prohibited solutions. Both violators and nonviolators verbally expressed their frustration at being bested by the puzzles, but the reactions of the former were uniquely characterized by aggressive, destructive explosions directed at the test, while the latter tended to blame themselves. Four weeks after the initial test, MacKinnon interviewed both the violators and nonviolators about their experience. Very few of those who admitted to violating the rules expressed any feeling of guilt about their actions, but 84 percent of nonviolators responded that they would have felt guilty about cheating. For MacKinnon, the paradox that the least liable felt the greater amount of guilt was addressed by Freud, who claimed that those who fail to gratify their aggressive impulses become an object of them through repression expressed as guilt.[89]

MacKinnon argued that his experiment, beyond providing a potential test for the psychoanalytic theory of repression, revealed the existence of two distinct personality types, identified by their emotional responses to the frustrating situation. Interviewing the subjects further, MacKinnon found that nonviolators tended to be emotionally disciplined as children by being made to feel unworthy while violators received physical punishment by their fathers. From a psychoanalytic perspective, "their violations of the prohibitions imposed by a male experimenter are not surprising but rather expected behavior of in the light of what is now known about their early discipline." Explicitly rejecting Hartshorne and May, MacKinnon argued for a "congruence of traits" and his experiment lending "support to a theory of the constituency of personality."[90] Deception and cheating were not opportunistic responses to specific stimulus situations, but derived from deep-rooted regularities within the individual's constitution. On the level of theory, MacKinnon broke with the behavioristic CEI. Such differences masked an important methodological continuity: while they had different interpretations of honesty as a consistent personality trait, both used deception as a method of eliciting an authentic response from subjects.

Other advocates of more psychoanalytic methods shared this conviction in the need for some level of deception to secure the validity of their tests. Murray made deception a central feature of his Thematic Apperception Test (TAT), a tool for assessing personality traits based on the narratives a subject generated upon viewing a series of evocative yet ambiguous illustrations. Grounded in the conviction that such talk would reveal unconscious tendencies and personal secrets, Murray insisted that the interpreter cloak the assessment's true purpose. The psychologist had to convince the patient that his or her imagination, a facet of intelligence, was being measured rather than knowing that the situation served to probe for potential neuroses or psychoses. Suspicion was a detriment to the test's success and it was the one of the examiner's primary roles to project a friendly, relaxing air. In general, Murray found that true confessions were not hard to achieve and "the subject leaves the tests unaware that he has presented the psychologist with what amounts to an X-Ray picture of his inner self."[91] Where he wanted patients to yield their privately held beliefs and feelings, the scientist's own true aim was not to be revealed.

This was certainly true of psychological research on sex. In 1936 Terman and Catharine Cox Miles published an influential scale for measuring masculinity and femininity. Terman claimed that he first became interested in the connection between sex and personality in 1922, through his encounter with gifted children. He believed that these young geniuses constituted a natural aristocracy fit to lead a truly meritocratic nation. The young geniuses were gifted in certain respects, but Terman also observed behaviors he found disturbing. Most famously, one nine-year-old boy, referred to simply as X, began dressing in girls' clothing. By the age of fifteen, "one of his favorite amusements was to dress himself as a stylish young woman, apply cosmetics liberally, and walk down the street to see how many men he could lure into flirtation."[92] Terman saw this as a deep perversion of the boy's natural intelligence. With the M-F scale, he aimed to detect such sexual maladjustment, to identify persons who did not adhere to normative sex roles as these persons risked leaving their role as reproductive elites unfulfilled through unhappy or nonexistent marital relations.

The psychologists contended that deception was necessary because the topic of sex was so emotionally charged that if a subject knew the true nature of the study, this awareness would inhibit a sincere response to the questions. Terman and Miles administered the M-F test to their subjects under the intentionally misleading name of "Attitude-Interest Analysis."[93] They, along with their associate E. Lowell Kelly, conducted experiments to determine the extent to which subjects could successfully "fake" responses to the test if they enjoyed knowledge about its aims. They found that both males and females could provide the responses appropriate for the opposite sex with great ease under such conditions. With these results, the three psychologists determined that faking scores on any undisguised personality test was an easy matter, so they insisted "that subjects should be kept in ignorance of the nature and purpose of the test."[94]

Deception was especially prominent in the psychologists' search for male and female homosexual test subjects, individuals ordinarily hidden from the scientist's view since they "live in constant fear of public exposure and consequent social ostracism."[95] Trying to locate female couples, Terman contacted the sex surveyor Katherine Bement Davis. He explained, "It would not be necessary to inform them of the purpose of the test unless you wished to. In fact it would perhaps be better if the subjects regarded it merely as a test of interests." Terman also recruited a pool of subjects through the wardens and inmates of San Quentin, Alcatraz, and other Bay Area correctional institutions. These prisoners "had been completely misled as to the purpose of the test and had no suspicion that they were known by the examiner to be sex inverts."[96]

From the outset, then, personality testers never trusted their subjects to reliably report their feelings and opinions, fearing what they understood as defense mechanisms would conceal the person's true nature. Perhaps the most striking aspect of this advocacy of deceit was its ubiquity among psychologists, regardless of their divergent politics and theories of human nature. For example, a Social Gospel progressive like Goodwin Watson assisted eugenicist Terman in securing subjects for his cloaked M-F test from the students attending the Union Theological Seminary.[97]

Hartshorne and May argued for the situational nature of personality, while Allport advocated its fixity over the lifespan of the individual. Despite these disagreements, they all converged on the conviction that the study of human beings required deceit on the scientist's part. What unified them was an investigative method. For them, human nature was a tricky and elusive object that required deception to ensnare it.

Not everyone conceded that deceit was the best means of achieving naturalistic responses. For example, Dorothy Swaine Thomas advocated other means of observing under the banner of "experimental sociology." By this she meant the careful study of interpersonal interactions in varying situations. Thomas distinguished her approach from mainstream behavioral research, which she felt aped too closely the methods of physics and chemistry, sciences that did not involve other humans. As she noted, "the very nature of the subject of investigation requires the development of techniques differing from those usually applied in psychology—not at present involving 'control' in the sense of setting up specially prepared situations but approaching the study of social behavior in a genuinely social environment."[98] Psychology experiments as traditionally conceived were inadequate for understanding the dynamics of human behavior because they placed the observed in artificial situations rather than examining lived experience. Rather than manipulate the research subject with deception, Thomas's project required the self-disciplining of the individual observer in order to produced standardized records.

As an intersubjective group, the investigators would decide upon what kinds of behavior constituted a response in the free-form play of the child. Using the preschool attached to the Teachers College, Thomas and her associates calibrated their measures by having multiple observers chart out a common situation simultaneously. Although involving a very different epistemological and ethical relationship between scientist and participant, Thomas's experimental sociology still operated within the ideal of the unobserved observer. Having the observers as a constant fixture of the playroom, while avoiding direct interaction with the children, would permit them to become part of the unnoticed scenery in the spaces where children operated. Furthermore, the intention of standardizing the units for measuring behavior was to efface the idiosyncrasies of the individual recorder, so that whoever recorded the information would be irrelevant. In such a schema, the spontaneous activity of young children could be watched carefully without the nearly invisible observer intervening in their activities.

Meanwhile, at the Yale Psycho-Clinic, Arnold Gesell developed a host of new means of secretly observing the behavior of infants. His concern was not deception per se, but rather in minimizing the artifice that laboratories inevitably introduced into the study of unadorned human nature. Gesell's major intellectual project was the mapping out of stages of mental maturation for supposedly normal children. This required that he and his fellow researchers carefully scrutinize the minute changes in the way infants responded to stimuli across specific temporal intervals. Gesell made several noted technological innovations, especially the use of film, to achieve these ends. He organized the design of apparatus and the architecture of his investigative space on the principle of having "the recording observer entirely out of the behavioral picture." Even in Gesell's more direct embrace of mechanical objectivity, the tropes of the magician and detective persisted. He conceded that "the cloak of invisibility of the fairy tale would best satisfy this acknowledged need."[99] In place of the magical device, having the psychologist observe the actions of the infants playing in a nursery from within a hidden alcove would suffice as a substitute. To this end, he introduced the "one-way screen," a technology that permitted the observer to study an unaware subject. Between the child and the scientist lay a mesh screen, a semipermeable barrier that permitted light to flow from the brighter nursery the child inhabited into the darkened room in which the psychologist worked. The scientific observer sat in a darkened corner of the nursery, masked by a sieve that permitted light and information to flow inside without disclosing the psychologist's presence.

Although the innovation of the one-way screen assisted in generating the fabled cloak of invisibility, the process still required bodily management on the part of the observer. Gesell encouraged researchers who sought to emulate his designs to wear dark clothing to maximize the effectiveness of the screen. Moreover, silence was necessary for obtaining proper results: although the observation room was darkened, it was not soundproof and a careless noise risked dissolving the arrangement's illusion. To minimize movement, he suggested that the alcove ought to be placed in such a fashion "that the observers' chairs do not have to be shifted to maintain uninterrupted observation of a child who may make swift excursions from one end of the nursery to another." Such arrangements permitted "an unseen observer to see" while still removing "the distorting and the disturbing influences of the observer." No longer were the virtues of intimate observation in a naturalistic setting and dispassionate, unobserved detachment irreconcilable; the technology was a guarantee of both.[100]

Even when it came to the study of the sensitive issue of sex, not everyone concurred that deceit was the best option for obtaining validity. The entomologist Alfred Kinsey developed an alternative to Terman's approach to cope with potentially reticent or deceptive subjects during his massive survey of sexual behavior. He rejected the possibility of deceiving his adult interviewees about the survey's aims, as each of them offered the possibility of recruiting further subjects. Within these confidential sessions, Kinsey argued for frankness and was not shy about directly confronting subjects he deemed deceptive about their histories. More important for securing a feeling of trust was the abstention from judgment and cultivation of empathy in his very select cohort of researchers. For Kinsey, sexology required a communion between the scientist and the subject rather than a struggle between the two.[101]

Hartshorne and May seemed aware that their own deceitfulness was controversial. May noted the greater difficulty in studying adults compared with the "more docile, easier controlled" child. The adult "is rather elusive and difficult to manage; he runs away from the laboratory, he refuses to take our tests unless he has first been 'sold' on the idea."[102] May's emphasis on the greater pliability of children as experimental subjects did not entail a commitment to presuming their innocence and vulnerability. Rather he and Hartshorne assumed that the children they studied were deceitful and fully capable of their own manipulations. Indeed, the deceitfulness that served as the particular object of the CEI became a generalized trait of every potential experimental subject. In 1934, a few years after *Studies in Deceit* was published, May addressed concerns about human experimentation as part of a radio address on the measurement of personality. There he noted the prevalence of "a popular prejudice against experimenting on human beings," but he assured his listeners that "most psychological experiments are 100 per cent harmless, most of them are interesting to the subject and some of them are good fun."[103] He emphasized how psychological experimentation constituted a partnership between researcher and subject from which both received enlightenment. Not just edifying, psychology was also highly amusing for all involved. Despite his reassurances, the ways in which experimental psychologists used deception remained morally ambiguous even as it became routine.

While worries about distress and harm were not prevalent in discussions of *Studies in Deceit*, deception as a scientific practice would become increasingly controversial during the 1960s. The first years of the decade saw the publication of a series of popular exposés condemning the activities of the "brain watchers" and emphasizing "the tyranny of testing," especially in the realm of vocational selection.[104] These critics focused on different issues than had the opponents of standardized testing in the 1920s. Privacy and consent were front and center. The deceptive nature of personality testing was a problematic practice because it violated another individual's civil rights.

Without doubt, Stanley Milgram's obedience studies remain the most notorious example of deception in research design. The experiment was presented as a study of the effects of applying electrical shocks on human learning, but the learner was an actor who faked painful responses to the stimulus. The goal was to determine how many volts a subject would apply to the learner at the experimenter's urging. The Milgram experiments brought to the fore concerns about the ways in which psychologists related to their experimental subjects. These experiments quickly gained notoriety within the psychological community and the culture at large as they simultaneously transformed Milgram into an intellectual celebrity and a controversial figure denied tenure. The debate initially particularized the obedience studies as an instance that seemingly confused experimentation with psychological torture. Newspapers and fellow psychologists questioned his procedure, which had subjects leaving the laboratory unsure of whether they had inflicted real physical harm onto others.[105]

Perhaps the pinnacle of the dramaturgical style of psychological experimentation initiated by studies of moral character, Milgram was far from an anomaly. Despite his notoriety, the character of psychological experimentation had been established long before Milgram unveiled his fake shock machine in the early 1960s. His experiments occurred during a period of heightened oversight into the practices of the social sciences. In part responding to the Nazi use of prisoners in medical research, both legislators and professional organizations sought increasingly to regulate what constituted legitimate research using human subjects. Milgram's capacity to produce distress in the laboratory generated uneasiness about the relationship between deception and scientific objectivity. Many of Milgram's peers felt that subjects ought to be able to refuse investigations into certain aspects of their lives, but, without full prior disclosure of an investigation's ends, this ideal was compromised. Within psychology specifically, the new code of ethics crafted in the late 1960s privileged debriefing, a term and practice derived from military procedure, as the technique for coping with deception during the course of experimentation.[106] There, the psychologist would effectively disassemble the preceding theatrical performance, outlining the fictions crafted to elicit the most authentic of responses.

As such, the adoption of deceit as crucial to the design of experiments and personality tests highlighted a paradox at the heart of the discipline, which became increasingly apparent at mid-century. The revelation of people's inner essences was achieved through deception on the investigator's part. On the one hand, psychologists served as leading spokespersons for the need of individuals to achieve authentic forms of self-expression. The therapeutic culture of mid-century centered on scripts of self-revelations of one's inner psychological depths. In its capacity to reveal the subject's hidden feelings and beliefs, psychological practice resonated with the Cold War era concerns with political and sexual self-revelation. Due to its deceptive methodologies, it also reflected the era's preoccupation with subterfuge, masquerades, and professional secrecy.[107]

Conclusion

Certainly attracting considerable attention during the political dramas of the early 1950s, this approach to psychology had its historical roots in the discipline's search for character in the 1920s. Patronage for research on this topic derived from Protestant worries about the changing morality of youth in an increasingly consumer-oriented society. In this regard, one aspect of *Studies in Deceit* given considerable mention by almost every commentator was the simultaneous omnipresence and inconsistency of deceit. Religiously minded experimenters initially hoped to curtail an undesirable behavior from the repertoire of America's youth. Deceitful behavior manifested itself in all kinds of situations, but persons were rarely dishonest in every imaginable situation. The CEI promoted the message that deception was not bred into one's fixed character or cultivated through development, but was called forth by particular situations. This led to Hartshorne and May's pessimistic verdict on the efficacy of character-building organizations in securing an honest corporate workforce in the next generation. They had deep reservations about what lessons could move beyond the light of the campfire. Hartshorne and May offered a picture of deception as a necessary and normal part of element of the everyday.

Indeed, one activity that required deceit for its regular operation was psychology as a scientific discipline. *Studies in Deceit* illuminated a transformation in the character of psychological experiments in terms of their content and moral valence. Initially, "character" in reference to psychological experiments pointed to those elements of morality that they could help make visible and measurable. Voelker, Cady, Hartshorne, May, and others sought to render vices, held privately in the mind, into publicly displayed behaviors. These same psychologists faced questions about the scruples and moral valence of their discipline as they came to develop novel forms of experimental design. They came to adopt the amoral, hard-boiled stance—previously reserved for psychics and criminals—for studying schoolchildren and other supposedly unremarkable persons in typical situations. Valid psychological measures required that individuals report the content of their inner life in a forthright manner. Yet, the greater visibility garnered by these tests through their administration in schools, clinics, and corporations risked generating a knowledgeable and hence hostile subject. These scientists claimed that, through their own minor deceits, they could generate deceivable selves amenable to psychological measurement. Deception did not circumvent the pursuit of objective science, but safeguarded it from the very persons under investigation.

CONCLUSION

Barnum's Ghost Gives an Encore Performance

"There's a sucker born every minute."
Barnum said it; there's sad truth in it
What burns me up, and turns me sour
Is that a crook is born every hour.

Joseph Jastrow penned the above poem while a resident of the Foundation Inn in Stockbridge, Massachusetts, about a year before his death in 1944.[1] Likely intended for and then rejected by a reputable literary magazine, the poem illuminates much about Jastrow's career in particular and the cultural place of psychology during his lifetime. There are a number of notable features in these lines of humorous verse, its literary merits and historical accuracy notwithstanding. Most obviously, the psychologist presented his own life's work as operating in a tradition established by P. T. Barnum. The showman was undoubtedly the most successful individual in making a personal fortune by directly confronting the uncertainty engendered by the novelty of an embryonic consumer culture. His career hinged upon his dual role as an educator about the dangers posed by a marketplace filled with deceitful people and deceptive goods and as an entrepreneur who profited from the promotion of these very same deceptive things to consumers. Throughout his life, he trafficked in the credulity of the public and its exposure. As Jastrow seemingly recognized, Barnum encapsulated a crucial aspect of the culture into which various strands of European psychology became indigenized. After the showman shifted his attention to the circus trade in 1869, these scientists, sometimes eagerly and sometimes reluctantly, took over much of his previous role as expert in the public's foibles and its propensity for deception. In this regard, the poem also reflected the psychologist's twin fascination with what I have been calling the "deceivable self" and "deceitful self." Jastrow etched out his theories of human nature through his encounters with both "suckers" and "crooks."

Jastrow was a particularly eloquent spokesperson on these issues, but he was far from an isolated figure. Many of his contemporaries had similarly established the public profile of psychology as a scientific discipline through their encounters with an array of various wonder-workers, cheats, and con men. These performances fulfilled a number of important functions. Exposing the activities of those they saw as charlatans helped demarcate the mind as the intellectual domain of the psychologist.[2] At the turn of the century, psychology was an ill-understood science, frequently indistinguishable in the court of public opinion from spiritualism, psychical research, or positive thinking. Beyond securing the boundaries of the discipline, this highly publicized work also promoted the psychologist as an individual largely knowledgeable about human deception, making it a notable feature of the American cultural landscape.

This became apparent at the dawn of the Cold War, when P. T. Barnum, of all people, made a surprising return to center stage. In 1949 Bertram Forer reported on an experiment he had recently conducted using his introductory psychology class. Perhaps unsurprisingly, it involved a certain level of deception on his part. Telling his students that he wanted to administer a standardized test designed to evaluate their personality traits, he dutifully distributed the questionnaires. The following week, with the students safely separated from one another due to an imminent quiz, he returned the results. Instead of providing the students with individualized profiles derived from their answers, Forer provided identical sketches he gleaned from a newsstand astrology book. When he asked the students whether they felt that their assessment accurately captured their individual personalities, virtually everyone concurred that they had. For Forer, his classroom demonstration illustrated the dangers posed by the "universal validity" of intentionally vague psychological diagnostics.[3] In a 1955 address, Paul Meehl suggested naming this phenomenon the "Barnum effect." He hoped that such a name might "stigmatize those pseudo-successful clinical procedures in which personality descriptions from tests are made to fit the patient largely or wholly by virtue of their triviality."[4] For Meehl, this certainly included astrological charlatans, but especially referred to accredited psychological professionals who relied upon their clinical judgment instead of statistical measures when it came to assessing personality traits.

In making these claims, Forer, Meehl, and a host of other psychologists made reference to a mimeographed character sketch circulated by Donald G. Paterson. It bore the telling title "Character Reading at Sight of Mr. X according to the System of Mr. P. T. Barnum." Paterson, who had started his career as an intelligence tester in the 1910s, emerged after the war as a founder of vocational psychology at a time when businesses were eager to obtain expert advice on rationalizing the hiring process. Yet, university-trained psychologists certainly did not possess a monopoly over this field. Since the 1920s, Paterson had sparred with the likes of Katherine Blackford, who claimed to be able to sort workers into appropriate vocations based on physical characteristics such as hair color.[5] As described by fellow vocational psychologist Ross Stagner, corporations purchased the majority of tools for personnel assessment because of "the deceptive effect of these glittering generalities incorporated in testing reports."[6] Paterson used his Barnum personality sketch over the years at Rotary clubs and businessmen luncheons to demonstrate the risks posed by this brand of charlatanry in the field of vocational assessment. It offered a pleasing yet vague personality sketch that could simultaneously apply to everyone and no one.[7] Psychologists saw Paterson's sketch as dramatically demonstrated the problem of "the gullibility of personnel managers." Its aim was to communicate to these organization men the inefficiency of their willingness to purchase any self-satisfying psychological service without inquiring into its experimental validity.

The Barnum effect—crafted by psychologists all born in the years following the showman's 1891 death—fused together the discipline's perennial boundary work against charlatans, suspicions about their own clinical judgment, and a rather unflattering opinion about human nature. Provide people with generic yet somewhat pleasing self-portraits and they will gladly consume your wares, regardless of accuracy. Where some suggested that these demonstrations materialized the clear boundary between archaic charlatanry and validated science, others argued that such distinctions were more difficult to maintain. Talk about the effect captured a dual concern: the psychologist's own erroneous self-confidence in his diagnostic abilities and the weakness of the "businessman" in search of a quick cure for his organizational troubles.[8] The Barnum effect portrayed deception as an integral element of the self, but who, exactly, was deceitful and who was deceivable? Was it the ambitious charlatan, the overconfident applied psychologist, the enthusiastic businessman bent on rational organization, the sincere undergraduate, or the employee receiving the results? The effect captured how deception was both normal and unsettling. This uncertainty about deception definitely resonated with Cold War political suspicions about unreliable and malleable psyches, but it had its formal roots in the encounter between psychological testing and America's expanding corporate organizations.

Such activity was not without precedent. Indeed, some historians have recently argued that psychology as a distinctly modern domain of inquiry and mode of explanation originated in the 1784 Royal Commission charged with investigating the nature of animal magnetism in prerevolutionary France. The threat of mesmerism was particularly salient in France on the eve of revolution due to its promotion of forms of authority outside the purview of the monarchy.[9] The commissioners—led by American ambassador Benjamin Franklin—determined that the dramatic bodily responses to the mesmerist's touch resulted merely from the influence of a distorted imagination and not, as Mesmer claimed, from the manipulation of Newtonian magnetic forces. The commissioners meant for this skeptical explanation to deflate the claims of those whom they perceived as commercial charlatans threatening the social order through their unruly public performances. By the mid-nineteenth century, this interpretation had been given a life of its own as a new generation of "mesmerists" embraced the commission's findings, now arguing that they offered health benefits due to their mastery of hypnotic suggestion. In 1872 Daniel Hack Tuke, great-grandson of the founder of the York Retreat, coined a new label to describe this healing power of the imagination, a discovery he traced back to the Royal Commission. For Tuke, the exposé of mesmerism constituted the birth of "psycho-therapeutics."[10]

Yet Franklin's work in France did not lead directly to the institutionalization of psychology as a scientific discipline in the United States. While subsequent generations of scholars described the commissioners as revealing the power of suggestion, this was not the language the commission used. Instead, they described mesmerism wholly in terms of the dangers posed by an individual's unruly imagination, one that generated improper associations from sensory experience. The imagination was particularly threatening on the eve of the French Revolution due to its associations with a highly individualized self unconstrained by the social hierarchies of the Ancien Régime.[11] Likewise, even though the commissioners used an assortment of tricks to keep those under investigation off guard, they did not describe their procedure in terms of deception. Mesmeric subjects may have been left ignorant of whether they had been magnetized or not, but they were not deceived about the aims and purposes of the experiments. In contrast to the writings of their late nineteenth-century successors, the moral qualms and uneasiness surrounding the use of deceit in a scientific inquiry were absent from the report to the king.[12] In important respects, the problem of deception was not really an issue for these men of science at the end of the eighteenth century.

A century after the commission delivered its verdict, a new generation recast this Enlightenment narrative about human nature. These late nineteenth-century scientists, working within a political framework where a person's imagination seemed less threatening, instead conceived of a self vulnerable to deception in the form of suggestion, expectations, and especially deception. The deceivable self operated in both the scientific sphere of psychophysics and the legal one of trademark infringement. This person was said to be governed by a law of economy: the individual possessed a frugal mind bent on the conservation of mental resources. A sudden confrontation with an abundance of experiences led this person into deception. I have argued that this economization of the mind within the discipline had much to do with the indigenization of scientific psychology into a culture captivated by the robber baron, the grafter, and the muckraker. Psychologists promoted their scientific wares in the public sphere as a means of navigating the uncertainty of commercial life. In the psychological laboratory this abundance was largely figurative, referring to novel and unexpected stimuli. As these scientists began applying their science to daily life, they entered into situations that literally dealt with a vast amount of newly branded commodities and purchasers selecting their goods in an instant. By engaging with preoccupations about the potential for fraud in both popular writings and practical applications, these scientists did much to seed the ground for the emergence of a psychological society in the twentieth century.

Psychologists helped facilitate the consolidation of a corporate order in the early twentieth century, but they were deeply ambivalent about the presence of such organizational structures in their own lives. Scholars have documented how through mental testing, psychologists helped sort an increasingly heterogeneous population into legible categories for commercial and corporate ends.[13] This extended beyond the measurement of intelligence and included interventions into trademark law, the development of various lie detectors, and the assessment of moral character.

The net result of this applied psychology was a hierarchically integrated social order. Alongside such activities, these same scientists reaffirmed notions of individuality and self-reliance in their popular writings. George Stratton's discussion of optical illusions in his *Experimental Psychology and its Bearing upon Culture* (1903) is a telling example. He insisted on the greater reliability of the inner development of the individual compared with succumbing to social influence.

Such a view resonated with the dominant political temper of these scientists. They understood themselves as a productive scientific elite, besieged as a class by both the leisured and working classes. James McKeen Cattell captured these views most clearly. He justified the creation of his Psychological Corporation on the following grounds: "A single advance in the applications of science, such as the Bessemer steel process, the electric motor or the internal combustion engine adds billions of dollars to the world's wealth, but the profits go to the millionaire who keeps a private yacht and to the laboring man who rides five miles for five cents."[14] Psychologists had to adopt a corporate form to protect members of the discipline from both kinds of parasites on their intellectual labors that flourished in a capitalist society. At the heart of twentieth-century American psychology, then, was a paradox. It was a discipline simultaneously dedicated to atomistic individualism and the mass administration of people. Throughout the century there was a perennial tension between the discipline's administrative function and the image of personhood it promoted.

Alongside a disciplinary identity and a cultural role, psychologists crafted an epistemology particularly suited for the human sciences, the ideal of "the unobserved but observing observer." This epistemic ideal had a number of components. It was not so much a "view from nowhere," but rather the perspective of a hidden or disguised undercover sleuth. Where the "mechanical objectivity" of the natural sciences demanded that the observer operate like a passive recording machine, this ideal required that the investigator engage in the elaborate stagecraft and dramaturgy of the magician. Finally, amorality was a prerequisite for this ideal. This was not value neutrality, but the demand of sacrificing one's personal scruples to best serve the truth. To combat a dangerous subjectivity on his own part as well as that of his subjects, the human scientist had to embrace a calculated insincerity, a variation of the ruthless, hard-boiled detective's masculinity.[15] First developed as a strategy to expose crafty spirit mediums, this approach toward the human subject soon migrated into other areas of psychology. By the 1930s, it had become an expected and unremarkable element of scientific practice when studying personality traits and individual behavior in social situations.

These scientists normalized deception in two, interrelated ways. They came to see deception as a necessary aspect of both human nature and scientific research into that human nature. The rise of behaviorism in the 1920s undoubtedly muted discussions about mentalistic phenomena such as deception, but the topic did not remain on the outskirts of the discipline for long. Applied psychologists seeking to delineate their approach to individual assessment from a host of challengers kept the topic very much alive. As explicit references to a Barnum effect surfaced in this literature, the specter of the marketplace and the uncertainty it engendered was never far from view in these discussions. Especially in the interrelated fields of personality and vocational testing, the specter of the showman was never far from view as Donald G. Paterson's "Character Reading at Sight of Mr. X according to the System of Mr. P. T. Barnum" indicated. Deception played a dual role in this arena. Psychologists were interested in both detecting dishonest persons and policing the boundaries of their discipline from commercial competitors. While they worked to screen potentially deceitful employees, they found deception a useful tool for exposing fraudulent counterfeiters of psychological expertise.

This transvaluation of deception among scientists had significant implications for therapeutic practice. The history of child adoption and the medical management of intersexuality are illustrative. The dominant approach taken toward adoption in the twentieth century focused on the ideal of "kinship by design." Hoping to conquer the dangers and uncertainty of adoption, policymakers and therapeutic experts sought to regulate, interpret, standardize, and naturalize the process. Through the paradigm of "matching," they sought to place children with parents who closely resembled them physically in the hopes of making the human intervention in the family invisible. This was accompanied by therapeutic interventions intended to socialize the child into their new family. Following the work of John Money in the 1950s, the medical management of children born with ambiguous genitalia focused on early surgical intervention followed by socialization into a single gender identity through psychological guidance.[16] The success of such enterprises hinged upon hiding the scientific intervention by shrouding the child in secrecy. In both cases, scientists argued that a knowledgeable therapeutic subject would impede the intended socialization. This equation of mental health and deception has been controversial among those living these therapeutic lives. In response to activists, there has also been a turn, beginning in the 1970s with adoption and much more recently in the case of sex assignment in infants, toward more openness and information in the name of long-term psychological health.

The psychologist's methods for identifying the deceivable and deceitful self certainly met with considerable resistance. In the legal sphere, judges rebuffed what they viewed as attempts by the psychological experts to expand the laws of evidence and replace the jury system. The rejection of the deception test in the Frye case was the most famous instance, and one that profoundly shaped the admission of expert testimony over the century. The lie detector's critics were not confined to the country's jurists, however. In her nationally syndicated advice column, Doris Blake skewered the fantasy of introducing the lie detector as an appliance designed to maintain truth in the household. Unlike the judges, Blake did not claim that a consensus was lacking within a community of experts about the machine's reliability. She protested because she was fairly convinced of the device's effectiveness. She saw the instrument as a threat to privacy and the small-time deceits that preserved marital happiness. Through her rejection of the particular machine, she promoted the validity of the psychologist's understanding of human nature.

Both of these attitudes featured prominently in legal cases addressing trademark infringement and deceptive advertising. Although overlooked by historians, these cases are critically important for understanding the nature of the psychological society in the twentieth century. They bring to the fore an essential tension: the rejection of psychologists' advice in the particular while championing a psychological understanding of the self more generally. By the 1910s a consensus had emerged that the law concerning trademarks had built within it an understanding of the purchaser that was profoundly psychological. Moreover, lawyers writing for law journals first advanced this interpretation, not psychologists. When these same lawyers called upon university-based scientists to design psychological scales to help rationalize legal decisionmaking, judges rejected their authority in the courts. Yet they did so without questioning the psychological definition of the self embedded in the law. Much like Doris Blake, these judges distrusted psychologists as experts, but embraced a psychological way of understanding the person as simultaneously deceivable and deceitful.

Our current historical moment has witnessed a renewal of interest in making the deceitful and deceivable self more legible. These contemporaneous sciences of deception are not grounded in the psychologist's law of mental economy, but in the language and technology of neuroscience. Behavioral economists and neuromarketers seek to predict the consumer's seemingly irrational choices. A host of scientific entrepreneurs offer new lie-detection technologies to the security apparatus of the state. Neuroscientists question whether the law's presumption of individual free will is an illusion. A new field of neuroethics has emerged to grapple with the societal impact of these endeavors. Are we on the cusp of a new era when it comes to understanding and manipulating the deceitful and deceivable self? Ending in the 1930s, I have eschewed directly confronting this question. This oblique approach was intentional because the dominant standpoint on these issues divorces contemporary concerns from their historical antecedents by stressing the novelty and uniqueness of our current knowledge.[17] The poverty of this neglect of history denies already existing resource for addressing these questions. This earlier, all too easily forgotten history, illustrates both a captivation with the deceptive components of human nature and the very real limits to influence of scientific approaches to making human deception more legible.

Notes

Introduction

1. See "Business Bureau Here Exempt from License," *New York Times*, April 14, 1931, 3.
2. Frank Dalton O'Sullivan, *Rackets, An Exposé of the Methods and Practices of the Better Business Bureaus Operating in Fifty Large Cities of the United States and of Other Outstanding Rackets* (Chicago: O'Sullivan Publishing, 1933), 17.
3. See also Edward J. Balleisen, "Private Cops on the Fraud Beat: The Limits of American Business Self-Regulation, 1895–1932," *Business History Review* 83 (2009): 113–60.
4. O'Sullivan, *Rackets*, 65–79.
5. He was particularly displeased that laboratory members held his textbook in low esteem. See Frank Dalton O'Sullivan, *Crime Detection* (Chicago: O'Sullivan Publishing, 1928).
6. On the development of the lie detector, see Ken Alder, *The Lie Detectors: The History of an American Obsession* (New York: Free Press, 2007).
7. By historicizing deception rather than uncovering historical examples of it, this project is distinct from the field of agnotology. See Robert N. Proctor and Londa Schiebinger, eds., *Agnotology: The Making and Unmaking of Ignorance* (Stanford, CA: Stanford University Press, 2008).
8. Michel de Montaigne, "Of Liars," in *The Works of Montaigne*, ed. William Hazlitt (London: C. Templeman, 1865), 15; Melville Bigelow, *A Treatise on Law of Fraud on Its Civil Side* (Boston: Little, Brown, 1890), 3.
9. Some have reserved the concept to describe solely intentional, verbal inversions of truth-telling. See Sissela Bok, *Lying: Moral Choice in Public and Private Life* (New York: Pantheon Books, 1978). Others note that such definitions fail to capture the variety of deceptions people practice, such as puffery, humbug, and bullshit. See Harry Frankfurt, *On Bullshit* (Princeton, NJ: Princeton University Press, 2005), 30–48.

10. I formulate my project in this way to signal my debts to Annemarie Mol, *The Body Multiple: Ontology in Medical Practice* (Durham, NC: Duke University Press, 2002), and Michelle Murphy, *Sick Building Syndrome and the Problem of Uncertainty: Environmental Politics, Technoscience, and Women Workers* (Durham, NC: Duke University Press, 2006).

11. See Albert O. Hirschman, *The Passions and the Interests: Political Arguments for Capitalism before Its Triumph* (Princeton, NJ: Princeton University Press, 1977); Theodore M. Porter, *Trust in Numbers: The Pursuit of Objectivity in Science and Public Life* (Princeton, NJ: Princeton University Press, 1995), 21–48; Harold J. Cook, *Matters of Exchange: Commerce, Medicine, and Science in the Dutch Golden Age* (New Haven, CT: Yale University Press, 2007).

12. Louis Galambos, "The Emerging Organizational Synthesis in Modern American History," *Business History Review* 44, no. 3 (1970): 280. See also Robert H. Wiebe, *The Search for Order, 1877–1920* (New York: Hill and Wang, 1967); Alan Trachtenberg, *The Incorporation of America: Culture and Society in the Gilded Age* (New York: Hill and Wang, 1982); Scott A. Sandage, *Born Losers: A History of Failure in America* (Cambridge, MA: Harvard University Press, 2005).

13. Patten, *New Basis of Civilization*, 11. For divergent interpretations of the significance of Patten's revolution, see T. J. Jackson Lears, "From Salvation to Self-Realization: Advertising and the Therapeutic Roots of Consumer Culture, 1880–1920," in *The Culture of Consumption: Critical Essays in American History, 1880–1980*, ed. Richard Wightman and T. J. Jackson Lears (New York: Pantheon, 1983), 1–38; James Livingston, *Pragmatism and the Political Economy of Cultural Revolution, 1850–1940* (Chapel Hill: University of North Carolina Press, 1994); Eric Rauchway, *Blessed among Nations: How the World Made America* (New York: Hill and Wang, 2006).

14. William Cronon, *Nature's Metropolis: Chicago and the Great West* (New York: Norton, 1991), 74–81.

15. On the nationally integrated consumer economy, see Alfred D. Chandler Jr., *The Visible Hand: The Managerial Revolution in American Business* (Cambridge, MA: Harvard University Press, 1977); Susan Strasser, *Satisfaction Guaranteed: The Making of the American Mass Market* (New York: Pantheon Books, 1989); Olivier Zunz, *Making America Corporate, 1870–1920* (Chicago: University of Chicago Press, 1990); T. J. Jackson Lears, *Fables of Abundance: A Cultural History of Advertising in America* (New York: Basic Books, 1994); Elspeth H. Brown, *The Corporate Eye: Photography and the Rationalization of American Commercial Culture, 1884–1929* (Baltimore: Johns Hopkins University Press, 2005).

16. Richard White, "Information, Markets, and Corruption: Transcontinental Railroads in the Gilded Age," *Journal of American History* 90, no. 1 (2003): 21.

17. Alessandro Stanziani, "Negotiating Innovation in a Market Economy: Foodstuffs and Beverage Adulteration in Nineteenth-Century France," *Enterprise and Society* 8, no. 2 (2007): 375–412.

18. James Harvey Young, *The Medical Messiahs: A Social History of Health Quackery in Twentieth-Century America* (Princeton, NJ: Princeton University Press, 1967).

19. Martin J. Sklar, *The Corporate Reconstruction of American Capitalism, 1890–1916: The Market, the Law and Politics* (Cambridge: Cambridge University Press, 1988).

20. On the paradoxical politics of the police powers, see William J. Novak, *The People's Welfare: Law and Regulation in Nineteenth-Century America* (Chapel Hill: University of North Carolina Press, 1996).

21. Consumer citizenship involves the binding together of political rights, civic responsibility, and an individual's purchasing power. It also constitutes a form of governance wherein the state serves as protector of consumer rights as part of its police powers. See Lizabeth Cohen, *A Consumer's Republic: The Politics of Mass Consumption in Postwar America* (New York: Knopf, 2003); Charles McGovern, *Sold American: Consumption and Citizenship, 1890–1945* (Chapel Hill: University of North Carolina Press, 2006); Lawrence B. Glickman, "The Strike in the Temple of Consumption: Consumer Activism and Twentieth-Century American Political Culture," *Journal of American History* 88 (2001): 99–128.

22. See Nancy Tomes, "The Great American Medicine Show Revisited," *Bulletin of the History of Medicine* 79 (2005): 627–63.

23. Kurt Danziger, "Universalism and Indigenization in the History of Modern Psychology," in *Internationalizing the History of Psychology*, ed. Adrian C. Brock (New York: New York University Press, 2006), 208–25. The nation's political landscape, namely the growth of segregation laws in the American South, significantly constrained the geographic reach of a psychological version of the self. See Anne C. Rose, *Psychology and Selfhood in the Segregated South* (Chapel Hill: University of North Carolina Press, 2009).

24. Kurt Danziger, *Constructing the Subject: Historical Origins of Psychological Research* (New York: Cambridge University Press, 1990). The Progressive Era was a period when much of the United States was particularly open and amenable to foreign intellectual influences. See Daniel T. Rodgers, *Atlantic Crossings: Social Politics in a Progressive Age* (Cambridge, MA: Harvard University Press, 1998).

25. My approach to the self is inspired by Charles Taylor, *Sources of the Self* (Cambridge, MA: Harvard University Press, 1989); Jonathan Crary, *Techniques of the Observer: On Vision and Modernity in the Nineteenth Century* (Cambridge, MA: MIT Press, 1990); Ian Hacking, *Rewriting the Soul: Multiple Personality and the Sciences of Memory* (Princeton, NJ: Princeton University Press, 1995); Nikolas Rose, *Inventing Ourselves: Psychology, Power, Personhood* (Cambridge: Cambridge University Press, 1996); Jan E. Goldstein, *The Post-Revolutionary Self: Politics and Psyche in France, 1750–1850* (Cambridge, MA: Harvard University Press, 2005).

26. On the forensic sciences of detection, see Sheila Jasanoff, *Science at the Bar: Law, Science, and Technology in America* (Cambridge, MA: Harvard University

Press, 1995); Geoff Bunn, "The Lie Detector, *Wonder Woman*, and Liberty: The Life and Work of William Moulton Marston," *History of the Human Sciences* 10, no. 1 (1997): 91–119; Simon A. Cole, *Suspect Identities: A History of Fingerprinting and Criminal Identification* (Cambridge, MA: Harvard University Press, 2001); Tal Golan, *Laws of Men and Laws of Nature: The History of Scientific Expert Testimony in England and America* (Cambridge, MA: Harvard University Press, 2004); Julia Rodriguez, "South Atlantic Crossings: Fingerprints, Science, and the State in Turn-of-the-Century Argentina," *American Historical Review* 109, no. 2 (2004): 387–416; Alison Winter, "The Making of 'Truth Serum,'" *Bulletin of the History of Medicine* 79, no. 3 (2005): 500–533; Alder, *Lie Detectors*. Important titles on visual culture and deceptive subjectivity include Tom Gunning, "An Aesthetic of Astonishment: Early Film and the (In)credulous Spectator," *Art and Text* 34 (1989): 31–45; Miles Orvell, *The Real Thing: Imitation and Authenticity in American Culture, 1880–1940* (Chapel Hill: University of North Carolina Press, 1989); Crary, *Techniques of the Observer*; Michael Leja, *Looking Askance: Skepticism and American Art from Eakins to Duchamp* (Berkeley: University of California Press, 2004).

27. Roger Smith, *The Norton History of the Human Sciences* (New York: Norton, 1997), 577.

28. Michael M. Sokal, ed., *Psychological Testing and American Society, 1890–1930* (New Brunswick, NJ: Rutgers University Press, 1987); John Carson, *The Measure of Merit: Talents, Intelligence, and Inequality in the French and American Republics, 1750–1940* (Princeton, NJ: Princeton University Press, 2007).

29. Nikolas Rose, *Governing the Soul: The Shaping of the Private Self* (London: Routledge, 1990); Ellen Herman, *The Romance of American Psychology: Political Culture in the Age of Experts* (Berkeley: University of California Press, 1995); James Capshew, *Psychologists on the March: Science, Practice, and Professional Identity in America, 1929–1969* (Cambridge: Cambridge University Press, 1999).

30. On the almost gleeful trickery in modern psychology, see Betty M. Bayer, "Between Apparatuses and Apparitions: Phantoms of the Laboratory," in *Reconstructing the Psychological Subject*, ed. Betty M. Bayer and John Shotter (London: Sage, 1998), 187–213.

31. See Glenn C. Altschuler and Stuart M. Blumin, "Limits of Political Engagement in Antebellum America: A New Look at the Golden Age of Participatory Democracy," *Journal of American History* 84, no. 3 (1997): 855–85.

32. Lawrence Levine, *Highbrow/Lowbrow: The Emergence of Cultural Hierarchy in America* (Cambridge, MA: Harvard University Press, 1988).

33. Neil Harris, *Humbug: The Art of P. T. Barnum* (Boston: Little Brown, 1973). Other key studies of Barnum include A. H. Saxon, *P. T. Barnum: The Legend and the Man* (New York: Columbia University Press, 1989); Bluford Adams, *E Pluribus Barnum: The Great Showman and the Making of U.S. Popular Culture* (Minneapolis: University of Minnesota Press, 1997); Andrea Stulman Dennett, *Weird and Wonderful: The Dime Museum in America* (New York: New York University Press,

1997); Benjamin Reiss, *The Showman and the Slave: Race, Death, and Memory in Barnum's America* (Cambridge, MA: Harvard University Press, 2001); James W. Cook, *The Arts of Deception: Playing with Fraud in the Age of Barnum* (Cambridge, MA: Harvard University Press, 2001); Steven Belletto, "Drink versus Printer's Ink: Temperance and the Management of Financial Speculation in *The Life of P. T. Barnum*," *American Studies* 46, no. 1 (2005): 45–65; Timothy J. Gilfoyle, "Barnum's Brothel: P. T.'s 'Last Great Humbug,'" *Journal of the History of Sexuality* 18, no. 3 (2009): 486–513. The marking of the "freak show" as unsavory venture was the production of a later historical epoch. See Robert Bogdan, *Freak Show: Presenting Human Oddities for Amusement and Profit* (Chicago: University of Chicago Press, 1988); Rosemarie Garland Thomson, ed., *Freakery: Cultural Spectacles of the Extraordinary Body* (New York: New York University Press, 1996); Rachel Adams, *Sideshow U.S.A.: Freaks and the American Cultural Imagination* (Chicago: University of Chicago Press, 2001).

34. See Christopher G. Sellers, *The Market Revolution: Jacksonian America, 1815–1846* (New York: Oxford University Press, 1991).

35. On the importance of contract law to the market revolution, see Morton J. Horwitz, *The Transformation of American Law, 1780–1860* (Cambridge, MA: Harvard University Press, 1977).

36. Janet Davis, "Freakishly, Fraudulently Modern," *American Quarterly* 55, no. 3 (2003): 536.

37. See Steven Shapin, *A Social History of Truth: Civility and Science in Seventeenth-Century England* (Chicago: University of Chicago Press, 1994); Porter, *Trust in Numbers*; Harry M. Marks, *Progress of Experiment: Science and Therapeutic Reform in the United States, 1900–1990* (Cambridge: Cambridge University Press, 1997); Mary Poovey, *A History of the Modern Fact: Problems of Knowledge in the Sciences of Wealth and Society* (Chicago: University of Chicago Press, 1998); Lorraine Daston and Peter Galison, *Objectivity* (New York: Zone Books, 2007).

38. Orvell, *The Real Thing*, xvii–xviii.

39. Lorraine Daston and Katherine Park, *Wonders and the Order of Nature, 1150–1750* (New York: Zone Books, 1998).

40. Daston and Galison, *Objectivity*, 115–90.

41. "From Superstition to Humbug," *Science* 2, no. 41 (November 16, 1883): 637.

42. See Gail Bederman, *Manliness and Civilization: A Cultural History of Gender and Race in the United States, 1880–1917* (Chicago: University of Chicago Press, 1995).

43. E. L. Youmans, "A Foreign Lesson and a Domestic Application," *Popular Science Monthly* 5 (May 1874): 112, 113.

44. Bernard Williams, *Truth and Truthfulness: An Essay in Genealogy* (Princeton, NJ: Princeton University Press, 2002), 172–204.

45. Mark Twain, "On the Decay of the Art of Lying," in *The Stolen White Elephant, Etc.* (Boston: James R. Osgood and Co., 1882), 217–25; Twain, "My First

Lie, and How I Got Out of It," in *The Man that Corrupted Hadleyburg, and Other Stories and Essays* (New York: Harper and Bros., 1902), 167–80.

46. See Jill G. Morawski and Gail A. Hornstein, "Quandary of the Quacks: The Struggle for Expert Knowledge in American Psychology, 1890–1940," in *The Estate of Social Knowledge*, ed. JoAnne Brown and David K. van Kuren (Baltimore: Johns Hopkins University Press, 1991), 106–33; Deborah J. Coon, "Testing the Limits of Sense and Science: American Experimental Psychologists Combat Spiritualism, 1880–1920," *American Psychologist* 47 (1992): 143–51; Thomas F. Gieryn, *Cultural Boundaries of Science: Credibility on the Line* (Chicago: University of Chicago Press, 1999). The examination of such boundary work was an important plank in the early research program of the social study of knowledge. See Roy Wallis, ed., *On the Margins of Science: The Social Construction of Rejected Knowledge* (Keele: University of Keele, 1979).

47. Joseph Jastrow, *The Psychology of Conviction* (Boston: Houghton Mifflin, 1918), 107.

48. See Donna Haraway, "Situated Knowledges: The Science Question in Feminism and the Privilege of Partial Perspective," *Feminist Studies* 14 (1988): 575–99.

49. On the tension between the hard-nosed demands of mechanical objectivity and the theatrics of public science, see Theodore M. Porter, "The Objective Self," *Victorian Studies* 50, no. 4 (2008): 641–47; Roderick D. Buchanan, *Playing with Fire: The Controversial Career of Hans J. Eysenck* (Oxford: Oxford University Press, 2010).

50. This was not the only way in which psychologists attempted to secure their objectivity. The use of statistics and a standardized language circulated through the American Psychological Association's Publication Manual point toward the importance of depersonalized technologies of trust in the discipline. See Porter, *Trust in Numbers*, 228.

51. Robert K. Merton, "Science and Social Order," *Philosophy of Science* 5, no. 3 (1938): 327.

52. In this regard, psychology followed a different "moral history" in the twentieth century than the natural sciences. See Steven Shapin, *The Scientific Life: A Moral History of a Late Modern Vocation* (Chicago: University of Chicago Press, 2008).

53. On the historical diverse setting of confidence games, see Karen Halttunen, *Confidence Men and the Painted Women: A Study of Middle-Class Culture in America, 1830–1870* (New Haven, CT: Yale University Press, 1982); Stephen Mihm, *A Nation of Counterfeiters: Capitalists, Con Men, and the Making of the United States* (Cambridge, MA: Harvard University Press, 2007); Ken Alder, "History's Greatest Forger: Science, Fiction, and Fraud along the Seine," *Critical Inquiry* 30 (2004): 702–16; Sheila Fitzpatrick, *Tear off the Masks! Identity and Imposture in Twentieth-Century Russia* (Princeton, NJ: Princeton University Press, 2005).

54. Edwin Boring, *A History of Experimental Psychology* (New York: Apple-

man, 1929), 623, and Boring, *Sensation and Perception in the History of Experimental Psychology* (New York: Appleman, 1942), 238–45.

Chapter One

1. J. P. Johnston, *Grafters I Have Met* (Chicago: Thompson and Thomas, 1906), 28–38. For Johnston's self-presentation as a hustling itinerate trader, see Johnston, *Twenty Years of Hus'ling* (Chicago: Hallet, 1888), and Johnston, *How to Hustle* (Chicago: Thompson and Thomas, 1905).

2. Sandage, *Born Losers*, 14.

3. On the entanglements between cultural psychology and confidence games at the heart of mass consumption, see Lears, *Fables of Abundance*.

4. In following such usage, I am defining *psychology* not as a body of theories espoused by a delineated cohort of academic specialists, but rather as a collection of techniques assembled around a configuration of the subject. On this usage, see Roger Smith, "Does the History of Psychology Have a Subject?" *History of the Human Sciences* 1, no. 2 (1988): 147–77.

5. E. G. Redmond, *The Frauds of America: How They Work and How to Foil Them* (Chicago: Francis Book Co., 1896), 168, 170.

6. Halttunen, *Confidence Men and the Painted Women*.

7. See Peter G. Filene, *Him/Her/Self: Gender Identities in Modern America*, 3rd ed. (1974; Baltimore: Johns Hopkins University Press, 1998), 75. On the ideology of the self-made man, see John G. Cawelti, *Apostles of the Self-Made Man: Changing Conceptions of Success in America* (Chicago: University of Chicago Press, 1964); Gary Lindberg, *The Confidence Man in American Literature* (New York: Oxford University Press, 1982), 73–89; Sandage, *Born Losers*. In contrast, Jackson Lears traces the cultural roots of the confidence man in the risky commercial ventures of the colonial era and a heterogeneous lineage of adventurers primarily from Native, African, and Catholic cultures. See Lears, *Something for Nothing: Luck in America* (New York: Viking, 2003).

8. See especially Donna J. Haraway, "Modest Witness: Feminist Diffractions in Science Studies," in *The Disunity of Science: Boundaries, Contexts, and Power*, ed. Peter Galison and David J. Stump (Stanford, CA: Stanford University Press, 1996), 428–41. The contemporaneous canonization of the African American trickster figure by folklorists is beyond the scope of this chapter. See Lawrence W. Levine, *Black Culture and Black Consciousness: Afro-American Folk Thought from Slavery to Freedom* (New York: Oxford University Press, 1977). In general, the confidence man genre tended to deny the ability of African Americans to successfully deceive.

9. On the racial underpinnings of manliness and masculinity, see Gail Bederman, *Manliness and Civilization: A Cultural History of Gender and Race in the United States, 1880–1917* (Chicago: University of Chicago Press, 1995).

10. For a pioneering attempt at tracing the social history of particular counterfeiters across jurisdictions, see Mihm, *A Nation of Counterfeiters*.

11. As Joan W. Scott has argued, historical experience is not something that is given in an unmediated form, but rather something that is constituted. See Scott, "The Evidence of Experience" *Critical Inquiry* 17, no. 4 (1991): 773–97.

12. Clifton R. Wooldridge, *Twenty Years a Detective in the Wickedest City in the World* (Chicago: Clifton R. Wooldridge, 1908), 52.

13. Johannes Dietrich Bergman, "The Original Confidence Man," *American Quarterly* 21, no. 3 (1969): 560–77.

14. See John F. Kasson, *Rudeness and Civility: Manners in Nineteenth-Century Urban America* (New York: Hill and Wang, 1990).

15. Steve Fraser, *Every Man a Speculator: A History of Wall Street in American Life* (New York: Basic Books, 2005), 30–32.

16. Halttunen, *Confidence Men and the Painted Women*, 1–55. On the American culture of gambling, see Ann Fabian, *Card Sharps, Dream Books, and Bucket Shops: Gambling in Nineteenth-Century America* (Ithaca, NY: Cornell University Press, 1990). On bunco schemes, see Redmond, *The Frauds of America*, 166–71, and Edward Smith, *Confessions of a Confidence Man: A Handbook for Suckers* (New York: Scientific American Publishing, 1923), 6–7. On the early history of games of chance see Lears, *Something for Nothing*, 10–54. The nineteenth-century police detectives Benjamin P. Eldridge and William B. Watts concur with the notion of an ancient, old world lineage. See Benjamin P. Eldridge and William B. Watts, *Our Rival, the Rascal: A Faithful Portrayal of the Conflict between the Criminals of This Age and the Defenders of Society—the Police* (Boston: Pemberton Publishing, 1896), vii.

17. On the gold brick scheme, see Eldridge and Watts, *Our Rival, the Rascal*, 191, and Smith, *Confessions of a Confidence Man*, 6–7.

18. On green goods schemes, see Eldridge and Watts, *Our Rival, the Rascal*, 198; David H. Leeper, *The Eye Opener or Man's Protector* (Kirkville, MO: Journal Printing, 1892), 46–50; Redmond, *The Frauds of America*, 57; Smith, *Confessions of a Confidence Man*, 7–8.

19. On the unstable cultural meanings of money and currency during this period, see Michael O'Malley, "Specie and Species: Race and the Money Question in Nineteenth-Century America," *American Historical Review* 99, no. 2 (1994): 369–95. On the increased federal powers to police money and counterfeiting, see David R. Robinson, *Illegal Tender: Counterfeiting and the Secret Service in Nineteenth-Century America* (Washington, DC: Smithsonian Institution Press, 1995).

20. Horwitz, *Transformation of American Law*, 263.

21. Laidlaw v. Organ, 15 U.S. 178 (1817). On the role of implicatures in deception, see Williams, *Truth and Truthfulness*, 97–110.

22. Rockafellow v. Baker, 41 Pa. 319 (1862).

23. Walton H. Hamilton, "The Ancient Maxim Caveat Emptor," *Yale Law Jour-*

nal 40, no. 8 (1931): 1186; W. Page Keeton, "Fraud: Misrepresentation of Opinion," *Minnesota Law Journal* 21 (1936): 650; Fowler V. Harper and Mary Coate McNeely, "A Synthesis of the Law of Misrepresentation," *Minnesota Law Journal* 22, no. 7 (1937): 956.

24. Such views could still be found in the Gilded Age in the polemics of certain crusading moral reformers like Anthony Comstock. For example, he berated the sharper for targeting the "unwary and credulous." See Comstock, *Frauds Exposed; or, How the People Are Deceived and Robbed, and Youth Corrupted* (New York: J. Howard Brown, 1882), 5.

25. Leeper, *The Eye Opener*, 51; J. B. Costello, ed., *Swindling Exposed: From the Diary of William B. Moreau, King of Fakirs* (Syracuse: J. B. Costello, 1907), 95. On the theme of complicity, see also Eldridge and Watts, *Our Rival, the Rascal*, 203; A. J. Greiner, *Swindles and Bunco Games in City and Country* (St. Louis: Sun Publishing, 1904), 337; Smith, *Confessions of a Confidence Man*, 8–10 and 153.

26. See *The Swindlers of America: Who They Are and How They Work* (New York: Frank M. Reed, 1875), 3. For similar arguments, see S.J.W. [S. James Weldon], *Twenty Years a Fakir* (Omaha, NE: Gate City Book and Novelty Co., 1899), 56; Greiner, *Swindles and Bunco Games in City and Country*, 9; Costello, ed., *Swindling Exposed*, 61.

27. Leeper, *The Eye Opener*, 10; Lawrence Friedman, "Crimes of Mobility," *Stanford Law Review* 43 (1991): 637–58.

28. People v. McCord, 46 N.Y. 470 (1871), 473.

29. "Notes on Cases," *Albany Law Journal* 6 (1872): 414.

30. People v. Williams, 4 Hill 9 (1842).

31. See "False Pretenses and Larceny—Dishonesty of Prosecutor," *Albany Law Journal* 61 (1900): 398.

32. People v. Tompkins, 186 N.Y. 413 (1906).

33. People v. McCord, 473–77.

34. For expression of these concerns, see J.R.R., "The Way of the Transgressor is Easy," *Michigan Law Review* 9, no. 5 (1911): 424–26.

35. Clarke J. Munn Jr., "Confidence Game—Conduct of Prosecuting Witness as a Defense," *Journal of Criminal Law and Criminology* 24, no. 4 (1933): 785.

36. Thomas Byrnes, *Professional Criminals in America* (New York: Cassell, 1886).

37. Zunz, *Making America Corporate*.

38. See Eric Foner, *The Story of American Freedom* (New York: Norton, 1998); Amy Dru Stanley, *From Bondage to Contract: Wage Labor, Marriage, and the Market in the Age of Slave Emancipation* (New York: Cambridge University Press, 1998); Gunther Peck, *Reinventing Free Labor: Padrones and Immigrant Workers in the North American West, 1880–1930* (New York: Cambridge University Press, 2000).

39. See Rose, *Inventing Ourselves*, and Rose, *Powers of Freedom: Reframing Political Thought* (Cambridge: Cambridge University Press, 1999).

40. On the role of habit in liberal self-governance, see Mariana Valverde, "'Despotism' and Ethical Liberal Government," *Economy and Society* 25, no. 3 (1996): 357–72.

41. Ben H. Kerns, *The Grafter* (Topeka, KS: Crane and Company, 1912), 16; Greiner, *Swindles and Bunco Games in City and Country*, 9; *How Fakers Fake: A Friendly Consideration of a Subject, Amusing in Its Types but Dangerous and Degrading in Its Possibilities* (Rochester, NY: Rochester Chamber of Commerce, 1912), 2; Eldridge and Watts, *Our Rival, the Rascal*, 187; Smith, *Confessions of a Confidence Man*, 8–10.

42. For example, see Greiner, *Swindles and Bunco Games in City and Country*, 337–38.

43. Alson Secor, ed., *Swindles* (Des Moines, IA: Successful Farming Publishing, 1910), 3.

44. Kerns, *The Grafter*, 45.

45. Greiner, *Swindles and Bunco Games in City and Country*, 55–57.

46. Kerns, *The Grafter*, 7; Judy Hilkey, *Character Is Capital: Success Manuals and Manhood in Gilded Age America* (Chapel Hill: University of North Carolina Press, 1997).

47. Walter A. Friedman, *Birth of a Salesman: The Transformation of Selling in America* (Cambridge, MA: Harvard University Press, 2004), 14–87.

48. Clifton R. Wooldridge, *Hands Up! In the World of Crime* (Chicago: Police Publishing, 1901), 88; S.J.W. [Weldon], *Twenty Years a Fakir*, 348.

49. On the economic failure as a particularly disparaged social type during this period, see Sandage, *Born Losers*.

50. Kerns, *The Grafter*, 45.

51. S.J.W. [Weldon], *Twenty Years a Fakir*, 12, 358.

52. Angel Kwolek-Folland, *Engendering Business: Men and Women in the Corporate Office, 1870–1930* (Baltimore: Johns Hopkins University Press, 1994), 77–89.

53. See Roland Marchand, *Advertising the American Dream: Making Way for Modernity, 1920–1940* (Berkeley: University of California Press, 1985).

54. Theodore Roosevelt, "The Man with the Muck Rake" (1906), in *The Muckrakers*, ed. Arthur and Lila Weinberg (Urbana: University of Illinois Press, 2001), 59.

55. Jean Marie Lutes, "Into the Madhouse with Nellie Bly: Girl Stunt Reporting in Late-Nineteenth-Century America," *American Quarterly* 54, no. 2 (2002): 217–53.

56. "Josiah Flynt's Biography," *New York Times*, September 6, 1903, 1.

57. Josiah Flynt, *Tramping with Tramps: Studies and Sketches of Vagabond Life* (New York: The Century Co., 1899), 1–27.

58. In the parlance of the day, a hobo was a migratory worker while a tramp refused work.

59. On the reification of class in this literary genre, see Mark Pittenger, "A World of Difference: Constructing the 'Underclass' in Progressive America,"

American Quarterly 49, no. 1 (1997): 26–65; Frank Tobias Higbie, *Indispensable Outcasts: Hobo Workers and Community in the American Midwest, 1880–1930* (Urbana: University of Illinois Press, 2003), 66–97.

60. Josiah Flynt, *My Life* (New York: Outing Publishing, 1908), 324, 326.

61. Flynt, *Tramping with Tramps*, ix.

62. Walter A. Wyckoff, *The Workers: An Experiment in Reality—The East* (New York: Charles Scribner's Sons, 1897), viii–ix; "A Study of Tramp Life," *Chicago Daily Tribune*, October 28, 1899, 13; A. M. Day, "Review: *The Workers: An Experiment in Reality. The East* and *The West*. By Walter A. Wyckoff," *Political Science Quarterly* 14, no. 4 (1899): 702 (emphasis in the original); review of *A Day with a Tramp, and Other Days*, *New York Daily Tribune*, October 26, 1901, 10.

63. For a contemporary description of Flynt's work along these lines, see the review "The Natural History of the Vagabond!," *New York Times*, March 31, 1900, Book Review section, 19.

64. Living among the Zuni for an extended period of time during the early 1880s, Cushing stressed how he had taken on their garb and customs in order to better understand their culture. Yet he remained a marginal figure within his discipline. Frank Hamilton Cushing, "My Adventures in Zuni," *Century Magazine* 25, no. 2 (December 1882): 191–208; 25, no. 4 (February 1883): 500–512; 26, no. 1 (May 1883): 28–48. For an assessment, see Brad Evans, "Cushing's Zuni Sketchbooks: Literature, Anthropology, and American Notions of Culture," *American Quarterly* 49, no. 4 (1997): 717–45.

65. Jennifer Platt, "The Development of the 'Participant Observation' Method in Sociology: Origin Myth and History," *Journal of the History of the Behavioral Sciences* 19, no. 4 (1983): 379–93.

66. On the epistemological status of this kind of research in anthropology, see Paul Rabinow, *Reflections on Fieldwork in Morocco* (Berkeley: University of California Press, 1977).

67. Flynt, *Tramping with Tramps*, 381.

68. "The Natural History of the Vagabond!," 19.

69. "Prof. Wyckoff as a Tramp," *The Sun*, October 5, 1902, 6.

70. See "Prof. Wyckoff Dies," *New York Times*, May 16, 1908, 1, and "'Tramp' Writer Who Died Sunday," *Chicago Daily Tribune*, January 22, 1907, 6.

71. On the theme of manly suffering in contemporaneous narratives of scientific exploration, see Rebecca Herzig, *Suffering for Science: Reason and Sacrifice in Modern America* (New Brunswick, NJ: Rutgers University Press, 2005).

72. Josiah Flynt, *The World of Graft* (New York: McClure, 1901).

73. Will Irwin, *Confession of a Con Man* (New York: B. W. Huebsch, 1909), 13.

74. Flynt, *The World of Graft*, 4, 11–12.

75. "Chicago Field for Thugs," *Chicago Daily Tribune*, January 26, 1901, 5; "Chicago as Crime Center," *Chicago Daily Tribune*, January 26, 1901, 12; "Josiah Flynt's Indictment of Chicago's Officials: Mayor Harrison and Chief Kipley Plead Not Guilty," *Chicago Daily Tribune*, January 27, 1901, 3.

76. "Mayor Harrison's Tardy Denial," *Chicago Daily Tribune*, January 28, 1901, 6, and "Kipley Replies to Flynt," *Chicago Daily Tribune*, January 31, 1901, 12.

77. On this strain of progressivism, see Richard L. McCormick, "The Discovery that Business Corrupts Politics: A Reappraisal of the Origins of Progressivism," *American Historical Review* 86, no. 2 (1981): 247–74.

78. Lincoln Steffens, *The Shame of the Cities* (New York: McClure, Phillips, 1904), 6–8.

79. In contrast, there existed significant transatlantic networks of reformers dedicated to the problem of urban government. See Rodgers, *Atlantic Crossing*, 112–59.

80. Steffens, *Shame of the Cities*, 11, 10.

81. Wooldridge, *Twenty Years a Detective*, 57.

82. See Fabian, *Card Sharps, Dream Books and Bucket Shops*, 153–202. On the transformation of nature that made this speculation possible, see Cronon, *Nature's Metropolis*, 97–147.

83. Josiah Flynt and Francis Walton, *The Powers that Prey* (New York: McClure, 1900), vi. On the Walton pseudonym, see "Alfred Hodder Dies at Forty," *New York Times*, March 4, 1907, 9.

84. "Concerning Three Articles in This Number of McClure's, and a Coincidence that May Set Us Thinking," *McClure's Magazine* 20, no. 3 (January 1903): 336.

85. Lincoln Steffens, "Enemies of the Republic," *McClure's Magazine* 22, no. 6 (April 1904): 599; Ida M. Tarbell, "Commercial Machiavellianism," *McClure's Magazine* 26, no. 4 (March 1906): 458, 456.

86. Costello, ed., *Swindling Exposed*, 215.

87. John Hill Jr., *Gold Bricks of Speculation* (Chicago: Lincoln Book Concern, 1904).

88. Fabian, *Card Sharps, Dream Books and Bucket Shops*, 198–200. See also Cawelti, *Apostles of the Self-Made Man*, 187–92; David Hochfelder, "'Where the Common People Could Speculate': The Ticker, Bucket Shops, and the Origins of Popular Participation in Financial Markets, 1880–1920," *Journal of American History* 93, no. 2 (2006): 335–58; Jonathan Ira Levy, "Contemplating Delivery: Futures Trading and the Problem of Commodity Exchange in the United States, 1875–1905," *American Historical Review* 111, no. 2 (2006): 307–35.

89. Thomas W. Lawson, *Frenzied Finance* (New York: Ridgway-Thayer, 1905), xiv.

90. On Lawson's multiple careers, see David A. Zimmerman, *Panic!: Markets, Crises, and Crowds in American Fiction* (Chapel Hill: University of North Carolina Press, 2006), 81–122.

Chapter Two

1. Lincoln Steffens, *The Autobiography of Lincoln Steffens* (New York: Harcourt, Brace, 1931), 146.

2. Boring, *History of Experimental Psychology*, 623; and Boring, *Sensation and Perception in the History of Experimental Psychology*, 238–45.

3. Jonathan Crary argues that the management of attention characterized the industrializing culture of the last quarter of the nineteenth century. See Crary, *Suspensions of Perception: Attention, Spectacle, and Modern Culture* (Cambridge, MA: MIT Press, 1999), 21–22.

4. See Christopher G. White, *Unsettled Minds: Psychology and the American Search for Spiritual Assurance, 1830–1940* (Berkeley: University of California Press, 2009), 105–12.

5. On the cultural work of these textbooks, see Jill G. Morawski, "There Is More to Our History of Giving: The Place of Introductory Textbooks in American Psychology," *American Psychologist* 47, no. 2 (1992): 161–69.

6. Paul Lucier, "The Professional and the Scientist in Nineteenth-Century America," *Isis* 100, no. 4 (2009): 699–732.

7. Arthur L. Blumenthal, "The Intrepid Joseph Jastrow," in *Portraits of Pioneers in Psychology*, vol. 1, ed. Gregory A. Kimble, Michael Wertheimer, and Charlotte L. White (Washington, DC: American Psychological Association, 1991), 84.

8. Joseph Jastrow, "Letter to the Editor," *New York Times*, June 9, 1935, Book Review section, 20. He wrote the letter in response to Isabel Proudfit, "More Confirmation of Barnum: Dr. Joseph Jastrow's *Wish and Wisdom* Shows How Our Desires to Be Fooled Sway Our Lives as of Old," *New York Times*, May 12, 1935, Book Review section, 5.

9. Patten, *A New Basis of Civilization*, 11.

10. On this shift, see Livingston, *Pragmatism and the Political Economy*, 67–70; Jeffrey Sklansky, *The Soul's Economy: Market Society and Selfhood in American Thought, 1820–1920* (Chapel Hill: University of North Carolina Press, 2002).

11. I am concerned with the emergence of a *mass* culture as opposed to simply a popular one. On this distinction, see Dwight MacDonald, "A Theory of Mass Culture," *Diogenes* 1, no. 1 (1953): 1–17. On the historical emergence of this distinction, see Levine, *Highbrow/Lowbrow*. On the turn-of-the-century mass culture, see John F. Kasson, *Amusing the Million: Coney Island at the Turn of the Century* (New York: Hill and Wang, 1978); Roy Rosenzweig, *Eight Hours for What We Will: Workers and Leisure in an Industrial City, 1870–1920* (Cambridge: Cambridge University Press, 1983); Kathy Lee Peiss, *Cheap Amusements: Working Women and Leisure in Turn-of-the-Century New York* (Philadelphia: Temple University Press, 1986); David Nasaw, *Going Out: The Rise and Fall of Public Amusements* (New York: Basic Books, 1993); Lauren Rabinowitz, *For the Love of Pleasure: Women, Movies, and Culture in Turn-of-the-Century Chicago* (New Brunswick, NJ: Rutgers University Press, 1998).

12. On the prominence of visual trickery in the mass culture of this era, see Gunning, "An Aesthetic of Astonishment," 31–45; Orvell, *The Real Thing*; Cook, *Arts of Deception*, 163–255; Leja, *Looking Askance*.

13. Crary, *Techniques of the Observer*.

14. Jutta Schickore, "Misperception, Illusion, and Epistemological Optimism: Vision Studies in Early Nineteenth-Century Britain and Germany," *British Journal for the History of Science* 39, no. 3 (2006): 383–405. See also Wendy Bellion, *Citizen Spectator: Art, Illusion, and Visual Perception in Early National America* (Chapel Hill: University of North Carolina Press, 2011).

15. Kurt Danziger, "The Positivist Repudiation of Wundt," *Journal for the History of the Behavioral Sciences* 15 (1979): 205–30.

16. On these national styles, see Danziger, *Constructing the Subject*, 49–67.

17. Crary, *Techniques of the Observer*, 97–98.

18. James Sully, *Illusions: A Scientific Study* (1881; New York: D. Appleton, 1891), 19.

19. On Peirce, see Joseph Brent, *Charles Sanders Peirce: A Life* (Bloomington: Indiana University Press, 1993).

20. Louis Menand, *The Metaphysical Club* (New York: Farrar, Straus and Giroux, 2001).

21. See Peter J. Behrens, "The Metaphysical Club at the Johns Hopkins University (1879–1885)," *History of Psychology* 8, no. 4 (2005): 331–46, and Christopher D. Green, "Johns Hopkins' First Professorship in Philosophy: A Critical Pivot Point in the History of American Psychology," *American Journal of Psychology* 120 (2007): 303–23.

22. See Bederman, *Manliness and Civilization*, 77–120.

23. See Joseph Jastrow, "Joseph Jastrow," in *A History of Psychology in Autobiography*, vol. 1, ed. Carl Murchison (Worcester, MA: Clark University Press, 1930), 138, 135–36.

24. Charles Sanders Peirce and Joseph Jastrow, "On Small Differences of Sensation," *Proceedings of the National Academy of Sciences* 3 (1884): 75–83.

25. Wilhelm Wundt, *Principles of Physiological Psychology*, trans. Edward Bradford Titchener (1874; New York: Macmillan, 1904), 8.

26. On the significance of this experiment for introducing randomization into experimental design, see Stephen M. Stigler, *The History of Statistics: The Measurement of Uncertainty before 1900* (Cambridge, MA: Belknap Press of Harvard University Press, 1986), 253–54.

27. Wundt opposed randomization in favor of studying the observer's expectation and habituation. See Trudy Dehue, "Deception, Efficiency, and Random Groups: Psychology and the Gradual Origination of Random Group Design," *Isis* 88 (1997): 656.

28. Peirce and Jastrow, "On Small Differences of Sensation," 83.

29. On Peirce and the cosmos of chance, see Ian Hacking, *The Taming of Chance* (Cambridge: Cambridge University Press, 1990).

30. Joseph Jastrow, "Some Particularities in the Age Statistics of the United States," *Science* 5, no. 122 (June 5, 1885): 461.

31. Ibid., 463, 464.

32. Antisemitism undoubtedly constrained his career opportunities. See Andrew S. Winston, "'As His Name Indicates': R. S. Woodworth's Letters of Reference and Employment for Jewish Psychologists in the 1930s," *Journal of the History of the Behavioral Sciences* 32 (1996): 36.

33. Jastrow, "Joseph Jastrow," 140.

34. I use the term "middlebrow" in a nonpejorative way to refer to ventures using mass-produced print material to circulate "high" culture and knowledge to a wider reading public. See Joan Shelley Rubin, *The Making of Middle-Brow Culture* (Chapel Hill: University of North Carolina Press, 1992). Rubin emphasizes the creation of the Book-of-the-Month Club in 1926 as the start of a true middlebrow culture, but the periodicals of the second half of the nineteenth century represented a crucial phase in the development of middlebrow tastes and styles.

35. Jastrow was far from isolated in such a trajectory. See George Cotkin, "Middle-Ground Pragmatists: The Popularization of Philosophy in American Culture," *Journal of the History of Ideas* 55, no. 2 (1994): 283–302.

36. Joseph Jastrow, "The Psychology of Deception," *Popular Science Monthly* 34 (December 1888): 145–57. An expanded version was included in his first collection of popular essays, *Fact and Fable in Psychology* (Boston: Houghton Mifflin, 1900).

37. On the widespread use of this rhetorical trope, see Maarten Derksen, "Are We Not Experimenting Then? The Rhetorical Demarcation of Psychology and Common Sense," *Theory and Psychology* 7, no. 4 (1997): 435–56.

38. Jastrow, "Psychology of Deception," 150 (emphasis in the original).

39. See William B. Carpenter, *Principles of Human Physiology* (Philadelphia: Blanchard and Lea, 1860), 594–96.

40. Jastrow, "Psychology of Deception," 147.

41. Charles S. Peirce to James McKeen Cattell, May 9, 1900, Box 35, James McKeen Cattell Papers, Manuscript Division, Library of Congress (hereafter JMCP). On Peirce's philosophy in connection to commercial publishing, see Paul Jerome Croce, *Science and Religion in the Era of William James: Eclipse of Certainty, 1820–1880* (Chapel Hill: University of North Carolina Press, 1995), 206–22.

42. Charles S. Peirce, "The Fixation of Belief," *Popular Science Monthly* 12 (November 1877): 1–15.

43. C. S. Peirce, "The Doctrine of Chances," *Popular Science Monthly* 12 (March 1878): 604–15, 610–11.

44. See Jastrow, "The Reconstruction of Psychology," *Psychological Review* 34, no. 3 (1927): 170.

45. Norman Triplett, "The Psychology of Conjuring Deceptions," *American Journal of Psychology* 11, no. 4 (1900): 440, 447.

46. Ibid., 439, 477, 505.

47. William James, *Talks to Teachers on Psychology and to Students on Some of Life's Ideals* (New York: Holt, 1899), 159.

48. William James, *The Principles of Psychology*, 2 vols. (New York: Holt, 1890), 2:13, 183–84, 240, 402, 476.

49. Lewis Mumford, *The Golden Day: A Study of American Literature and Culture* (1926; New York Dover, 1968), 97.

50. William James, *Pragmatism: A New Name for Some Old Ways of Thinking* (New York: Longmans, Green, 1907), 207–8. For the reflection of historians on the metaphor, see George B. Cotkin, *William James, Public Philosopher* (Baltimore: Johns Hopkins University Press, 1990), 164–65; Livingston, *Pragmatism and the Political Economy*, 208–9; Levy, "Contemplating Delivery," 332–34.

51. James, *Principles of Psychology*, 1:122–23.

52. On the Darwinian law of economy, see Silvan S. Schweber, "Darwin and the Political Economists: Divergence of Character," *Journal of the History of Biology* 13, no. 2 (1980): 195–289. On Mach's principle of economy and its influence on American psychologists, including James, see Andrew S. Winston, "Cause Into Function: Ernest Mach and the Reconstruction of Explanation in Psychology," in *The Transformation of Psychology: Influences of Nineteenth-Century Philosophy, Technology, and Natural Science*, ed. Christopher D. Green, Thomas Teo, and Marlene Shore (Washington, DC: American Psychological Association, 2001), 107–31.

53. Due to his increasing concern about the fate of ethnic minorities within the polity, Jastrow came to distance himself from this racialist thinking. Jastrow built his defense of the male immigrant's place in the republic upon his own demarcations of social difference. He held immigrants up as bastions of rationality as opposed to the superstitious primitivism of the supposedly lower races and rejection of a feminized, Protestant culture of hysteria and mass consumption. See Joseph Jastrow, "Heredity and Mental Traits," *Science* 40 (October 16, 1914): 555–56; Jastrow, *Psychology of Conviction*; Jastrow, "Delusion, Mass-Suggestion, and the War: A Dream and the Awakening," *Scientific Monthly* 8, no. 5 (1919): 427–32.

54. Joseph Jastrow, "The Problems of 'Psychic Research,'" *Harper's Monthly Magazine* (1889): 76–82, 76.

55. Jackson Lears argues that many Americans of the period prized such superstitions explicitly understood as primitive as a salve against the stress of overcivilization through a recapturing of childlike innocence. See Lears, *Something for Nothing*, 201–15.

56. G. Stanley Hall, "Children's Lies," *American Journal of Psychology* 3, no. 1 (1890): 62; Jastrow, "Reconstruction of Psychology," 184.

57. Jastrow, "Psychology of Deception," 157.

58. See Triplett, "Psychology of Conjuring Deceptions," 505.

59. Struggling against one's own will also characterized nineteenth-century scientific objectivity; Daston and Galison, *Objectivity*, 191–252.

60. Simon During, *Modern Enchantments: The Cultural Power of Secular Magic* (Cambridge, MA: Harvard University Press, 2002), 135, 107–34; Peter Lamont, "Magician as Conjuror: A Frame Analysis of Victorian Mediums," *Early Popular Visual Culture* 4, no. 1 (2006): 21–33.

61. On illusions at American fairs, see Leja, *Looking Askance*, 153–83.

62. Robert Rydell, *All the World's a Fair: Visions of Empire at American International Expositions, 1876–1916* (Chicago: University of Chicago Press, 1984), 38–71; James Gilbert, *Perfect Cities: Chicago's Utopias of 1893* (Chicago: University of Chicago Press, 1991), 75–130; Rabinowitz, *For the Love of Pleasure*, 47–67.

63. It was in this capacity that he first made contact with Hugo Münsterberg. See Joseph Jastrow to Hugo Münsterberg, November 13, 1892, Mss. Acc. 1837, Hugo Münsterberg Papers, Boston Public Library (hereafter HMP).

64. See M. F. Washburn, "Some Apparatus for Cutaneous Stimulation," *American Journal of Psychology* 6, no. 3 (1894): 422–26.

65. Hall had conducted a detailed study of another famed deaf-blind woman, Laura Bridgman, several years beforehand. See G. Stanley Hall, "Laura Bridgman," *Mind* 4 (1879): 149–72.

66. Joseph Jastrow, "Psychological Notes on Helen Kellar [sic]," *Psychological Review* 1 (1894): 361. Since his time at Johns Hopkins, Jastrow empirically studied the mental life of the visually impaired. See Jastrow, "The Perception of Space by Disparate Senses," *Mind* 11 (1886): 539–54; Jastrow, "The Dreams of the Blind," *New Princeton Review* 5 (1888): 18–34.

67. See "When Men's Senses Are Tested," *Chicago Daily Tribune*, September 10, 1893, 35; "Experimental Psychology at the Fair," *Chicago Daily Tribune*, November 5, 1893, 28; Jastrow, "Joseph Jastrow," 142.

68. "At the Midwinter Fair," *Sacramento Daily Record-Union*, February 7, 1894, 6; Albert A. Hopkins, ed., *Magic, Stage Illusions, and Scientific Diversions* (1897; New York: Benjamin Blom, 1967), 91–94; "Amusements," *Omaha Daily Bee*, June 12, 1898, 13. See also "The Haunted Swing," *Salt Lake Herald*, June 7, 1894, 4; "Scenes at Atlanta's Show," *Omaha Daily Bee*, December 12, 1895, 8; "Most Wonderful Amusement Resort on Earth," *New York Sun*, June 12, 1904, 8.

69. See R. W. Wood, "The 'Haunted Swing' Illusion," *Psychological Review* 2 (1895): 277–78.

70. See Rydell, *All the World's a Fair*; Raymond Corbey "Ethnographic Showcases, 1870–1930," *Cultural Anthropology* 8, no. 3 (1993): 338–69; Shawn Michelle Smith, *American Archives: Gender, Race, and Class in Visual Culture* (Princeton, NJ: Princeton University Press, 1999), 157–86.

71. Joseph Jastrow, review of *Magic Stage Illusions and Scientific Diversions, Including Trick Photography*, *Science*, n.s., 6, no. 153 (December 3, 1897): 851.

72. Sofie Lachapelle, "From the Stage to the Laboratory: Magicians, Psychologists, and the Science of Illusion," *Journal of the History of the Behavioral Sciences* 44, no. 4 (2008): 319–34; E. W. Scripture, "Accurate Work in Psychology," *American Journal of Psychology* 6, no. 3 (1894): 429; G. Stanley Hall, "Spooks and Telepathy," *Appleton's Magazine* 12 (1908): 678; Norman Triplett, "Communication," *American Journal of Psychology* 12 (1900): 144.

73. On Herrmann and Kellar's contributions to this reformed magic, see Cook, *Arts of Deception*, 205–13.

74. See Alexander Herrmann, "New Light on the Black Art," *Cosmopolitan* 14, no. 2 (1892): 208–13; "Kellar, Magician, Dead," *New York Times*, May 11, 1922, 11.

75. "Magician Herrmann Dead," *New York Times*, December 18, 1896, 5.

76. Alexander Herrmann, "The Art of Magic," *North American Review* 153, no. 416 (1891): 95–96.

77. Joseph Jastrow, "Psychological Notes upon Sleight-of-Hand Experts," *Science*, n.s., 3, no. 7 (May 8, 1896): 685–89.

78. Ibid., 688, 686–87.

79. Ibid., 689.

80. See Augustin Charpentier, "Analyse experimentale de quelque element de la sensation de poids," *Archives de physiologie normale and pathologique* 3 (1891): 122–35; Edouard Claparède, "Expériences sur la vitesse de soulèvement des poids de volumes différents," *Archives de Psychologie* 1 (1902): 69–94.

81. Such ventures met with considerable criticism from his peers, who leveled charges of vulgarization and the plagiarism of a recent translation of Wundt's textbook. See Margaret Washburn, review of *Thinking, Feeling, Doing*, *The Philosophical Review* 4, no. 6 (1895): 659–61; Frank Angell, review of *Thinking, Feeling, Doing*, *Mind* 5 (1896): 272–73.

82. A heroic self-experimenter, Scripture declined to pursue this particular line of research because of the "disagreeable after-effects of the drug on my organism." See Scripture, "Consciousness under the Influence of *Cannabis Indica*," *Science* 22 (October 27, 1893): 233–34. Although his was the earliest published account by an American psychologist, Cattell recorded his experiences under the influence of various substances in the diary kept during his graduate studies. See Michael M. Sokal, ed., *An Education in Psychology: James McKeen Cattell's Journal and Letters from Germany and England, 1880–1888* (Cambridge, MA: MIT Press, 1981).

83. See Michael M. Sokal, "Biographical Approach: The Psychological Career of Edward Wheeler Scripture," in *Historiography of Modern Psychology: Aims, Resources, Approaches*, ed. Josef Broszek and Ludwig J. Pongratz (Toronto: C. J. Hogrefe, 1980), 255–78.

84. E. W. Scripture and C. E. Seashore, "On the Measurement of Hallucinations," *Science* 22 (December 29, 1893): 353.

85. C. E. Seashore, "Measurement of Illusions and Hallucinations in Normal Life," *Studies from the Yale Psychological Laboratory* 3 (1895): 64–65, 5, 24–29.

86. Henry Sidgwick et al., "Report of the Census of Hallucinations," *Proceedings of the Society for Psychical Research* 10 (1894): 25–422. William James was the committee's North American coordinator. See James, "Census of Hallucinations," *Science* 15 (May 16, 1890): 304.

87. E. W. Scripture, "The Law of Size-Weight Suggestion," *Science* 5 (February 5, 1897): 227.

88. E. W. Scripture, "Measuring Hallucinations," *Science* 3 (May 22, 1896): 762; Seashore, "Measurement of Illusions and Hallucinations in Normal Life," 2.

89. See Horst Gunlach, "What Is a Psychological Instrument?" in *Psychology's Territories: Historical and Contemporary Perspectives from Different Disciplines*, ed. Mitchell Ash and Thomas Strum (Mahwah, NJ: Lawrence Erlbaum Associates, 2007), 203–5.

90. See E. W. Scripture, *New Psychology* (New York: Scribner, 1898), 280.

91. "Carl Emil Seashore," in *History of Psychology in Autobiography*, ed. Murchison, 1:249.

92. J. Albert Gilbert, "Research on the Mental and Physical Development of School-Children," *Studies from the Yale Psychological Laboratory* 2 (1894): 59–63.

93. Seashore, "Measurement of Illusions and Hallucinations in Normal Life," 65; E. W. Scripture, *Thinking, Feeling, Doing* (1895; New York: Putnam, 1907), 217–22.

94. For an overview of Münsterberg's public science, see Matthew Hale Jr., *Human Science and Social Order: Hugo Münsterberg and the Origins of Applied Psychology* (Philadelphia: Temple University Press, 1980).

95. Milton Bradley to Hugo Münsterberg, October 29, 1895, Mss. Acc. 1575, HMP.

96. Edward Bradford Titchener, "A Psychological Laboratory," *Mind* 7 (1898): 311–31.

97. E. W. Scripture, "Notes and News: *Thinking, Feeling, Doing*," *Mind* 5, no. 20 (1896): 580.

98. See Peter Galison, "Image of Self," in *Things that Talk: Object Lessons from Art and Science*, ed. Lorraine Daston (New York: Zone Books, 2004), 257–94.

99. On the duck-rabbit as a psychological figure, see Peter Brugger, "One Hundred Years of an Ambiguous Figure: Happy Birthday, Duck/Rabbit!" *Perceptual and Motor Skills* 89 (1999): 973–77.

100. Joseph Jastrow, "The *Mind's Eye*," *Popular Science Monthly* 54 (1899): 295.

101. W.H.R. Rivers, "Vision," in *Reports of the Anthropological Expedition to Torres Strait*, vol. 2, *Physiology and Psychology*, ed. A. C. Haddon (Cambridge: Cambridge University Press, 1901), 130.

102. Ibid., 99, 126–27; Joseph Jastrow, review of *Reports of the Cambridge Anthropology Expedition to Torres Strait*, vol. 2, *Physiology and Psychology*, *Science*, n.s., 15, no. 384 (May 9, 1902): 742–44.

103. On the self-fashioning of early advertising professionals as cultural missionaries of modern culture, see Marchand, *Advertising the American Dream*.

104. Walter Dill Scott, *Theory and Practice of Advertising* (1903; New York: Small, Maynard, 1916), 168, 175–81.

105. Ibid., 191. On Scott's consulting career, see Friedman, *Birth of a Salesman*, 172–89.

106. James, *Principles of Psychology*, 1:293; George H. Mead, "Social Psychology as Counterpart to Physiological Psychology," *Psychological Bulletin* 6, no. 12

(1909): 401–8; Livingston, *Pragmatism and the Political Economy*, 158–80. On the Progressive Era intellectual's revolt against "bigness," see Michael Tavel Clarke, *These Days of Large Things: The Culture of Size in America, 1865–1930* (Ann Arbor: University of Michigan Press, 2007).

107. See Deborah J. Coon, "'One Moment in the World's Salvation': Anarchism and the Radicalization of William James," *Journal of American History* 83, no. 1 (1996): 70–99.

108. George M. Stratton, "Some Preliminary Experiments on Vision without Inversion of the Retinal Image," *Psychological Review* 3 (1896): 611–17, and Stratton, "Vision without Inversion of the Retinal Image," *Psychological Review* 4 (1897): 341–60.

109. George Stratton, *Experimental Psychology and Its Bearing upon Culture* (London: Macmillan, 1903), 157, 158 (emphasis in the original).

110. Seashore, "Measurement of Illusions and Hallucinations in Normal Life," 67; Triplett, "Psychology of Conjuring Deceptions," 507.

111. Joseph Jastrow to Ferris Greenslet, December 16, 1926, bMS Am 1925 (949), Houghton Mifflin Company Correspondence, Houghton Library, Harvard University.

Chapter Three

1. "Medium Screams at Dark Lantern," *Los Angeles Herald*, November 1, 1906, 9.

2. Bessie Beatty, "How Medium Fakers Dupe Innocent," *Los Angeles Sunday Herald*, November 4, 1906, sec. 3, 1, 6.

3. Bessie Beatty, "Truth Makes Fake Mediums Squirm," *Los Angeles Herald*, November 18, 1906, 6–7; Bessie Beatty, "Fakers Prove Easy Marks When Dealing with Bright Persons," *Los Angeles Herald*, November 25, 1906, 4.

4. See Harry Houdini, *The Miracle Mongers and Their Methods* (New York: Dutton, 1920).

5. George M. Beard, "The Psychology of Spiritism," *North American Review* 129 (July 1879): 67.

6. For an overview, see Robert C. Fuller, *Spiritual, but Not Religious: Understanding Unchurched America* (Oxford: Oxford University Press, 2001).

7. See Ann Taves, *Fits, Trances, and Visions: Experiencing Religion and Explaining Experience from Wesley to James* (Princeton, NJ: Princeton University Press, 1999); Leigh Eric Schmidt, *Hearing Things: Religion, Illusion, and the American Enlightenment* (Cambridge, MA: Harvard University Press, 2000); Robert S. Cox, *Body and Soul: A Sympathetic History of American Spiritualism* (Charlottesville: University of Virginia Press, 2003); Alex Owen, *The Place of Enchantment: British Occultism and the Culture of the Modern* (Chicago: University of Chicago Press, 2004); Corinna Treitel, *A Science for the Soul: Occultism and the Genesis of the*

German Modern (Baltimore: Johns Hopkins University Press, 2004); John Warne Monroe, *Laboratories of Faith: Mesmerism, Spiritism, and Occultism in Modern France* (Ithaca, NY: Cornell University Press, 2008).

8. See R. Laurence Moore, *In Search of White Crows: Spiritualism, Parapsychology, and American Culture* (New York: Oxford University Press, 1977); Wallis, ed., *On the Margins of Science*; Janet Oppenheim, *The Other World: Spiritualism and Psychical Research in England, 1850–1914* (Cambridge: Cambridge University Press, 1985); Morawski and Hornstein, "Quandary of the Quacks," 106–33; Coon, "Testing the Limits of Sense and Science," 143–51; Trudy Dehue, "From Deception Trials to Control Reagents: The Introduction of the Control Group about a Century Ago," *American Psychologist* 55, no. 2 (2000): 264–68.

9. During this very period, the scientist came to be seen as the moral equivalent of any other worker as opposed to an exalted interpreter of Nature. See Shapin, *The Scientific Life*, 21–46.

10. Ian Hacking, "Telepathy: Origins of Randomization in Experimental Design," *Isis* 79 (1988): 427–51; Richard Noakes, "'Instruments to Lay Hold of Spirits': Technologising the Bodies of Victorian Spiritualism," in *Bodies/Machines*, ed. Iwan Rhys Morus (Oxford: Berg Publishers, 2002), 125–63.

11. See Christopher Breu, *Hard-Boiled Masculinities* (Minneapolis: University of Minnesota Press, 2005); Allan Pinkerton, *The Detective and the Somnambulist* (Chicago: W. B. Keen, Cooke, 1875). The great exemplar of the hardboiled detective, Dashiell Hammett's Continental Op, encountered the California "spook racket" in *The Dain Curse* (1929). While Sherlock Homes may have also provided a model, these psychologists verbally sparred with his creator, Arthur Conan Doyle, over his advocacy of spiritualism.

12. On this aspect of Pinkerton's narratives, see J. K. Van Dover, *You Know My Method: The Science of the Detective* (Bowling Green, OH: Bowling Green State University Popular Press, 1994), 58.

13. In this respect, my interpretation of the relationship among psychologists, magicians, and detectives differs from that offered by Francesca Bordogna. She suggests that psychologists invoked the twin figures of the detective and stage magician as a means of avoiding William James's challenge to engage in psychical research. See Francesca Bordogna, *William James at the Boundaries: Philosophy, Science, and the Geography of Knowledge* (Chicago: University of Chicago Press, 2008), 103–6. Her account overlooks how psychologists continued to engage in such activities long after James's 1910 death. In contrast, I suggest that they offered the magician and detective as an alternative epistemological and moral model in their dealings with this particular class of psychological subjects. While both accounts stress how psychical phenomena challenged the psychologist's dependence on the ideal of mechanical objectivity, the reasons for the rejection differ.

14. William James to G. Stanley Hall, July 8, 1879, B1-3-2, Box 25, Folder 1, G. Stanley Hall Papers, Clark University (hereafter GSHP).

15. Eric Caplan, *Mind Games: American Culture and the Birth of Psychotherapy* (Berkeley: University of California Press, 1998), 37–60.

16. Spiritualism was not a movement particular to wealthy whites. On the role of spiritualism in other ethnic communities largely ignored by these psychologists, see Hans A. Baer, *The Black Spiritual Movement: A Religious Response to Racism* (Knoxville: University of Tennessee Press, 1984).

17. George M. Beard, "The Delusions of Clairvoyance," *Scribner's Monthly* 18, no. 3 (July 1879): 433.

18. Beard, "Psychology of Spiritism," 68.

19. Beard, "Delusions of Clairvoyance," 440.

20. Edward M. Brown, "Neurology and Spiritualism in the 1870s," *Bulletin of the History of Medicine* 57 (1983): 563–77.

21. George M. Beard, "Experiments with Living Human Beings," *Popular Science Monthly* 14, no. 44 (March 1879): 619–20, and no. 45 (April 1879): 751. Beard was not without precedent. In 1843 a trio of Baltimore physicians "humbugged" the city's Mesmerists by having an African American youth fake the somnambulistic trance on stage and then reveal his trickery before an audience of several hundred. See "Mesmerism in Baltimore," *Boston Medical and Surgical Journal* 28, no. 18 (June 7, 1843): 359–62. Alison Winter notes the staging of these anti-Mesmerist shows beginning in the 1840s, but also observes that spirit mediums rather than Mesmerists became the main target of these demonstrations and they were increasingly prevalent in the final quarter of the nineteenth century into the early years of the twentieth. See Winter, *Mesmerized: Powers of Mind in Victorian Britain* (Chicago: University of Chicago Press, 1998), 125–28.

22. G. Stanley Hall to Joseph Jastrow, February 24, 1920, B1-3-2, Box 25, Folder 2, GSHP; G. Stanley Hall, "Introduction," in *Studies in Spiritism*, ed. Amy E. Tanner (New York: Appleton, 1910), xv; Louis N. Wilson, *G. Stanley Hall: A Sketch* (New York: G. E. Stechert, 1914), 109–13.

23. Joseph Jastrow and George Nuttall, "On the Existence of a Magnetic Sense," *Proceedings of the American Society for Psychical Research* 1 (1885–89): 116–26, and Joseph Jastrow, "The Existence of a Magnetic Sense," *Science* 9 (July 2, 1886): 7–9; G. Stanley Hall to William James, December 21, 1886, MS Am 1092.9 (171), William James Papers, Houghton Library, Harvard University.

24. William James to Hugo Münsterberg, October 22, 1909, in *The Correspondence of William James*, vol. 12, ed. Ignas K. Skrupskelis and Elizabeth M. Berkeley (Charlottesville: University of Virginia Press, 2004). Future references to this particular volume are hereafter cited as CWJ.

25. See Edward L. Youmans's preface to Wilhelm Wundt, "Spiritualism as a Scientific Question," *Popular Science Monthly* 15 (1879): 577–78. On the distinction between American spiritualism and the more austere spiritism of continental Europe, see Monroe, *Laboratories of Faith*, 95–149.

26. Blewett Lee, "Spiritualism and Crime," *Columbia Law Review* 22, no. 5 (1922): 439–49; "Restricting False Prophets," *Chicago Tribune*, November 28, 1907, 10.

27. David Rasch, "Wills as Affected by Belief in Spiritualism," *New York University Law Review* 4, no. 1 (1927): 65.

28. McClary v. Stull, 62 N.W. 501 (1895)

29. Harry Hibschman, "Spooks and Wills," *United States Law Review* 64, no. 9 (1930): 471–80; Christopher J. Buccafusco, "Spiritualism and Will(s) in the Age of Contract," University of Illinois Legal Working Paper 91 (2008).

30. See Courtenay Grean Raia, "From Ether Theory to Ether Theology: Oliver Lodge and the Physics of Immortality," *Journal of the History of the Behavioral Sciences* 43, no. 1 (2007): 19–43.

31. Hereward Carrington to Hugo Münsterberg, October 10, 1909, Mss. Acc. 1611, HMP, and Hereward Carrington to G. Stanley Hall, October 10, 1909, B1-3-4, Box 27, Folder 8, GHSP.

32. Hereward Carrington, "Notes for Sitters," B1-3-4, Box 27, Folder 8, GSHP.

33. Charles L. Dana to G. Stanley Hall, November 17, 1909, B1-3-4, Box 27, folder 8, GSHP. See also Charles L. Dana to Hugo Münsterberg, November 17, 1909, Mss. Acc. 1645. HMP.

34. G. Stanley Hall to Hereward Carrington, November 17, 1909, B1-3-4, Box 27, Folder 8, GSHP.

35. Dickinson S. Miller to G. Stanley Hall, November 29, 1909, B1-3-4, Box 27, Folder 8, GSHP.

36. On the distinct meanings of "scientist" and "man of science," see Paul White, *Thomas Huxley: Making the "Man of Science"* (Cambridge: Cambridge University Press, 2003).

37. "Difficult to Detect Her," *New York Times*, May 12, 1910, 2.

38. Ibid., 3.

39. Carrington, "Notes for Sitters."

40. Joseph Jastrow, "The Unmasking of Pal[l]adino," *Collier's Weekly* 45, no. 8 (May 14, 1910): 21. Press accounts of the medium often misspelled her last name as "Paladino."

41. "Medium Tells How He Tricked Dupes," *New York Times*, June 5, 1910, 9.

42. Jastrow, "Unmasking of Pal[l]adino," 21; William James, "The Confidences of a 'Psychical Researcher'" (1909), in *Essays in Psychical Research* (Cambridge: Cambridge University Press, 1986), 372.

43. Jastrow, "Unmasking of Pal[l]adino," 21.

44. Jastrow, *Fact and Fable in Psychology*, 54.

45. Jon Palfreman, "Between Scepticism and Credulity: A Study of Victorian Scientific Attitudes to Modern Spiritualism," in *On the Margins of Science*, ed Wallis, 215–17.

46. Dickinson Miller, "Pal[l]adino's Tricks All Laid Bare," *New York Times*, May 12, 1910, 1; Dickinson Miller to James McKeen Cattell, January 8, 1910, Box 30, JMCP (emphasis in the original).

47. W. S. Davis, "The New York Exposure of Eusapia Palladino," *Journal of the American Society for Psychical Research* 4, no. 8 (1910): 402.

48. Joseph Jastrow, "The Case of Pal[l]adino," in Jastrow, *Psychology of Conviction*, 107.

49. For a contemporary explanation of the "Black Art" illusion, see Albert A. Hopkins, ed., *Magic, Stage Illusions and Scientific Diversions* (New York: Munn and Co., 1897), 67.

50. Jastrow, "The Case of Pal[l]adino," 107.

51. Jastrow, "Unmasking of Pal[l]adino," 22.

52. Miller, "Pal[l]adino's Tricks All Laid Bare," 1.

53. Jastrow, "Unmasking of Pal[l]adino," 21.

54. Davis, "New York Exposure," 411–12.

55. Ibid., 422; Jastrow, "Unmasking of Pal[l]adino," 42.

56. William James to Hereward Carrington, June 10, 1910, CWJ.

57. Davis, "New York Exposure," 424.

58. Jastrow, "The Case of Pal[l]adino," 102; G. Stanley Hall to Joseph Jastrow, February 9, 1920, B1-3-2, Box 25, Folder 2, GHSP.

59. John Jay Chapter to William James, June 19, 1910, CWJ.

60. Miller, "Pal[l]adino's Tricks All Laid Bare," 1.

61. Jastrow, "Unmasking of Pal[l]adino," 21.

62. "Palladino Test Won't Take Place," *New York Times*, May 27, 1910, 1–2.

63. R. W. Wood, "Palladino's Methods," *New York Times*, May 17, 1910, 8.

64. R. W. Wood, "Report of an Investigation of the Phenomena Connected with Eusapia Palladino," *Science* 31 (May 20, 1910): 779–80.

65. [Joseph Jastrow], "Chips from a Psychologist's Workshop," *The Dial*, May 16 1914, 426–27; Hugo Münsterberg to Doubleday, Page and Co., January 8, 1914, Mss. Acc. 2315(2), HMP; Hugo Münsterberg, "Spiritualism and Fraud," *Christian Observer* 98 (February 16, 1910): 9.

66. Hugo Münsterberg, "My Friend, the Spiritualists: Some Theories and Conclusion Concerning Eusapia Palladino," *Metropolitan Magazine* 31, no. 5 (February 1910): 571; Joseph Jastrow, "The Investigation of Palladino," *New York Times*, March 17, 1910, 10.

67. Margaret Münsterberg claimed that Edgar Scott, "who at the time did not wish to have his name known," served as the confederate. See Margaret Münsterberg, *Hugo Münsterberg: His Life and Work* (New York: Appleton, 1922), 181.

68. "The Foot of Palladino," *Washington Post*, February 7, 1910, 6.

69. Hugo Münsterberg to Hereward Carrington, November 16, 1909, Mss. Acc. 2301(2), HMP.

70. Hugo Münsterberg to Joseph Jastrow, June 10, 1914, Mss. Acc. 2343(2), HMP; Joseph Jastrow to Hugo Münsterberg, June 8, 1914, Mss. Acc. 1837, HMP.

71. Tanner, ed., *Studies in Spiritism*, 9–10.

72. Hall, cited in ibid., 179.

73. Ibid., 188.

74. Wundt rejected experimental misdirection, but it had become somewhat more prevalent following the imageless thought controversy. See Martin Kusch,

"Recluse, Interlocutor, Interrogator: Natural and Social Order in Turn-of-the-Century Psychological Research Schools," *Isis* 86 (1995): 419–39.

75. Oliver Lodge to G. Stanley Hall, November 9, 1909, B1-3-4, Box 27, Folder 9, GSHP.

76. Tanner, *Studies in Spiritism*, 214–15. She is the sole author, with a preface by Hall.

77. G. Stanley Hall to Mrs. Piper, May 18, 1909, B1-3-4; Alta L. Piper to George Dorr, n.d., ca. May 15, 1909; George Dorr to G. Stanley Hall, May 16, 1909; and Mrs. Piper to G. Stanley Hall, May 20, 1909, B1-3-4, Box 27, Folder 9, GSHP. According to her daughter, the physical strains of these tests caused a temporary lapse in Piper's ability. See Alta L. Piper, *The Life and Work of Mrs. Piper* (London: Kegan Paul, 1929), 173–75.

78. G. Stanley Hall to George Dorr, May 28, 1909, B1-3-4, Box 27, Folder 9, GSHP.

79. G. Stanley Hall to William James, September 13, 1909, and William James to Oliver Lodge, December 29, 1909, CWJ.

80. G. Stanley Hall to Oliver Lodge, October 15, 1909, B1-3-4, Box 27, Folder 9, GSHP.

81. Ann Braude, *Radical Spirits: Spiritualism and Women's Rights in Nineteenth-Century America* (Bloomington: University of Indiana Press, 2001); Alex Owen, *The Darkened Room: Women, Power, and Spiritualism in Late Victorian England* (1989; Philadelphia: University of Pennsylvania Press, 1990); Judith Walkowitz, *City of Dreadful Delight: Narratives of Sexual Danger in Late Victorian London* (Chicago: University of Chicago Press, 1992), 171–90; Beryl Satter, *Each Mind a Kingdom: American Women, Sexual Purity, and the New Thought Movement, 1875–1920* (Berkeley: University of California Press, 1999); John F. Kasson, *Houdini, Tarzan, and the Perfect Man: The While Male Body and the Challenge of Modernity in America* (New York: Hill and Wang, 2001), 77–155.

82. Kasson, *Houdini, Tarzan, and the Perfect Man*, 147–53.

83. Hugo Münsterberg to James McKeen Cattell, January 29, 1913, Mss. Acc. 2303(6), HMP.

84. Amy E. Tanner to G. Stanley Hall, February 10, 1907, B1-2-10, Box 12, Folder 5, GSHP.

85. Tanner, *Studies in Spiritism*, 166.

86. See Martha H. Patterson, *Beyond the Gibson Girl: Reimagining the American New Woman, 1895–1915* (Chicago: University of Illinois Press, 2005).

87. On this strain of feminist Progressivism, see Robyn Muncy, *Creating a Female Dominion in American Reform, 1890–1935* (New York: Oxford University Press, 1990).

88. "$20,000 to Girl Reporter," *New York Times*, February 8, 1914, 1.

89. Amy E. Tanner, "Glimpses at the Mind of a Waitress," *American Journal of Sociology* 13, no. 1 (1907): 48–55.

90. Ishbel Ross, *Ladies of the Press: The Story of Women in Journalism by an*

Insider (New York: Harper and Brothers, 1936), 581–83; Tanner, *Studies in Spiritism*, vi.

91. See "Prof. Quackenbos Silent," *New York Times*, March 27, 1893, 9; "Hypnotism the Cure-All," *New York Times*, April 30, 1899, 12; "Dr. Quackenbos Dies in 77th Year," *New York Times*, August 2, 1926, 17; "Child Is Psychic Marvel: Tests Show Uncanny Power," *Chicago Daily Tribune*, March 15, 1913, 1; "Girl Has Remarkable Power," *Washington Post*, April 27, 1913, MS2; "Girl Amazes Experts," *Washington Post*, February 23, 1913, 12; "Girl with X-Ray Eyes," *Washington Post*, March 16, 1913, ES4.

92. Hugo Münsterberg, "Thought Transference," in *Psychology and Social Sanity* (Garden City, NY: Doubleday, Page, 1914), 144. See also "Girl Prodigy Meets Tests," *New York Times*, February 16, 1913, 9.

93. Münsterberg, "Thought Transference," 145, 147, 148; "Beulah No Marvel, Says Münsterberg," *New York Times*, April 23, 1913, 20.

94. Münsterberg, "Thought Transference," 159–62; "Prof. Münsterberg of Harvard Lectures on Beulah Miller," *New York Times*, April 20, 1913, 8.

95. W. S. Davis, "Psychic Marvels," *New York Times*, March 19, 1913, 12.

96. "Claim Girl Is Fake," *Washington Post*, April 25, 1913, 3.

97. "Credulity Alone Miraculous," *New York Times*, March 20, 1913, 10.

98. "She Doesn't Even Mind Read," *New York Times*, April 24, 1913, 10.

99. Fred L. Holmes, "Girl, Blind-Deaf, Is Called World Wonder," *The Silent Worker* 34, no. 4 (January 1922): 121–23; "Helen Keller Meets Wonder Girl of Janesville School in Half-Hour Talk Here," *Wisconsin State Journal*, January 21, 1922, http://www.wisconsinhistory.org/wlhba/ (accessed March 7, 2010).

100. "Extraordinary Development of Tactile and Olfactory Senses Compensatory for the Loss of Sight and Hearing," *Official Bulletin of the Chicago Medical Society* 21 (June 3, 1922): 29.

101. Joseph Jastrow, "The Will to Believe," *Journal of the American Medical Association* 78 (June 17, 1922): 1891–93.

102. "Scientists at Odds Whether or Not Girl Sees," *Chicago Daily Tribune*, June 17, 1922, 12; Robert H. Gault, "An Unusual Case of Olfactory and Tactile Sensitivity," *Journal of Abnormal Psychology* 17, no. 4 (1923): 395–401. See also "Proves Girl Smells Color," *New York Times* (November 21, 1922): 9; "Can We See with Our Noses and Hear with Our Fingers?" *Popular Science Monthly* 102, no. 4 (April, 1923): 30–31; "Can We See with Our Noses and Hear with Our Fingers?" *Current Opinion* 75 (July 1, 1923): 91–92; "Deaf and Blind Girl Pronounced Cured," *New York Times*, January 30, 1924, 4.

103. See William James, *The Will to Believe and Other Essays in Popular Philosophy* (1896; Cambridge, MA: Harvard University Press, 1979). Although they held different views on the nature of spiritualism, Jastrow remained an admirer of James and dedicated his *Psychology of Conviction* to James, the "master psychologist."

104. Jastrow, "Will to Believe," 1892.

Chapter Four

1. FTC v. Standard Education Society, 86 Fed. 2d, 692 (2d Cir., 1936).
2. FTC v. Standard Education Society, 302 U.S. 112 (1937).
3. Lewis Wood, "Black Lines Up with Three Liberals in Divided Court," *New York Times*, November 9, 1937, 1. The "divided court" referred to another decision delivered on the same day; tellingly the court decision in *Standard Education Society* was unanimous.
4. On the law's practices for generating and using "common knowledge," see Mariana Valverde, *The Law's Dream of a Common Knowledge* (Princeton, NJ: Princeton University Press, 2003). See also Alan Hyde, *Bodies of Law* (Princeton, NJ: Princeton University Press, 1997); Susanna Blumenthal, "The Mind of a Moral Agent: Scottish Common Sense and the Problem of Responsibility in Nineteenth-Century American Law," *Law and History Review* 26, no. 1 (2008): 99–159.
5. See Earl Kintner, *A Primer on the Law of Deceptive Practice: A Guide for Businessmen* (New York: Macmillan, 1971), 7, and Ivan L. Preston, *The Great American Blowup: Puffery in Advertising* (Madison: University of Wisconsin Press, 1996), 27–46.
6. See Cohen, *Consumer's Republic*.
7. Florence Manufacturing Co. v. J. C. Dowd and Co., 178 Fed. 73 (1910), 75.
8. See Peter Miller and Nikolas Rose, "Mobilizing the Consumer: Assembling the Subject of Consumption," *Theory, Culture and Society* 14, no. 1 (1997): 1–36.
9. David P. Kuna, "The Concept of Suggestion in the Early History of Advertising Psychology," *Journal of the History of the Behavioral Sciences* 12 (1976): 347–53; Marchand, *Advertising the American Dream*, 52–87; Lears, *Fables of Abundance*; Brown, *The Corporate Eye*, 159–216.
10. Michel Callon, ed., *The Laws of Markets* (London: Blackwell, 1998).
11. Strasser, *Satisfaction Guaranteed*, 29–57.
12. J. R. Lunsford Jr., "The Unwary Purchaser in Unfair Trade Cases," *Mercer Law Review* 1 (1949): 48–68.
13. Trademark Cases, 100 U. S. 82 (1879), 94; Coats v. Merrick Thread Co., 149 U.S. 566 (1893), 567.
14. Young, *Medical Messiahs*; Nancy Tomes, "Merchants of Health: Medicine and Consumer Culture in the United States, 1900–1940," *Journal of American History* 88 (2001): 519–47; Joseph M. Gabriel, "A Thing Patented Is a Thing Revealed: Francis E. Stewart, George S. Davis, and the Legitimization of Patent Rights in Pharmaceutical Manufacturing, 1879–1911," *Journal of the History of Medicine and Allied Sciences* 64, no. 2 (2009): 135–72.
15. "Circular," *Journal of Materia Medica* 1, no. 1 (1858): 16; George H. Thurston, *Pittsburgh and Allegheny in the Centennial Year* (Pittsburgh: Anderson and Son, 1876), 240.
16. McLean v. Fleming, 96 U.S. 245 (1878), 256.

17. Wirtz v. Eagle Bottling Co., 50 NJ Eq. 164 (1892). See also Centaur Co. v. Robinson, 91 Fed. 899 (1899).

18. Fairbank v. Bell, 77 Fed. 869 (1896); P. Lorillard Co. v. Peper, 86 Fed. 956 (1898), 959.

19. "Judge Jenkins Much Affected," *New York Times*, July 14, 1893, 1. The charges were nullified a few months later. See "Indictments Null and Void," *New York Times*, November 7, 1893, 5.

20. Farmers' Loan and Trust Co. v. Northern Pacific Railroad Co., 60 Fed. 803 (1894).

21. Stephen George, *Enterprising Minnesotans: 150 Years of Business Pioneers* (Minneapolis: University of Minnesota Press, 2003), 27–30.

22. Pillsbury v. Pillsbury-Washburn Flour Co., 64 Fed. 841 (1894), 848.

23. On the notion of the purchaser's habitual behavior, see Jenkins Bros. v. Kelly and Jones Co., 227 Fed. 211 (1915).

24. See Charles Rosenberg, *The Trial of the Assassin Guiteau: Psychiatry and the Law in the Gilded Age* (Chicago: University of Chicago Press, 1968), and Susanna Blumenthal, "The Deviance of the Will: Policing the Bounds of Testamentary Freedom in Nineteenth-Century America," *Harvard Law Review* 119 (2006): 959–1034.

25. See Hilson Co. v. Foster, 80 Fed 896 (1897).

26. Pat O'Malley, "Risk and Responsibility," in *Foucault and Political Reason*, ed. Andrew Barry, Thomas Osborne, and Nikolas Rose (London: UCL Press, 1996), 189–207.

27. Fairbank v. Bell, 77 Fed. 869 (1896), 876.

28. In this sense, these decisions are a more extreme version of diminished personal responsibility embodied in the law during the Progressive Era. See Barbara Young Welke, *Recasting American Liberty: Gender, Race, Law, and the Railroad Revolution, 1865–1920* (New York: Cambridge University Press, 2001) and Michael Willrich, *City of Courts: Socializing Justice in Progressive Era Chicago* (New York: Cambridge University Press, 2003).

29. On the mental differences between judges and ordinary purchasers, see A. W. Barbar, "The Psychology of Trade-Marks," *Bulletin of the United States Trade-Mark Association* 6 (1911): 152–65.

30. On the concept of the "implicated actor" see Adele E. Clarke, *Disciplining Reproduction: Modernity, American Life Science, and "the Problem of Sex"* (Berkeley: University of California Press, 1998).

31. McCann v. Anthony, 21 Mo. App. 83 (1886); Britton v. White Manufacturing 61 Fed 93 (1894).

32. Oliver Wendell Holmes Jr., *The Common Law* (Boston: Little, Brown 1881), 53–58.

33. "Psychology and Trade Marks," *New York Times*, June 12, 1911, 10.

34. This was true when it came to product liability cases dealing with similarly inexpensive goods. There the burden was placed on the purchaser to prove that

the manufacturer had acted negligently with an explicit knowledge of a defect; the purchaser's unwariness and vulnerability were not presumed. Such was certainly the case in *Hudgins v. Coca-Cola Bottling Company* (1905), when a grocery merchant was unable to collect damages inflicted upon his person by an exploding cola bottle. See Hudgins v. Coca-Cola Bottling Company, 122 Ga. 695 (1905).

35. Williams v. Brooks, 50 Conn. 278 (1882), 283. A similar opinion was offered in Colman v. Crump, 70 NY 573 (1877) and Battle v. Finlay, 45 Fed. 796 (1891).

36. On the timing of this shift in advertising strategies, see Merle Curti, "The Changing Concept of 'Human Nature' in the Literature of American Advertising," *Business History Review* 41, no. 4 (1967): 335–57.

37. Bernard C. Steiner, "The Ordinary and the Ultimate Purchaser," *Yale Law Journal* 16 (1907): 112–25.

38. McLean v. Fleming, 251.

39. See Porter, *Trust in Numbers*. Porter does note (on pages 89–93) that judges as a professional group have been particularly hostile to this brand of objectivity.

40. While much legal scholarship has focused on the liberal rule of law tradition in America, the long-standing police powers in American governance proliferated in the Progressive Era. Federal, state, and municipal governments passed a plethora of statues constraining or prohibiting a variety of objects, practices, and behaviors in the name of the common good. The nature of these police powers was paradoxical: promoting social welfare in the name of the common good on the one hand while enforcing coercive state power against individuals on the other. These traditional, common law powers provided the legal basis for regulatory state of the early twentieth century. See Ernst Freund, *The Police Power: Public Policy and Constitutional Rights* (Chicago: Callahan, 1904); Christopher Tomlins, *Law, Labor, and Ideology in the Early Republic* (New York: Cambridge University Press, 1993), 19–59; Novak, *The People's Welfare*.

41. Markus Dirk Dubber, *The Police Power: Patriarchy and the Foundations of American Government* (New York: Columbia University Press, 2005).

42. "American Industries: No. 40: Oleomargarine—How It Is Made," *Scientific American* 42, no. 17 (April 24, 1880): 255–58.

43. Richard A. Ball and J. Robert Lilly, "The Menace of Margarine: The Rise and Fall of a Social Problem," *Social Problems* 29, no. 5 (1982): 488–98.

44. Geoffrey P. Miller, "Public Choice at the Dawn of the Special Interest State: The Story of Butter and Margarine," *California Law Review* 77, no. 1 (1989): 83–131; Henry C. Bannard, "The Oleomargarine Law: A Study in Congressional Politics," *Political Science Quarterly* 2, no. 4 (1887): 545–55.

45. James Harvey Young, "This Greasy Counterfeit: Butter versus Oleomargarine in the U.S. Congress, 1886," *Bulletin of the History of Medicine* 53, no. 3 (1979): 392–414; Powell v. Pennsylvania, 127 US 678 (1888).

46. Plumley v. Massachusetts, 155 US 461 (1894).

47. James Harvey Young, *Pure Food: Securing the Federal Food and Drugs Act of 1906* (Princeton, NJ: Princeton University Press, 1989).

48. Florence Kelley, *Some Ethical Gains through Legislation* (New York: Macmillan, 1905), 210–11, 212–13.

49. For Wiley's political biography, see Clayton A. Coppin and Jack C. High, *The Politics of Purity: Harvey Washington Wiley and the Origins of Federal Food Policy* (Ann Arbor: University of Michigan Press, 1999).

50. H. L. Hollingworth, "The Influence of Caffeine on Mental and Motor Efficiency," *Archives of Psychology* 20 (1912): iii–166. For a thoughtful discussion of Harry and Leta S. Hollingworth's innovations in experimental design aimed at isolating himself from the influence of their powerful patrons, see Ludy T. Benjamin Jr., Anne M. Rogers, and Angela Rosenbaum, "Coca-Cola, Caffeine, and Mental Deficiency: Harry Hollingworth and the Chattanooga Trial of 1911," *Journal of the History of the Behavioral Sciences* 27 (1991): 42–55.

51. United States v. Johnson, 221 U.S. 488 (1911).

52. Samuel V. Kennedy III, *Samuel Hopkins Adams and the Business of Writing* (Syracuse, NY: Syracuse University Press, 1999), 76–78.

53. Young, *Medical Messiahs*.

54. On this novel aspect of the commission, see Allyn A. Young, "The Sherman Act and the New Anti-Trust Legislation: III," *Journal of Political Economy* 23, no. 5 (1915): 417–36.

55. Sears, Roebuck Co. v. FTC, 258 Fed 307 (1919), 311.

56. See *Annual Report of the Federal Trade Commission for the Fiscal Year Ended June 30, 1931* (Washington, DC: U.S. Government Printing Office, 1931), 66–69.

57. See *Annual Report of the Federal Trade Commission for the Fiscal Year Ended June 30, 1928* (Washington, DC: U.S. Government Printing Office, 1928), 63.

58. Edward S. Rogers, "Comments on the Modern Law of Unfair Trade," *Illinois Law Review* 3 (1909): 551–64, and Rogers, "The Unwary Purchaser: A Study in the Psychology of Trade-Mark Infringement," *Michigan Law Review* 8 (1910): 613–22.

59. For a rare case that dismissed the claim based on a lack of actually deceived purchasers, see Pfeiffer v. Wilde, 102 Fed. 658 (1900). For the contrasting protection of the potentially careless, see Scriven v. North, 134 Fed. 366 (1904).

60. Rogers, "The Unwary Purchaser," 621–22. For an account of his grocery store haunting, see Edward S. Rogers, *Good-will, Trade-Marks, and Unfair Trading* (Chicago: Shaw, 1914), 66.

61. On the social investigators' production of the underclass as a natural kind, see Pittenger, "A World of Difference," 26–65. On the social scientist's shift to a mass public conceptualized in terms of consumer behavior, see Sarah E. Igo, *The Averaged American: Surveys, Citizens, and the Making of a Mass Public* (Cambridge, MA: Harvard University Press, 2007).

62. It was most likely John Wigmore that made this connection since he shared Rogers's concerns about the laxity and inconsistency of American trademark

law. See John H. Wigmore, "Justice, Commercial Morality, and the Federal Supreme Court: The Waterman Pen Case," *Illinois Law Review* 10 (1915): 178–89. The choice was not accidental: Münsterberg had just published his *On the Witness Stand* (1908), his collection of essays applying psychological techniques to legal situations. See chapter 5.

63. Edward S. Rogers to Hugo Münsterberg, February 19, 1909, Mss. Acc. 2088, HMP.

64. Münsterberg, "Psychology and the Market," *McClure's Magazine* 34, no. 1 (November 1909): 87–93.

65. Münsterberg, "Experiments with Reference to Illegal Imitation," in *Psychology and Industrial Efficiency* (1913; New York: Arno Press, 1973), 283.

66. Ibid., 286–90.

67. Gustave A. Feingold, "Recognition and Discrimination," *Psychological Monographs* 18 (1915): 1–128; Harold E. Burtt, "Measurement of Confusion between Similar Trade Names," *Illinois Law Review* 19 (1925): 320–36; Burtt, *Legal Psychology* (New York: Prentice-Hall, 1931).

68. Feingold, "Recognition and Discrimination," 34.

69. Feingold, "The Influence of Environment on the Identification of Persons and Things," *Journal of the American Institute of Criminal Law and Criminology* 5 (1914): 39–51.

70. On the importance of standardization and factory-like discipline in the psychological laboratory, see Deborah J. Coon, "Standardizing the Subject: Experimental Psychologists, Introspection, and the Quest for a Technoscientific Ideal," *Technology and Culture* 34 (1993): 757–83.

71. Feingold, "Recognition and Discrimination," 9–10.

72. Edward S. Rogers, "An Account of Some Psychological Experiments on the Subject of Trade-Mark Infringement," *Michigan Law Review* 18, no. 2 (1919): 75–103.

73. On the marketing of Coca-Cola, see Richard S. Tedlow, *New and Improved: The Story of Mass Marketing in America* (New York: Basic Books, 1990), 34 (emphasis in original).

74. Coca-Cola Co. v. Koke Co. of America, 254 U.S. 143 (1920), 146.

75. The best overview of the Coca-Cola's corporate history is Mark Pendergrast, *For God, Country, and Coca-Cola: The Unauthorized History of the Great American Soft Drink and the Company that Makes It* (New York: Scribner's, 1993).

76. "Stipulations," In the United States Patent Office, Opposition No. 1662, The Coca-Cola Company vs. Chero-Cola Company, 497, Coca-Cola Archives, Atlanta, Georgia (hereafter CCA).

77. Richard H. Paynter, "A Psychological Investigation of the Likelihood of Confusion between the Words 'Coca-Cola' and 'Chero-Cola,'" *Journal of Applied Psychology* 3 (1919): 329–51, and Paynter, "A Psychological Study of Trade-Mark Infringement," *Archives of Psychology* 42 (1920): 1–72.

78. Rogers, "Some Psychological Experiments," 78–82.
79. Ibid., 82–83.
80. Ibid.; Paynter, "Psychological Study of Trade-Mark Infringement," 71.
81. Rogers, "Some Psychological Experiments," 91–92.
82. "Applicant's Objections to Paynter's Report," In the United States Patent Office, Opposition No. 1662, The Coca-Cola Company vs. Chero-Cola Company, 528, 530, CCA; "Stipulations," 497.
83. Coca-Cola Co. v. Chero-Cola Co., 273 Fed. 755 (1921), 756.
84. Golan, *Laws of Men and Laws of Nature*, 211–53.
85. See Harry L. Hollingworth, "Characteristic Differences between Recall and Recognition," *American Journal of Psychology* 24 (1913): 532–44.
86. See Edward K. Strong Jr., "The Effect of Length of Series upon Recognition Memory," *Psychological Review* 19 (1912): 447–62; Feingold, "Recognition and Discrimination," 113.
87. See Strong, "Effect of Length of Series"; Strong, "The Effect of the Time-Interval upon Recognition Memory," *Psychological Review* 20 (1913): 339–72; Edith Mulhall, "Experimental Studies in Recognition and Recall," *American Journal of Psychology* 26 (1915): 217–28.
88. Feingold, "Recognition and Discrimination," 47.
89. It also became an important object of educational psychology in the form of word-recognition tests.
90. See Susan Ohmer, *George Gallup in Hollywood* (New York: Columbia University Press, 2006), 13–50. For a contemporary of review of Gallup's recognition test, see D. B. Lucas, "A Rigid Technique for Measuring the Impression Values of Specific Magazine Advertisements," *Journal of Applied Psychology* 24, no. 6 (1940): 778–90.
91. Henry C. Link, "A New Method of Testing Advertising Effectiveness," *Harvard Business Review* 11, no. 2 (1933): 165–77; Link, "A New Method for Testing Advertising and a Psychological Sales Barometer," *Journal of Applied Psychology* 18, no. 1 (1934): 1–26.
92. Münsterberg, *Psychology and Industrial Efficiency*, 286; Feingold, "Recognition and Discrimination," 117; Paynter, "Psychological Study of Trade-Mark Infringement," 6, 72.
93. See *Annual Report of the Federal Trade Commission for the Fiscal Year Ended June 30, 1938* (Washington, DC: U.S. Government Printing Office, 1938), 2–4, and C. H. Sandage, *Advertising: Theory and Practice* (1936; Chicago: Richard D. Irwin, 1945), 47–48.
94. On the importance of the consumer as a target of New Deal policy, see "The Ultimate Consumer: A Study in Economy of Illiteracy," *Annals of the American Academy of Political and Social Science* 173 (1934): 1–197. On the demographically diverse consumer movement of the period, see Cohen, *Consumer's Republic*, 19–61; Lawrence Glickman, *Buying Power: A History of Consumer Activism in America* (Chicago: University of Chicago Press, 2009), 189–218.

95. J. Philip Wernette, "Grade Labeling versus Descriptive Labeling," *National Marketing Review* 1, no. 2 (1935): 131–40.

96. On the New Deal grading system and its critics, see Gwen Kay, "Healthy Public Relations: The FDA's 1930s Legislative Campaign," *Bulletin of the History of Medicine* 75, no. 3 (2001): 446–87; Meg Jacobs, *Pocketbook Politics: Economic Citizenship in Twentieth-Century America* (Princeton, NJ: Princeton University Press, 2005), 124–26; McGovern, *Sold American*, 166–82, 255–58.

97. Testimony of Harrison S. Lyman, *Hearings before the Committee on Patents Subcommittee on Trade-Marks*, U.S. House of Representatives, 75th Cong., 3rd sess., H.R. 9041, March 15, 16, 17, 18, 1938 (Washington, DC: U.S. Government Printing Office, 1938), 36, 38.

98. Testimony of Milton Handler, *Hearings before the Committee on Patents Subcommittee on Trade-Marks*, U.S. House of Representatives, 77th Cong., 1st sess., H.R. 102, H.R. 5461, and S. 895, November 4, 12, 13, and 14, 1941 (Washington, DC: U.S. Government Printing Office, 1941), 106.

99. Testimony of Elliott H. Moyer, *Hearings before the Committee on Patents Subcommittee on Trade-Marks*, U.S. Senate, 77th Cong., 2nd sess., S. 895, December 11, 1942 (Washington, DC: U.S. Government Printing Office, 1942), 24

100. Testimony of Edward S. Rogers, *Hearings before the Committee on Patents Subcommittee on Trade-Marks*, U.S. House of Representatives, 75th Cong., 3rd sess., H.R. 9041, March 15, 16, 17, 18, 1938 (Washington, DC: U.S. Government Printing Office, 1938), 13.

101. Testimony of Edward S. Rogers, *Hearings before the Committee on Patents Subcommittee on Trade-Marks*, U.S. House of Representatives, 75th Cong., 3rd sess., H.R. 9041, March 15, 16, 17, 18, 1938 (Washington, DC: U.S. Government Printing Office, 1938), 50.

102. Ibid., 51.

103. Prestonettes, Inc. v. Coty, 264 US 359 (1924), 368.

104. Mishawaka Rubber and Woolen Manufacturing Co. v. S. S. Kresege Co. 316 U.S. 203 (1942), 205.

105. California Fruit Growers Exchange v. Sunkist Baking Co., 166 Fed. 2d. 971 (1949), 973.

106. Coca-Cola Co. v. Carlisle Bottling Works, 43 Fed. 2d 101 (1929), 114.

Chapter Five

1. A. A. Lewis, "Looking for an Honest Man," *Scientific Monthly* 49, no. 3 (1939): 267, 270.

2. In this regard, I depart from the excellent histories of the lie detector, namely Geoffrey C. Bunn, "The Hazards of the Will to Truth: A History of the Lie Detector," PhD diss., York University, 1998; Bunn, "Spectacular Science: The Lie Detector's Ambivalent Powers," *History of Psychology* 10, no. 2 (2007): 156–78;

Golan, *Laws of Men and Laws of Nature*, 211–53; Sean That O'Donnell, "Courting Science, Binding Truth: A Social History of *Frye v. U.S.*," PhD diss., Harvard University, 2007; Alder, *Lie Detectors*. The story of Lewis's "the statistical" lantern is central to chapter 6.

3. See Allan Sekula, "The Body and the Archive," *October* 39 (1986): 3–64; Ronald R. Thomas, *Detective Fiction and the Rise of Forensic Science* (New York: Cambridge University Press, 1999), 21–39; Jennifer L. Mnookin and Michelle West, "Theaters of Proof: Visual Evidence and the Law in *Call Northside 777*," *Yale Journal of Law and Humanities* 13 (2001): 329–402.

4. On the emergence of emotions as objects of scientific investigation, see Otniel E. Dror, "The Affect of Experiment: The Turn to Emotions in Anglo-American Physiology, 1900–1940," *Isis* 90 (1999): 205–37.

5. On psychoanalysis as a detective narrative, see Anne Harrington, *The Cure Within: A History of Mind-Body Medicine* (New York: Norton, 2008).

6. On the persistence of fixed types in "environmental" accounts of deviancy, see Pittenger, "A World of Difference," 26–65.

7. On the mental hygiene movement, see Elizabeth Lunbeck, *The Psychiatric Persuasion: Knowledge, Gender, and Power in Modern America* (Princeton, NJ: Princeton University Press, 1994); Jack D. Pressman, "Psychiatry and Its Origins," *Bulletin of the History of Medicine* 71 (1997): 129–39; Joseph M. Gabriel, "Mass Producing the Individual: Mary C. Jarrett, Elmer E. Southard, and the Industrial Origins of Psychiatric Social Work," *Bulletin of the History of Medicine* 79, no. 3 (2005): 430–58.

8. As the jurisdiction that first devoted a distinct legal procedure for youthful offenders, Chicago appears prominently in the literature on juvenile delinquency. Some accounts emphasize the theme of social control and the expansion of state power into a novel domain. See Anthony M. Platt, *The Child Savers: The Invention of Delinquency* (Chicago: University of Chicago Press, 1969); Anne Meis Knupfer, *Reform and Resistance: Gender, Delinquency, and America's First Juvenile Court* (New York: Routledge, 2001). Others remain more sympathetic to the court's program. See Victoria Getis, *The Juvenile Court and the Progressives* (Urbana: University of Illinois Press, 2000); David S. Tanenhaus, *Juvenile Justice in the Making* (Oxford: Oxford University Press, 2005). The most sophisticated account of the Progressive Era transformation of the law in Chicago focuses exclusively on the Municipal Court. See Willrich, *City of Courts*. Earlier accounts of Healy's work have failed to discuss *Pathological Lying* and its place in his own intellectual development as well as the field as a whole.

9. On the polygraph as placebo technology, see Alder, *Lie Detectors*, xiv.

10. Annette Mülberger, "Teaching Psychology to Jurists: Initiatives and Reactions Prior to World War I," *History of Psychology* 12, no. 2 (2009): 60–86.

11. See Carl G. Jung, "The Association Method," *American Journal of Psychology* 21 (1910): 219–40, and William Stern, "Abstracts of Lectures on the Psychology

of Testimony and on the Study of Individuality," *American Journal of Psychology* 21, no. 2 (1910): 270–82. For defenses of the reaction-time methodology in the wake of the turn toward physiological measures, see Eva R. Goldstein, "Reaction Times and the Consciousness of Deception," *American Journal of Psychology* 34 (1923): 562–81; H. R. Crosland, *The Psychological Methods of Word-Association and Reaction-Time as Tests of Deception* (Eugene: University of Oregon Press, 1929).

12. See Hugo Münsterberg, "Nothing but the Truth," *McClure's Magazine* 29 (1907): 532–36; Münsterberg, "The Third Degree," *McClure's Magazine* 29 (1907): 614–22; Münsterberg, "Hypnotism and Crime," *McClure's Magazine* 30 (1908): 317–22; Münsterberg, "The Prevention of Crime," *McClure's Magazine* 30 (1908): 750–56; Münsterberg, "Traces of Emotion and the Criminal," *Cosmopolitan* 44 (1908): 524–29; Münsterberg, *On the Witness Stand: Essays on Psychology and Crime* (1908; New York: Clark Boardman Co., 1925); "Machine to Detect Liars," *Washington Post*, March 17, 1909, 6.

13. Golan argues that it was for this very reason that jurists were much more resistant to psychological expertise compared with the physical and medical sciences. See Golan, *Laws of Men and Laws of Nature*, 212–13.

14. Münsterberg, *On the Witness Stand*, 15.

15. "Battering His Story: Orchard Is Put to the Test," *Boston Globe*, June 26, 1907, 1.

16. See J. Anthony Lukas, *Big Trouble: A Murder in a Small Western Town Sets Off a Struggle for the Soul of America* (New York: Simon and Schuster, 1997), 584–601.

17. Münsterberg, "Experiments with Harry Orchard" (1907), unpublished manuscript in Mss. Acc. 2450, 6, HMP.

18. Clarence Darrow to Hugo Münsterberg, August 16, 1907, Mss. Acc. 1648, HMP; "Moyer Follows W. D. Haywood on the Stand," *Washington Times*, July 5, 1907, 2.

19. Marlis Schweitzer, "'The Mad Search for Beauty': Actresses' Testimonials, the Cosmetics Industry, and the 'Democratization of Beauty,'" *Journal of the Gilded Age and Progressive Era* 4, no. 3 (2005): 255–92.

20. See "Finds Our Culture in Women's Hands: Business Man Too Tired to Do More than See Comic Opera, says Prof. Muensterberg," *New York Times*, January 29, 1910, 5.

21. Hugo Münsterberg to Samuel S. McClure, July 14, 1907, Mss. Acc. 2358, and Hugo Münsterberg to Clarence Darrow, July 14, 1907, Mss. Acc. 2311, HMP.

22. Charles C. Moore, "Yellow Psychology," *American Law Review* 42 (1908): 437–45; John H. Wigmore, "Professor Münsterberg and the Psychology of Testimony," *Illinois Law Review* 3, no. 7 (1909): 399–445; Wigmore to Hugo Münsterberg, November 11, 1908, and January 3, 1913, Mss. Acc. 2244, HMP; "When Scales of Justice Wobble," *New York Daily Tribune*, June 6, 1909, 3.

23. Lightner Witmer, "Mental Healing and the Emmanuel Movement," *The Psychological Clinic* 2, no. 8 (1909): 241.

24. Morton Prince to A. A. Roback, January 22, 1927, A. A. Roback Papers, Houghton Library, Harvard University (hereafter Roback Papers).

25. Caplan, *Mind Games*, 117–48. For Münsterberg's own rejection of the movement, see "In Lecture to Students Attacks Em[m]anuel Movement," *Richmond Times-Dispatch*, July 18, 1909, 2.

26. Samuel Hopkins Adams, *The Great American Fraud* (New York: Collier's, 1905).

27. On the role of diagnosis technologies in Cabot's own clinical practice, see Christopher Crenner, *Private Practice: In the Early Twentieth-Century Medical Office of Dr. Richard Cabot* (Baltimore: Johns Hopkins University Press, 2005).

28. Richard C. Cabot, "The Physician's Responsibility for the Nostrum Evil," *Journal of the American Medical Association* 47 (September 29, 1906): 982–84.

29. Cabot distinguished between deceitful placebos and "hypnotism and any other form of frankly mental therapeutics." See Richard C. Cabot, *Social Service and the Art of Healing* (New York: Moffat, Yard, and Co., 1909), 159.

30. Max Eastman, "Exploring the Soul and Healing the Body," *Everybody's Magazine* 32 (1915): 741; Peter Clark Macfarlane, "Diagnosis by Dreams," *Good Housekeeping* 60 (1915): 126.

31. Goddard, cited in "Electric Machine to Tell Guilt of Criminals," *New York Times*, September 19, 1911, 6.

32. Anton Delbrück, *Die Pathologische Lüge und die Psychisch Abnormen Schwindler* (Stuttgart: Enke, 1891).

33. William in a Shaw Dunn, "Pseudologia Phantastica, or Pathological Lying, in a Case of Hysteria with Moral Defect," *Journal of Mental Science* 62 (1916): 595–99.

34. Eugen Bleuler, *Textbook of Psychiatry*, trans. A. A. Brill (New York: Macmillan, 1924), 587.

35. "The Disease of Lying," *The Dial* 59 (November 11, 1915): 426. On Healy's ambivalence toward the exact tenets of psychoanalysis, see William Healy, *Mental Conflicts and Misconduct* (Boston: Little, Brown, 1915), 18–19.

36. The link between the social settlement workers and Healy is further discussed in Getis, *Juvenile Court and the Progressives*, 67–78.

37. On the social survey, see Martin Bulmer, Kevin Bales, and Kathryn Kish Sklar, eds., *The Social Survey in Historical Perspective, 1880–1940* (New York: Cambridge University Press, 1991).

38. The argument that Healy introduced a revolution in American criminological practice was widespread in the 1910s and 1920s. The clearest articulation can be found in Arthur E. Fink, "Causes of Crime: Biological Theories in the United States, 1800–1915," PhD diss., University of Pennsylvania, 1938, viii–ix, 149–50. Fink's closing year of 1915 is due to the publication of Healy's *The Individual Delinquent*.

39. See David G. Horn, *The Criminal Body: Lombroso and the Anatomy of De-

viance (New York: Routledge, 2003); William Healy, "The Individual Study of the Young Criminal," *Journal of the American Institute for Criminal Law and Criminology* (hereafter *JAICLC*) 1, no. 1 (1910): 50–62; William Healy, *The Individual Delinquent: A Text Book of Diagnosis and Prognosis for All Concerned in Understanding Offenders* (Boston: Little, Brown, 1915). For another discussion of Healy's investigative methodology, see Kathleen W. Jones, *Taming the Troublesome Child: American Families, Child Guidance, and the Limits of Psychiatric Authority* (Cambridge, MA: Harvard University Press, 1999), 38–61.

40. See W. Leslie MacKenzie, review of *The Individual Delinquent, Mind* 28, no. 111 (1919): 354–58; Henry M. Hyde, "Expert Laughs at Theory that Crime Is a Disease," *Chicago Tribune*, January 16, 1915, 13.

41. Healy, *Individual Delinquent*, 5–8; John C. Burnham, "Interviews of William Healy and Augusta Bonner, January 1960," 236–37. A transcript of the interview is available at the Chicago Historical Society. The interviewees requested that it not be quoted directly. This retrospective assessment may not reflect his actual historical development.

42. Christopher Lawrence, "Incommunicable Knowledge: Science, Technology and the Clinical Art in Britain, 1850–1914," *Journal of Contemporary History* 20, no. 4 (1985): 503–20.

43. Healy, *Individual Delinquent*, 34, 36–37. Healy married Bonner after his first wife died of cancer in 1932. See "Healy Interview," 78.

44. William Healy and Mary Healy, *Pathological Lying, Accusation, and Swindling: A Study in Forensic Psychology* (Boston: Little, Brown, 1915). Healy's interpretation of pathological lying first appeared in his *Individual Delinquent*, 729–52. The book-length study of the pathological liar was coauthored with his wife Mary. From existing records it is difficult to determine each author's relative contribution to the monograph. When referring to findings from *Pathological Lying* I will refer to the Healys in the plural, but refer to William Healy in the singular when information is derived from texts solely authored by him.

45. Healy and Healy, *Pathological Lying*, 1.

46. For obvious reasons, the Healys used pseudonyms and altered some identifying details in their published accounts. I retain the names they gave. The account of the case of Inez M. is from ibid., 54–81.

47. Ibid., 60.

48. Ibid., 93, 43.

49. Ibid., 107.

50. Ibid., 266.

51. See Burnham, "Interviews," 105.

52. On the innocent face of the young thief, see Healy, "Individual Study of the Young Criminal," 52–53.

53. "Menace of the Lying Child: Investigation Reveals Odd Instances of Apparently Normal Juveniles with Queer Traits," *Washington Post Sunday Magazine*, November 7, 1915, 2.

54. Healy and Healy, *Pathological Lying*, 250–52, 266.

55. On the proliferation of diagnostic technologies between 1900 and 1925, see Joel D. Howell, *Technology in the Hospital: Transforming Patient Care in the Early Twentieth Century* (Baltimore: Johns Hopkins University Press, 1995).

56. William Healy and Grace Maxwell Fernald, "Tests for Practical Mental Classification," *Psychological Monographs* 13, no. 2 (1911): 22; William Healy to John Wigmore, December 4, 1914, Box 64, Folder 21, John Henry Wigmore Papers, University Archives, Northwestern University (hereafter JHWP); Healy and Healy, *Pathological Lying*, 12.

57. See Augusta F. Bonner, William Healy, Gladys M. Lowe, and Myra E. Shimberg, *A Manual of Individual Mental Tests and Testing* (Boston: Little, Brown, 1927), 66.

58. Healy and Healy, *Pathological Lying*, 253–54.

59. Ibid., 258, 265, 263. In an interview conducted with Healy in 1960, he discussed his growing fascination with Freud although he was discouraged from calling on him during his 1900 grand tour of European psychiatric sites, which included Vienna. Burnham, "Interviews," 54.

60. Healy and Healy, *Pathological Lying*, 277–78.

61. "Wherein the Pathological Liar Is Superior to the Truthful Man," *Current Opinion* 59, no. 5 (1916): 325.

62. Richard C. Cabot, *Honesty* (New York: Macmillan, 1938), 30–31.

63. Herman Adler, Review of *Pathological Lying*, *Harvard Law Review* 29 (1915): 347.

64. On the significance of "thinking with cases," see John Forrester, "If *p*, then what? Thinking in Cases," *History of the Human Sciences* 9, no. 3 (1996): 1–25. Other readers were more sympathetic. For example, a number of reviewers stressed the value of the case-based narrative style and agreed on the importance of training forensic experts through the reading of cases rather than presenting the information abstractly. See the unsigned review in *Columbia Law Review* 16 (1916): 364–65.

65. For retrospective criticisms, see Abraham P. Ordover, "Admissibility of Patterns of Similar Sexual Conduct: The Unlamented Death of Character for Chastity," *Cornell Law Review* 63 (1978): 120–21, and Leigh B. Bienen, "A Question of Credibility: John Henry Wigmore's Use of Scientific Authority in Section 924a of the Treatise on Evidence," *California Western Law Review* 19 (1983): 235–68.

66. Philip Kinsley, "Jury Acquits Wolfrum of Girls' Charge," *Chicago Tribune*, May 25, 1927, 1. Papers would occasionally report a defendant receiving clemency after having been identified as a pathological liar. For example, see "Lie Preventive Needed," *Washington Post*, June 21, 1923, 6.

67. See "Sexual Immorality as Affecting Female Credibility," *American Law Review* 61 (1927): 273–75, and C. W. Hall, "Impeachment by Evidence of a Witness's Bad Character," *North Carolina Law Review* 5 (1927): 340–45.

68. See Elizabeth S. Abelson, *When Ladies Go A-Thieving: Middle-Class Shoplifters in the Victorian Department Store* (New York: Oxford University Press, 1989).

69. Bernard Glueck, *Studies in Forensic Psychiatry* (Boston: Little, Brown, 1916), 241, 163.

70. Nathan G. Hale, *The Rise and Crisis of Psychoanalysis in the United States: Freud and the Americans, 1917–1985* (New York: Oxford University Press, 1995).

71. William Moulton Marston, "Studies on Testimony," *JAICLC* 15, no. 1 (1924): 5–31; John Larson, *Lying and Its Detection: A Study of Deception and Deception Tests* (Chicago: University of Chicago Press, 1932), 35; Fred E. Inbau, "Scientific Evidence in Criminal Cases. II: Methods of Detecting Deception," *JAICLC* 24, no. 6 (1934): 1140–58.

72. For an overview of Reeve's work, see Van Dover, *You Know My Method*, 159–89.

73. Arthur B. Reeve, "In Defense of the Detective Story," *The Independent* 75 (July 10, 1913): 92–93; Reeve, *The Dream Doctor* (New York: Harper and Bros., 1913), 10, 66–67, 213–18; Reeve, *The Silent Bullet* (New York: Harper and Bros., 1910), 53–58.

74. W. B. Cannon, "The Interrelations of Emotions as Suggested by Recent Physiological Researches," *American Journal of Psychology* 25, no. 2 (1914): 256–82; William M. Marston, "Systolic Blood Pressure Symptoms of Deception," *Journal of Experimental Psychology* 2 (1917): 117–63.

75. Christopher W. Crenner, "Introduction of the Blood Pressure Cuff into U.S. Medical Practice: Technology and Skilled Practice," *Annals of Internal Medicine* 128, no. 6 (March 15, 1998): 488–93.

76. William Moulton Marston, *The Lie Detector Test* (New York: Richard Smith, 1938), 23.

77. Marston, "Systolic Blood Pressure Symptoms of Deception," 124, 126.

78. William M. Marston, "Psychological Possibilities in the Deception Tests," *JAICLC* 11 (1921): 551–70. With few exceptions, all test subjects in this report were women.

79. E. A. Kreger to Surgeon General, September 2, 1919, Projects Proposed File, Division of Anthropology and Psychology, 1919–1939, National Academies Archives, Washington, DC.

80. Marston, "Systolic Blood Pressure Symptoms of Deception," 162 (emphasis in original).

81. William M. Marston, "Reaction-Time Symptoms of Deception," *Journal of Experimental Psychology* 3, no. 1 (1920): 73.

82. See Anne Roller, "Vollmer and His College Cops," *Survey* 62 (1929): 304–7; Claire Bond Potter, *War on Crime: Bandits, G-Men, and the Politics of Mass Culture* (New Brunswick, NJ: Rutgers University Press, 1998), 31–56. Potter's analysis pertains to J. Edgar Hoover's reforms of the Federal Bureau of Investigation starting

in 1924. Vollmer's nationally recognized work in Berkeley influenced this trend at the federal level.

83. See Ken Alder, "A Social History of Untruth: Lie Detection and Trust in Twentieth-Century America," *Representations* 80 (2002): 1–33.

84. August Vollmer to Leonarde Keeler, April 4, 1934, Box 11, Folder 20, Charles Augustus Keeler Papers, Bancroft Library, University of California, Berkeley (hereafter Keeler Papers).

85. John A. Larson, "Modification of the Marston Deception Test," *JAICLC* 12, no. 3 (1921): 390–99.

86. Rollin Perkins, review of Larson, *Lying and Its Detection*, *Iowa Law Review* 18 (1933): 575.

87. J. A. Larson, "The Cardio-Pneumo-Psychogram and Its Use in the Study of Emotions, with Practical Applications," *Journal of Experimental Psychology* 5 (1922): 323–28.

88. William Marston to John Wigmore, July 30, 1922, Box 90, Folder 12, JHWP.

89. Frye v. United States, 293 Fed 1013 (1923). On objectivity as disciplinary consensus rather than mechanical procedures, see Porter, *Trust in Numbers*, 3–4.

90. O'Donnell, "Courting Science, Binding Truth," 99–141, helpfully situates the case in relationship to the African American community in Washington and the racial politics of policing in the city.

91. Marston, *Lie Detector Test*, 70–73.

92. Ibid., 67; "Blondes Lose in Film Love Test," *New York Times*, January 31, 1928, 25.

93. The skepticism of psychologists was famously reported in C. T. McCormick, "Deception-Tests and the Law of Evidence," *California Law Review* 15, no. 6 (1927): 484–504.

94. John Wigmore to Leonard[e] Keeler, August 8, 1929, Box 78, Folder 31, JHWP; Leonarde Keeler to August Vollmer, December 17, 1929, Box 7, Folder 1, Keeler Papers.

95. Wigmore explained the origins of the laboratory in a letter to the director of the Bureau of Social Hygiene. See John Wigmore to Lawrence B. Dunham, December 27, 1929, Box 103, Folder 6, JHWP.

96. Memorandum, November 23, 1929, Box 103, Folder 6, JHWP.

97. Memorandum, "The Massee Crime Detection Laboratory Project," June 1, 1929, Box 103, Folder 6, JHWP.

98. John Wigmore to Walter Dill Scott, January 14, 1937, Box 103, Folder 9, JHWP (emphasis in original).

99. "The World's Fair Police Exhibit," *Police 13–13* (June 1933): 8–11; "Science Puts an End to 'Perfect Crime,'" *World's Fair Weekly*, June 17 1933, 9, 10.

100. For example, see "Lie Test Fails Man Who Paid for Taking It," *Chicago Defender*, March 26, 1938, 6.

101. Edwin W. Sims, "Fighting Crime in Chicago: The Crime Commission," *JAICLC* 11, no. 1 (1920): 21, 27–28; "Inventor Tells How His Device Picks Out Liars," *Chicago Daily Tribune*, August 30 1930, 4; John Larson to Donald L. Kooken, April 18, 1933, Box 1, Folder 10, John A. Larson Papers, Bancroft Library, University of California, Berkeley (hereafter Larson Papers).

102. "'Secret Six' Open Warfare on Rackets as Crime Cure," *Chicago Daily Tribune*, February 22, 1930, 2; Richard Lindberg, *To Serve and Collect: Chicago Politics and Police Corruption from the Lager Beer Riot to the Summerdale Scandal* (New York: Praeger, 1991), 206–7.

103. John Larson to August Vollmer, May 27, 1930, Box 1, Folder 18, BANC MSS 78/160 cz, Larson Papers.

104. John Larson to Leonarde Keeler, December 21, 1931, Box 1, Folder 9, Larson Papers; Larson to Kooken, April 18, 1933.

105. Larson, "Cardio-Pneumo-Psychogram," 326.

106. John A. Larson and Herman M. Adler, "A Study of Deception in the Penitentiary: A Preliminary Communication" (1925), quote 5, Carton 1, Folder 16, Larson Papers.

107. Herman Adler and John Larson, "Deception and Self-Deception," *Journal of Abnormal Psychology and Social Psychology* 22 (1928): 364–71. Despite this collaboration, Larson held Adler in particularly low regard. See Alder, *Lie Detectors*, 98.

108. Walter G. Summers, "Science Can Get the Confession," *Fordham Law Review* 8 (1939): 351.

109. Marston, *Lie Detector Test*, 15, 8–9, 16.

110. Ibid., 17.

111. William Moulton Marston to John A. Larson, June 30, 1936, Box 1, Folder 13, Larson Papers; Fred E. Inbau, review of *The Lie Detector*, *JAICLC* 29, no. 2 (1938): 305.

112. A. A. Roback, "The Psychology of Confession," *University Magazine* 16, no. 2 (1917): 265–84; Münsterberg, "The Third Degree," 621; Paul Schilder, "Comments," in Larson, *Lying and its Detection*, xviii–xx; William Healy, review of *The Lie Detector Test*, *Psychoanalytic Quarterly* 7 (1938): 400–403.

113. Doris Blake, "White Lies or Fact," *Chicago Daily Tribune*, October 19, 1924, part 6, 9. "Doris Blake" was the pen name used by Antoinette Donnelly, who also wrote advice columns under her own name. See "Doris Blake Advises the Lovelorn," *Life*, September 15, 1941, 65–72.

114. For Blake's use of the lie detector evidence, see Doris Blake, "Lie Detector Shows Women 'Fib' Little," *Chicago Daily Tribune*, April 14, 1938, 14. For a sample of her advice to wives to deceive their husbands in the name of marital happiness, see Blake, "The Fib and the Lie," *Chicago Daily Tribune*, December 5, 1913, 11; Blake, "How to Be Happy though Married," *Chicago Daily Tribune*, September 21, 1923, 22; Blake, "There Are Worse Lies in Life than Those Told in

Courtship," *Chicago Daily Tribune*, August 29, 1925, 14; Blake, "This Business of Deception Offers Food for Thought," *Chicago Daily Tribune*, January 29, 1927, 15; Blake, "Can Woman Love Her Mate and Lie, Too? Wotta Question!" *Chicago Daily Tribune*, December 27, 1930, 11.

115. She was equally incredulous of psychoanalysis. See Blake, "Think Twice Before You Take That Child to a Psychoanalyst," *Chicago Daily Tribune*, November 18, 1924, 26.

Chapter Six

1. William H. Whyte, Jr., *The Organization Man* (New York: Simon and Schuster, 1956), 405, 406.

2. Catherine Lutz, "Epistemology of the Bunker: The Brainwashed and Other New Subjects of Permanent War," in *Inventing the Psychological: Toward a Cultural History of Emotional Life in America*, ed. Joel Pfister and Nancy Schnog (New Haven, CT: Yale University Press, 1997), 245–67.

3. L. L. Thurstone, "Attitudes Can Be Measured," *American Journal of Sociology* 33, no. 4 (1928): 532.

4. Hugh Hartshorne and Mark A. May, *Studies in the Nature of Character*, vol. 1, *Studies in Deceit* (New York: Macmillan, 1928). The subsequent volumes were Hugh Hartshorne, Mark A. May, and Julius B. Maller, *Studies in Service and Self-Control* (New York: Macmillan, 1929); Hartshorne, May, and Frank K Shuttleworth, *Studies in the Organization of Character* (New York: Macmillan, 1930).

5. On the Lewinian origins of dramaturgical and deceptive experiments, see James H. Korn, *Illusions of Reality: A History of Deception in Social Psychology* (Albany: SUNY University Press, 1997), 38–54.

6. Historian Lynn Dumenil borrowed Walter Lippmann's phrase to capture "the wide-ranging forces that were disrupting the certainty and stability of modern men and women." See Dumenil, *The Modern Temper: American Culture and Society in the 1920s* (New York: Hill and Wang, 1995), 148.

7. The best history of the movements remains David I. Macleod, *Building Character in the American Boy: The Boy Scouts, YMCA, and Their Forerunners, 1870–1920* (Madison: University of Wisconsin Press, 1983).

8. Paula S. Fass, *The Damned and the Beautiful: American Youth in the 1920s* (Oxford: Oxford University Press, 1977), 18–25.

9. "Home Training Comes First," *Los Angeles Times*, September 9, 1922, sec 2, p. 4; Edward F. Roberts, "Boy Crooks Steal Billions Yearly," *Los Angeles Times*, December 7, 1924, sec. J, p. 10; "Teaching Honesty," *The Outlook*, October 24, 1923, 296.

10. Ellis W. Hawley, "Herbert Hoover, the Commerce Secretariat, and the Vision of an 'Associative State,' 1921–1928," *Journal of American History* 61, no. 1 (1974): 116–40.

11. Cited in Dumenil, *The Modern Temper*, 38.

12. "Reports Average Man Still Honest," *Boston Daily Globe*, June 22, 1923, 3.

13. Roberts, "Boy Crooks Steal Billions Yearly."

14. William Byron Forbush, "The National Honesty Bureau," *Journal of Social Forces* 1, no. 2 (1923): 127–28.

15. *Honesty Book: A Handbook for Teachers, Parents and Other Friends of Children* (New York: National Honesty Bureau, 1923), iii.

16. William Byron Forbush, *The Boy Problem: A Study in Social Pedagogy* (Boston: Pilgrim, 1901); Heather A. Warren, "Character, Public Schooling, and Religious Education, 1920–1934," *Religion and American Culture* 7, no. 1 (1997): 61–80.

17. Mark May, "A Summary of the Work of the Character Education Inquiry" [1929], 2, MS 1447, Box 8, Folder 9, Mark Arthur May Papers, Yale University Libraries (hereafter May Papers).

18. John Carson, *The Measure of Merit: Talents, Intelligence, and Inequality in the French and American Republics, 1750–1940* (Princeton, NJ: Princeton University Press, 2007), 159–270.

19. Paul M. Dennis, "The Edison Questionnaire," *Journal of the History of the Behavioral Sciences* 20, no. 1 (1984): 23–37.

20. "Value of Mental Tests in Hiring," *New York Times*, September 19, 1920, sec. E, p. 19.

21. Ibid.

22. Nicholas Pastore, "The Army Intelligence Tests and Walter Lippmann," *Journal of the History of the Behavioral Sciences* 14, no. 4 (1978): 316–27; Stephen Leacock, "Manual of the New Mentality," *Harper's Monthly* 148 (March 1924): 473.

23. "Educators Outline Duty of Teachers," *New York Times*, May 15, 1926, 15; C. Leslie Updegraph, "Present Status of Moral Education in the Public Schools," *The Phi Delta Kappan* 9, no. 5 (1927): 137–40; "Schoolroom Ability," *New York Times*, January 31, 1926, sec. E, p. 12.

24. Percival M. Symonds, *Diagnosing Personality and Conduct* (New York: Century, 1931), v.

25. Lewis M. Terman, *The Intelligence of School Children: How Children Differ in Ability, the Use of Mental Tests in School Grading, and the Proper Education of Exceptional Children* (Boston: Houghton Mifflin, 1919), 10; Lewis M. Terman, "Foreword," to Vernon M. Cady, "The Estimation of Juvenile Incorrigibility," *Journal of Delinquency Monographs* 2 (1923): 4.

26. See Edward L. Thorndike, "Intelligence and Its Uses," *Harper's Monthly* 140 (January 1920): 227–35.

27. On this connection, see Emily C. Davis, "Psychologists Look for Honest Children," *Science News-Letter* 13 (January 16, 1928): 371–72, 381–82.

28. Retha E. Breeze, "What Constitutes Good Citizenship," *The School Review* 32, no. 7 (1924): 535.

29. Paul Frederick Voelker, *The Function of Ideals and Attitudes in Social Education: An Experimental Study* (New York: Teachers College, Columbia University, 1921), 5.

30. Ibid., 64.

31. John Dewey, "The Chaos in Moral Training," *Popular Science Monthly* 45 (August 1894): 433–43.

32. Voelker, *Function of Ideals and Attitudes*, 65.

33. Ibid., 100, 74–75.

34. Jones, *Taming the Troublesome Child*, 120–21. More broadly, see Fass, *The Damned and the Beautiful*, and John Modell, *Into One's Own: From Youth to Adulthood in the United States, 1920–1975* (Berkeley: University of California Press, 1989), 67–120.

35. Eunice Fuller Barnard, "Moral Tests Reveal Children's Standards," *New York Times Magazine*, March 20, 1927, 15.

36. Mark A. May and Hugh Hartshorne, "Personality and Character Tests," *Psychological Bulletin* 23, no. 7 (1926): 395–411; Mark A. May, Hugh Hartshorne, and Ruth E. Welty, "Personality and Character Tests," *Psychological Bulletin* 24, no. 7 (1927): 418–35; Mark A. May, Hugh Hartshorne, and Ruth E. Welty, "Personality and Character Tests," *Psychological Bulletin* 25, no. 7 (1928): 422–43; A. A. Roback, *A Bibliography of Character and Personality* (Cambridge, MA: Sci-Art Publishers, 1927).

37. Vernon M. Cady, "The Estimation of Juvenile Incorrigibility," *Journal of Delinquency Monograph* 2 (1923).

38. Terman, "Were We Born That Way?" *World's Work* 44 (1922): 655–60.

39. Lewis Terman to Commonwealth Fund, April 10, 1922, SC 038, Box 10, Folder 2, and Terman, "The Physical and Mental Traits of Gifted Children," 12, 14, Box 10, Folder 36, Lewis Madison Terman Papers, Stanford University Archives, Stanford, California (hereafter Terman Papers).

40. Cady, "Estimation of Juvenile Incorrigibility," 50–53.

41. Albert Sydney Raubenheimer, "An Experimental Study of Some Behavior Traits of the Potentially Delinquent Boy," *Psychological Monographs* 34, no. 6 (1925): 8–13.

42. A. A. Roback, *The Psychology of Character* (New York: Harcourt, Brace, 1927), 360.

43. Katharine Murdoch, "A Study of Differences Found Between Races in Intellect and in Morality," *School and Society* 22 (1925): 628–32, 659–64.

44. See S. L. Pressey and L. C. Pressey, "Results of Certain Honesty Tests Given to a Group of Rural White Children and to Two Groups of Indian Children," *Journal of Applied Psychology* 17, no. 2 (1933): 120–29.

45. Hugh Hartshorne, "A Research Extraordinary: The Character Education Inquiry at Teachers College," *Religious Education* 19, no. 6 (1924): 397–98.

46. Charles E. Harvey, "John D. Rockefeller, Jr., and the Interchurch World

Movement of 1919–1920: A Different Angle on the Ecumenical Movement," *Church History* 51, no. 2 (1982): 198–209.

47. All in Raymond Fosdick Papers, Rockefeller Archive Center, Tarrytown, New York: "Minutes of the Committee on Social and Religious Surveys, Twelfth Meeting, April 21, 1922," Folder 10; Raymond B. Fosdick to John D. Rockefeller Jr., June 27, 1922, and John D. Rockefeller Jr. to John R. Mott, January 5, 1923, Folder 13; "Minutes of the Committee on Social and Religious Surveys, Nineteenth Meeting, June 29th, 1923," Folder 10.

48. Robert S. Lynd and Helen Merrell Lynd, *Middletown: A Study in American Culture* (New York: Harcourt, Brace, 1929).

49. See Richard Wightman Fox, "Epitaph for Middletown: Robert S. Lynd and the Analysis of Consumer Culture," in *The Culture of Consumption: Critical Essays in American History, 1880–1980*, ed. Fox and T. J. Jackson Lears (New York: Pantheon, 1983), 101–41.

50. "Minutes of the Committee on Social and Religious Surveys, Nineteenth Meeting, June 29th, 1923."

51. "Minutes of the Institute of Social and Religious Research, Twenty-Fourth Meeting, May 2, 1924," Folder 10, and Galen M. Fisher to Ernest D. Burton and Raymond B. Fosdick, March 15, 1924, Folder 13, Fosdick Papers.

52. In 1924 Lynd's Small Industrial City survey received $15,033.84. See "Minutes of the Institute of Social and Religious Research, Twenty-Second Meeting, January 7, 1924," Folder 10, Fosdick Papers.

53. Galen M. Fisher to Raymond M. Fosdick, June 23, 1926, Folder 15, and "Minutes of the Meeting of the Board of Directors, Institute of Social and Religious Research, held at the Hotel Commodore, New York City, June 26, 1926," Folder 11, Fosdick Papers.

54. On the role of the psychology of religion in the generational shift from evangelical to liberal Protestantism, see Ian A. M. Nicholson, "Academic Professionalization and Protestant Reconstruction, 1890–1902: George Albert Coe's Psychology of Religion," *Journal of the History of the Behavioral Sciences* 30, no. 4 (1994): 348–68; White, *Unsettled Minds*, 134–57.

55. Lewis Terman to E. Lowell Kelly, August 11, 1931, Box 14, Folder 29, Terman Papers; Mark A. May, "The Psychological Examination of Conscientious Objectors," *American Journal of Psychology* 31, no. 2 (1920): 152–65; Hugh Hartshorne to Mark A. May, February 6, 1924, Box 3, Folder 37, May Papers.

56. Mark A. May to Hugh Hartshorne, January 24, 1924, and Hartshorne to May, February 6, 1924, Box 3, Folder 37, May Papers. On this pathological style of scientific thought, see Carson, *The Measure of Merit*, 126–130.

57. Hugh Hartshorne, "Sociological Implications of the Character Education Inquiry," *American Journal of Sociology* 36, no. 2 (1930): 253.

58. See Hartshorne and May, *Studies in Deceit*, 47, but especially the discussion in Symonds, *Diagnosing Personality and Conduct*, 299–300.

59. Hugh Hartshorne, *Worship in the Sunday School: A Study in the Theory and Practice of Worship* (New York: Teachers College, Columbia University, 1913), 118, 3 (emphasis in the original).

60. Hugh Hartshorne to Mark May, April 12, 1924, Box 3, Folder 37, May Papers.

61. Ibid.

62. Hartshorne and May, *Studies in Deceit*, 15, 379.

63. A. A. Roback, "Personality Tests—Wither?" *Character and Personality* 1, no. 3 (1933): 217.

64. Hartshorne and May, *Studies in Deceit*, 10.

65. Ibid., 89.

66. Symonds, *Diagnosing Personality and Conduct*, 298.

67. Barnard, "Moral Tests Reveal Children's Standards," 15.

68. "Wealthy Pupils Steal Most, Tests Indicate," *Washington Post*, December 19, 1926, M1; "Says Schools Teach Lying," *New York Times*, July 12, 1928, 10; Hartshorne and May, *Studies in Deceit*, 410.

69. Hartshorne and May, *Studies in Deceit*, 409–10, 412. See also "Tests Show Home Is Greatest Moral Force, Friends Next, Sunday Schools of Little Aid," *New York Times*, February 5, 1927, 7. On the ways in which the Lynds obscured ethnic and class differences in designing the *Middletown* study, see Igo, *The Averaged American*, 54–60.

70. Hartshorne and May, *Studies in Deceit*, 402–6.

71. Hugh Hartshorne, in "Stenographic Report of the Character Education Research Conference Held at the Chicago Theological Seminary under auspices of the Religious Education Association, September 14–16, 1928," 4, Archives of the Religious Education Association, RG 74, Series vii, Box 89, Folder 1057, Archives of the Religious Education Association, Yale Divinity School Library.

72. Mark A. May, "The Adult in the Community," in *The Foundations of Experimental Psychology*, ed. Carl Murchison (Worcester, MA: Clark University Press, 1929), 766 (emphasis in original); May, "The Foundations of Personality," in *Psychology at Work* (New York: McGraw-Hill, 1932), 89.

73. Henry A. Murray, "Introduction," in *Explorations in Personality*, ed. Murray (New York: Oxford University Press, 1938), 9.

74. See Dale Carnegie, *How to Win Friends and Influence People* (New York: Simon and Schuster, 1936).

75. "Says Boy Scouts Cheat in the Classroom, Showing that Situations Govern Honesty," *New York Times*, November 19, 1927, 6.

76. Henry Pratt Fairchild, *Conduct Habits of Boy Scouts* (New York: Boy Scouts of America, 1931).

77. Ruth Benedict, *Patterns of Culture* (Boston: Houghton Mifflin, 1934), 236.

78. Roback, *Psychology of Character*, 362; Roback, "Personality Tests," 218.

79. Gordon W. Allport, *Personality: A Psychological Interpretation* (New York: Holt, 1937), 248–55.

80. Gordon Allport to A. A. Roback, October 30, 1929, Ac Ms 2518 file 13, Roback Papers.

81. Ian A. M. Nicholson, *Inventing Personality: Gordon Allport and the Science of Selfhood* (Washington, DC: APA Press, 2003).

82. Gordon W. Allport, "The Study of Undivided Personality," *Journal of Abnormal Psychology and Social Psychology* 19 (1924): 132–41; Allport, *Personality*, 250–55.

83. Katherine Pandora, *Rebels within the Ranks: Psychologists' Critique of Scientific Authority and Democratic Realities in New Deal America* (Cambridge: Cambridge University Press, 1997), 42–44.

84. Gardner Murphy, Lois Barclay Murphy, and Theodore M. Newcomb, *Experimental Social Psychology: An Interpretation of Research upon the Socialization of the Individual* (New York: Harper, 1937), 661–62.

85. Goodwin M. Watson, "Some Accomplishments in Summer Camps" (1928), Box 20, Folder 16, Hugh Hartshorne Papers, Group 30, Miscellaneous Personal Papers Collections, Yale Divinity School Library; Watson, "A Supplementary Review of Measures of Personality Traits," *Journal of Educational Psychology* 28, no. 2 (1927): 87.

86. For example, see Saul Rosenzweig, "The Experimental Situation as Psychology Problem," *Psychological Review* 40 (1933): 337–54; Harry Charles Steinmetz, "Measuring Ability to Fake Occupation Interest," *Journal of Applied Psychology* 16 (1932): 123–30; Gordon Hendrickson, "Some Assumptions Involved in Personality Measurement," *Journal of Experimental Education* 2, no. 3 (1934): 243–49; Douglas Spencer, "The Frankness of Subjects on Personality Measures," *Journal of Educational Psychology* 29 (1938): 26–35.

87. Allport, *Personality*, 250; Gordon W. Allport, "A Test for Ascendance-Submission," *Journal of Abnormal and Social Psychology* 23, no. 2 (1928): 133.

88. Donald W. MacKinnon, "Violation of Prohibitions in Solving Problems," in *Explorations in Personality*, ed. Murray, 491–501, quote 494. His work was part of larger project that sought to subject psychoanalysis to experimental validation. See Gail A. Hornstein, "The Return of the Repressed: Psychology's Problematic Relations with Psychoanalysis, 1909–1960," *American Psychologist* 47 (1992): 254–63.

89. MacKinnon, "Violation of Prohibitions in Solving Problems," 492–96.

90. Ibid., 500, 501.

91. Henry A. Murray, *Thematic Apperception Test Manual* (Cambridge, MA: Harvard University Press, 1943), 1.

92. Lewis M. Terman and Catharine Cox Miles, *Sex and Personality* (New York: McGraw-Hill, 1936), 14. On the relationship between Terman's work on intelligence and sex, see Peter Hegarty, "From Genius Inverts to Gendered Intelligence: Lewis Terman and the Power of the Norm," *History of Psychology* 10, no. 2 (2007): 132–55.

93. Terman and Miles, *Sex and Personality*, 4. For a critical overview of these

methods, J. G. Morawski, "The Measurement of Masculinity and Femininity: Engendering Categorical Realities," *Journal of Personality* 53 (1985): 207–11.

94. E. Lowell Kelly, Catharine Cox Miles, and Lewis M. Terman, "Ability to Influence One's Score on a Typical Pencil-and-Paper Test of Personality," *Character and Personality* 4, no. 3 (1936): 215.

95. Terman and Miles, *Sex and Personality*, 239. Although the book as a whole is attributed to Terman and Miles, Terman and Kelly are named as the authors of the chapter on homosexuality.

96. Terman to Katherine Bement Davis, March 18, 1929, Box 12, Folder 31, Terman Papers; Terman and Miles, *Sex and Personality*, 244–45. See Peter Hegarty, "'More Feminine than 999 Men Out of 1,000': Measuring Sex Roles and Gender Nonconformity in Psychology," in *Gender Non-conformity, Race, and Sexuality: Charting the Connections*, ed. Toni Lester (Madison: University of Wisconsin Press, 2003), 62–83.

97. Goodwin Watson to Terman, April 25, 1929, Box 12, Folder 31, Terman Papers.

98. Dorothy Swaine Thomas, "The Methodology of Experimental Sociology," in *Some Techniques for Studying Social Behavior*, ed. Thomas (New York: Teachers College, Columbia University, 1929), 5.

99. Arnold Gesell, *Infancy and Human Growth* (New York: Macmillan, 1928), 30, 31.

100. Ibid., 37–38; Arnold Gesell and Frances L. Ilg, *Infant and Child in the Culture of Today* (New York: Harper, 1943), 370; Ellen Herman, "Families Made by Science: Arnold Gesell and the Technologies of Modern Child Adoption," *Isis* 92, no. 4 (2001): 684–715.

101. Alfred C. Kinsey, Wardell B. Pomeroy, and Clyde Martin, *Sexual Behavior in the Human Male* (Philadelphia: W. B. Saunders, 1948), 41–43. My analysis of Kinsey draws on Igo, *The Averaged American*, 208–15.

102. May, "The Adult in the Community," 738.

103. The address was published as Mark A. May, "Measurements of Personality," *Scientific Monthly* 38, no. 1 (1934): 75.

104. Roderick D. Buchanan, "On Not 'Giving Psychology Away': The Minnesota Multiphasic Personality Inventory and Public Controversy Over Testing in the 1960s," *History of Psychology* 5, no. 3 (2002): 284–309.

105. Ian Nicholson, "'Shocking' Masculinity: Stanley Milgram, 'Obedience to Authority,' and the 'Crisis of Manhood' in Cold War America," *Isis* 102, no. 2 (2011): 238–68.

106. Ben Harris, "Key Words: A History of Debriefing in Social Psychology," in *The Rise of Experimentation in American Psychology*, ed. J. G. Morawski (New Haven, CT: Yale University Press, 1988), 188–212; Laura Stark, "The Science of Ethics: Deception, the Resilient Self, and the APA Code of Ethics, 1966–1973," *Journal of the History of the Behavioral Sciences* 46, no. 4 (2010): 337–70; Nicholson, "'Shocking' Masculinity," 238–68.

107. On the Cold War confessional scripts of political and sexual sins, see David Serlin, *Replaceable You: Engineering the Body in Postwar America* (Chicago: University of Chicago Press, 2004).

Conclusion

1. Found in a binder of typewritten poems likely composed in 1943. Joseph Jastrow Papers, Rare Books, Manuscript, and Special Collections Library, Duke University.

2. Gieryn, *Cultural Boundaries of Science*.

3. Bertram R. Forer, "The Fallacy of Personal Validation: A Classroom Demonstration of Gullibility," *Journal of Abnormal Psychology* 44, no. 1 (1949): 118–23.

4. Paul E. Meehl, "Wanted—A Good Cookbook," *American Psychologist* 11, no. 6 (1956): 266.

5. On the battles between these two kinds of vocational experts, see Brown, *The Corporate Eye*, 23–64. For an example of his particular contribution, see Donald G. Paterson, and Katherine E. Ludgate, "Blond and Brunette Traits: A Quantitative Study," *Journal of Personnel Research* 1 (1922): 122–27.

6. For a discussion of the Barnum effect in this regard, see Ross Stagner, "The Gullibility of Personnel Managers," *Personnel Psychology* 11, no. 3 (1958): 351.

7. For a reproduction of Paterson's sketch and his description of its use, see Milton L. Blum and Benjamin Balinsky, *Counselling and Psychology: Vocational Psychology and Its Relation to Educational and Personal Counseling* (New York: Prentice-Hall, 1951), 46–47.

8. See Marvin D. Dunnette, "Use of the Sugar Pill by Industrial Psychologists," *American Psychologist* 12 (1957): 223–25.

9. Robert Darnton, *Mesmerism and the End of the Enlightenment* (Cambridge, MA: Harvard University Press, 1968).

10. See Sonu Shamdasani, "'Psychotherapy': The Invention of a Word," *History of the Human Sciences* 18 (2005): 1–22. See also Adam Crabtree, *From Mesmer to Freud: Magnetic Sleep and the Roots of Psychological Healing* (New Haven, CT: Yale University, 1993); Ted J. Kaptchuk, "Intentional Ignorance: A History of Blind Assessment and Placebo Controls in Medicine," *Bulletin of the History of Medicine* 72, no. 3 (1998): 389–433; Harrington, *The Cure Within*.

11. On the problem of imagination in sensationalist psychology, see Jan E. Goldstein, *The Post-Revolutionary Self: Politics and Psyche in France, 1750–1850* (Cambridge, MA: Harvard University Press, 2005).

12. *Report of Dr. Benjamin Franklin, and other Commissioners, Charged by the King of France with the Examination of the Animal Magnetism, as now practised at Paris* (London: J. Johnson, 1785).

13. For a helpful overview of psychology's "marketable methods," see Danziger, *Constructing the Subject*, 101–17.

14. James McKeen Cattell, "A Corporation for the Advancement of Psychology," circa 1921, Box 185, JMCP.

15. Without doubt this crafty persona in the laboratory was tempered by the self-discipline introduced by statistical analysis. See Porter, *Trust in Numbers*.

16. Ellen Herman, *Kinship by Design: A History of Adoption in the Modern United States* (Chicago: University of Chicago Press, 2008); Anne Fausto-Sterling, *Sexing the Body: Gender Politics and the Construction of Sexuality* (New York: Basic Books, 2000).

17. For a critique of how the field of neuroethics traffics in (rather than challenges) the neurosciences' unsubstantiated claims to novelty, see Fernado Vidal, "Brainhood, Anthropological Figure of Modernity," *History of the Human Sciences* 22, no. 1 (2009): 5–36.

Bibliography

Abelson, Elizabeth S. *When Ladies Go A-Thieving: Middle-Class Shoplifters in the Victorian Department Store.* New York: Oxford University Press, 1989.
Adams, Bluford. *E Pluribus Barnum: The Great Showman and the Making of U.S. Popular Culture.* Minneapolis: University of Minnesota Press, 1997.
Adams, Rachel. *Sideshow U.S.A.: Freaks and the American Cultural Imagination.* Chicago: University of Chicago Press, 2001.
Alder, Ken. "History's Greatest Forger: Science, Fiction, and Fraud along the Seine." *Critical Inquiry* 30 (2004): 702–16.
———. *The Lie Detectors: The History of an American Obsession.* New York: Free Press, 2007.
———. "A Social History of Untruth: Lie Detection and Trust in Twentieth-Century America." *Representations* 80 (2002): 1–33.
Altschuler, Glenn C., and Stuart M. Blumin. "Limits of Political Engagement in Antebellum America: A New Look at the Golden Age of Participatory Democracy." *Journal of American History* 84, no. 3 (1997): 855–85.
Baer, Hans A. *The Black Spiritual Movement: A Religious Response to Racism.* Knoxville: University of Tennessee Press, 1984.
Ball, Richard A., and J. Robert Lilly. "The Menace of Margarine: The Rise and Fall of a Social Problem." *Social Problems* 29, no. 5 (1982): 488–98.
Balleisen, Edward J. "Private Cops on the Fraud Beat: The Limits of American Business Self-Regulation, 1895–1932." *Business History Review* 83 (2009): 113–60.
Bederman, Gail. *Manliness and Civilization: A Cultural History of Gender and Race in the United States, 1880–1917.* Chicago: University of Chicago Press, 1995.
Behrens, Peter J. "The Metaphysical Club at the Johns Hopkins University (1879–1885)." *History of Psychology* 8, no. 4 (2005): 331–46.
Belletto, Steven. "Drink versus Printer's Ink: Temperance and the Management of Financial Speculation in *The Life of P. T. Barnum.*" *American Studies* 46, no. 1 (2005): 45–65.

Bellion, Wendy. *Citizen Spectator: Art, Illusion, and Visual Perception in Early National America.* Chapel Hill: University of North Carolina Press, 2011.

Benjamin, Ludy T., Jr., Anne M. Rogers, and Angela Rosenbaum. "Coca-Cola, Caffeine, and Mental Deficiency: Harry Hollingworth and the Chattanooga Trial of 1911." *Journal of the History of the Behavioral Sciences* 27 (1991): 42–55.

Bergman, Johannes Dietrich. "The Original Confidence Man." *American Quarterly* 21, no. 3 (1969): 560–77.

Bienen, Leigh B. "A Question of Credibility: John Henry Wigmore's Use of Scientific Authority in Section 924a of the Treatise on Evidence." *California Western Law Review* 19 (1983): 235–68.

Blumenthal, Arthur L. "The Intrepid Joseph Jastrow." In *Portraits of Pioneers in Psychology,* ed. Gregory A. Kimble, Michael Wertheimer, and Charlotte L. White, 1:75–87. Washington, DC: APA Press, 1991.

Blumenthal, Susanna. "The Deviance of the Will: Policing the Bounds of Testamentary Freedom in Nineteenth-Century America." *Harvard Law Review* 119 (2006): 959–1034.

———. "The Mind of a Moral Agent: Scottish Common Sense and the Problem of Responsibility in Nineteenth-Century American Law." *Law and History Review* 26, no. 1 (2008): 99–159.

Bogdan, Robert. *Freak Show: Presenting Human Oddities for Amusement and Profit.* Chicago: University of Chicago Press, 1988.

Bok, Sissela. *Lying: Moral Choice in Public and Private Life.* New York: Pantheon Books, 1978.

Bordogna, Francesca. *William James at the Boundaries: Philosophy, Science, and the Geography of Knowledge.* Chicago: University of Chicago Press, 2008.

Braude, Ann. *Radical Spirits: Spiritualism and Women's Rights in Nineteenth-Century America.* Bloomington: University of Indiana Press, 2001.

Brent, Joseph. *Charles Sanders Peirce: A Life.* Bloomington: University of Indiana Press, 1993.

Breu, Christopher. *Hard-Boiled Masculinities.* Minneapolis: University of Minnesota Press, 2005.

Brown, Edward M. "Neurology and Spiritualism in the 1870s." *Bulletin of the History of Medicine* 57 (1983): 563–77.

Brown, Elspeth H. *The Corporate Eye: Photography and the Rationalization of American Commercial Culture, 1884–1929.* Baltimore: Johns Hopkins University Press, 2005.

Brugger, Peter. "One Hundred Years of an Ambiguous Figure: Happy Birthday, Duck/Rabbit!" *Perceptual and Motor Skills* 89 (1999): 973–77.

Buccafusco, Christopher J. "Spiritualism and Will(s) in the Age of Contract." University of Illinois Legal Working Paper 91. 2008.

Buchanan, Roderick D. "On Not 'Giving Psychology Away': The Minnesota Multiphasic Personality Inventory and Public Controversy Over Testing in the 1960s." *History of Psychology* 5, no. 3 (2002): 284–309.

———. *Playing with Fire: The Controversial Career of Hans J. Eysenck.* Oxford: Oxford University Press, 2010.

Bulmer, Martin, Kevin Bales, and Kish Sklar, eds. *The Social Survey in Historical Perspective, 1880–1940.* New York: Cambridge University Press, 1991.

Bunn, Geoffrey C. "The Hazards of the Will to Truth: A History of the Lie Detector." PhD diss., York University, 1998.

———. "The Lie Detector, *Wonder Woman,* and Liberty: The Life and Work of William Moulton Marston." *History of the Human Sciences* 10, no. 1 (1997): 91–119.

———. "Spectacular Science: The Lie Detector's Ambivalent Powers." *History of Psychology* 10, no. 2 (2007): 156–78.

Callon, Michel, ed. *The Laws of Markets.* London: Blackwell, 1998.

Caplan, Eric. *Mind Games: American Culture and the Birth of Psychotherapy.* Berkeley: University of California Press, 1998.

Capshew, James. *Psychologists on the March: Science, Practice, and Professional Identity in America, 1929–1969.* Cambridge: Cambridge University Press, 1999.

Carson, John. *The Measure of Merit: Talents, Intelligence, and Inequality in the French and American Republics, 1750–1940.* Princeton, NJ: Princeton University Press, 2007.

Cawelti, John G. *Apostles of the Self-Made Man: Changing Conceptions of Success in America.* Chicago: University of Chicago Press, 1964.

Chandler, Alfred D., Jr. *The Visible Hand: The Managerial Revolution in American Business.* Cambridge, MA: Harvard University Press, 1977.

Clarke, Adele E. *Disciplining Reproduction: Modernity, American Life Science, and "the Problem of Sex."* Berkeley: University of California Press, 1998.

Clarke, Michael Tavel. *These Days of Large Things: The Culture of Size in America, 1865–1930.* Ann Arbor: University of Michigan Press, 2007.

Cohen, Lizabeth. *A Consumer's Republic: The Politics of Mass Consumption in Postwar America.* New York: Knopf, 2003.

Cole, Simon A. *Suspect Identities: A History of Fingerprinting and Criminal Identification.* Cambridge, MA: Harvard University Press, 2001.

Cook, Harold J. *Matters of Exchange: Commerce, Medicine, and Science in the Dutch Golden Age.* New Haven, CT: Yale University Press, 2007.

Cook, James W. *The Arts of Deception: Playing with Fraud in the Age of Barnum.* Cambridge, MA: Harvard University Press, 2001.

Coon, Deborah J. "'One Moment in the World's Salvation': Anarchism and the Radicalization of William James." *Journal of American History* 83 (1996): 70–99.

———. "Standardizing the Subject: Experimental Psychologists, Introspection, and the Quest for a Technoscientific Ideal." *Technology and Culture* 34 (1993): 757–83.

———. "Testing the Limits of Sense and Science: American Experimental Psychologists Combat Spiritualism, 1880–1920." *American Psychologist* 47 (1992): 143–51.

Coppin, Clayton A., and Jack C. High. *The Politics of Purity: Harvey Washington Wiley and the Origins of Federal Food Policy.* Ann Arbor: University of Michigan Press, 1999.
Corbey, Raymond. "Ethnographic Showcases, 1870–1930." *Cultural Anthropology* 8, no. 3 (1993): 338–69.
Cotkin, George B. "Middle-Ground Pragmatists: The Popularization of Philosophy in American Culture." *Journal of the History of Ideas* 55, no. 2 (1994): 283–302.
———. *William James, Public Philosopher.* Baltimore: Johns Hopkins University Press, 1990.
Cox, Robert S. *Body and Soul: A Sympathetic History of American Spiritualism.* Charlottesville: University of Virginia Press, 2003.
Crabtree, Adam. *From Mesmer to Freud: Magnetic Sleep and the Roots of Psychological Healing.* New Haven, CT: Yale University Press, 1993.
Crary, Jonathan. *Suspensions of Perception: Attention, Spectacle, and Modern Culture.* Cambridge, MA: MIT Press, 1999.
———. *Techniques of the Observer: On Vision and Modernity in the Nineteenth Century.* Cambridge, MA: MIT Press, 1990.
Crenner, Christopher W. "Introduction of the Blood Pressure Cuff into U.S. Medical Practice: Technology and Skilled Practice." *Annals of Internal Medicine* 128, no. 6 (March 15, 1998): 488–93.
———. *Private Practice: In the Early Twentieth-Century Medical Office of Dr. Richard Cabot.* Baltimore: Johns Hopkins University Press, 2005.
Croce, Paul Jerome. *Science and Religion in the Era of William James: Eclipse of Certainty, 1820–1880.* Chapel Hill: University of North Carolina Press, 1995.
Cronon, William. *Nature's Metropolis: Chicago and the Great West.* New York: Norton, 1991.
Curti, Merle. "The Changing Concept of 'Human Nature' in the Literature of American Advertising." *Business History Review* 41, no. 4 (1967): 335–57.
Danziger, Kurt. *Constructing the Subject: Historical Origins of Psychological Research.* New York: Cambridge University Press, 1990.
———. "The Positivist Repudiation of Wundt." *Journal of the History of the Behavioral Sciences* 15 (1979): 205–30.
———. "Universalism and Indigenization in the History of Modern Psychology." In *Internationalizing the History of Psychology,* ed. Adrian C. Brock, 208–25. New York: New York University Press, 2006.
Darnton, Robert. *Mesmerism and the End of the Enlightenment.* Cambridge, MA: Harvard University Press, 1968.
Daston, Lorraine, and Peter Galison. *Objectivity.* New York: Zone Books, 2007.
Daston, Lorraine, and Katherine Park. *Wonders and the Order of Nature, 1150–1750.* New York: Zone Books, 1998.
Davis, Janet. "Freakishly, Fraudulently Modern." *American Quarterly* 55, no. 3 (2003): 525–38.

Dehue, Trudy. "Deception, Efficiency, and Random Groups: Psychology and the Gradual Origination of Random Group Design." *Isis* 88 (1997): 653–73.

———. "From Deception Trials to Control Reagents: The Introduction of the Control Group about a Century Ago." *American Psychologist* 55, no. 2 (2000): 264–68.

Dennett, Andrea Stulman. *Weird and Wonderful: The Dime Museum in America.* New York: New York University Press, 1997.

Dennis, Paul M. "The Edison Questionnaire." *Journal of the History of the Behavioral Sciences* 20, no. 1 (1984): 23–37.

Derksen, Maarten. "Are We Not Experimenting Then? The Rhetorical Demarcation of Psychology and Common Sense." *Theory and Psychology* 7, no. 4 (1997): 435–56.

Dror, Otniel E. "The Affect of Experiment: The Turn to Emotions in Anglo-American Physiology, 1900–1940." *Isis* 90 (1999): 205–37.

Dubber, Markus Dirk. *The Police Power: Patriarchy and the Foundations of American Government.* New York: Columbia University Press, 2005.

Dumenil, Lynn. *The Modern Temper: American Culture and Society in the 1920s.* New York: Hill and Wang, 1995.

During, Simon. *Modern Enchantments: The Cultural Power of Secular Magic.* Cambridge, MA: Harvard University Press, 2002.

Evans, Brad. "Cushing's Zuni Sketchbooks: Literature, Anthropology, and American Notions of Culture." *American Quarterly* 49, no. 4 (1997): 717–45.

Fabian, Ann. *Card Sharps, Dream Books, and Bucket Shops: Gambling in Nineteenth-Century America.* Ithaca, NY: Cornell University Press, 1990.

Fass, Paula S. *The Damned and the Beautiful: American Youth in the 1920s.* Oxford: Oxford University Press, 1977.

Fausto-Sterling, Anne. *Sexing the Body: Gender Politics and the Construction of Sexuality.* New York: Basic Books, 2000.

Filene, Peter G. *Him/Her/Self: Gender Identities in Modern America.* 3rd ed. Baltimore: Johns Hopkins University Press, 1998.

Fitzpatrick, Sheila. *Tear Off the Masks! Identity and Imposture in Twentieth-Century Russia.* Princeton, NJ: Princeton University Press, 2005.

Foner, Eric. *The Story of American Freedom.* New York: Norton, 1998.

Forrester, John. "If *p*, then what? Thinking in Cases." *History of the Human Sciences* 9, no. 3 (1996): 1–25.

Fox, Richard Wightman. "Epitaph for Middletown: Robert S. Lynd and the Analysis of Consumer Culture." In *The Culture of Consumption: Critical Essays in American History, 1880–1980,* ed. Richard Wightman Fox and T. J. Jackson Lears, 101–41. New York: Pantheon Books, 1983.

Frankfurt, Harry. *On Bullshit.* Princeton, NJ: Princeton University Press, 2005.

Fraser, Steve. *Every Man a Speculator: A History of Wall Street in American Life.* New York: Basic Books, 2005.

Friedman, Lawrence. "Crimes of Mobility." *Stanford Law Review* 43 (1991): 637–58.

Friedman, Walter A. *Birth of a Salesman: The Transformation of Selling in America.* Cambridge, MA: Harvard University Press, 2004.

Fuller, Robert C. *Spiritual, but Not Religious: Understanding Unchurched America.* Oxford: Oxford University Press, 2001.

Gabriel, Joseph M. "Mass Producing the Individual: Mary C. Jarrett, Elmer E. Southard, and the Industrial Origins of Psychiatric Social Work." *Bulletin of the History of Medicine* 79, no. 3 (2005): 430–58.

———. "A Thing Patented Is a Thing Revealed: Francis E. Stewart, George S. Davis, and the Legitimization of Patent Rights in Pharmaceutical Manufacturing, 1879–1911." *Journal of the History of Medicine and Allied Sciences* 64, no. 2 (2009): 135–72.

Galambos, Louis. "The Emerging Organizational Synthesis in Modern American History." *Business History Review* 44, no. 3 (1970): 279–90.

Galison, Peter. "Image of Self." In *Things that Talk: Object Lessons from Art and Science,* ed. Lorraine Daston, 257–94. New York: Zone Books, 2004.

Getis, Victoria. *The Juvenile Court and the Progressives.* Urbana: University of Illinois Press, 2000.

Gieryn, Thomas F. *Cultural Boundaries of Science: Credibility on the Line.* Chicago: University of Chicago Press, 1999.

Gilbert, James. *Perfect Cities: Chicago's Utopias of 1893.* Chicago: University of Chicago Press, 1991.

Gilfoyle, Timothy J. "Barnum's Brothel: P.T.'s 'Last Great Humbug.'" *Journal of the History of Sexuality* 18, no. 3 (2009): 486–513.

Glickman, Lawrence B. *Buying Power: A History of Consumer Activism in America.* Chicago: University of Chicago Press, 2009.

———. "The Strike in the Temple of Consumption: Consumer Activism and Twentieth-Century American Political Culture." *Journal of American History* 88 (2001): 99–128.

Golan, Tal. *Laws of Men and Laws of Nature: The History of Scientific Expert Testimony in England and America.* Cambridge, MA: Harvard University Press, 2004.

Goldstein, Jan E. *The Post-Revolutionary Self: Politics and Psyche in France, 1750–1850.* Cambridge, MA: Harvard University Press, 2005.

Green, Christopher D. "Johns Hopkins' First Professorship in Philosophy: A Critical Pivot Point in the History of American Psychology." *American Journal of Psychology* 120 (2007): 303–23.

Gunlach, Horst. "What Is a Psychological Instrument?" In *Psychology's Territories: Historical and Contemporary Perspectives from Different Disciplines,* ed. Mitchell Ash and Thomas Strum, 193–224. Mahwah, NJ: Lawrence Erlbaum Associates, 2007.

Gunning, Tom. "An Aesthetic of Astonishment: Early Film and the (In)credulous Spectator." *Art and Text* 34 (1989): 31–45.

Hacking, Ian. *Rewriting the Soul: Multiple Personality and the Sciences of Memory.* Princeton, NJ: Princeton University Press, 1995.
———. *The Taming of Chance.* Cambridge: Cambridge University Press, 1990.
———. "Telepathy: Origins of Randomization in Experimental Design." *Isis* 79 (1988): 427–51.
Hale, Matthew, Jr. *Human Science and Social Order: Hugo Münsterberg and the Origins of Applied Psychology.* Philadelphia: Temple University Press, 1980.
Hale, Nathan G. *The Rise and Crisis of Psychoanalysis in the United States: Freud and the Americans, 1917–1985.* New York: Oxford University Press, 1995.
Halttunen, Karen. *Confidence Men and the Painted Women: A Study of Middle-Class Culture in America, 1830–1870.* New Haven: Yale University Press, 1982.
Haraway, Donna J. "Modest Witness: Feminist Diffractions in Science Studies." In *The Disunity of Science: Boundaries, Contexts, and Power,* ed. Peter Galison and David J. Stump, 428–41. Stanford, CA: Stanford University Press, 1996.
———. "Situated Knowledges: The Science Question in Feminism and the Privilege of Partial Perspective." *Feminist Studies* 14 (1988): 575–99.
Harrington, Anne. *The Cure Within: A History of Mind-Body Medicine.* New York: Norton, 2008.
Harris, Ben. "Key Words: A History of Debriefing in Social Psychology." In *The Rise of Experimentation in American Psychology,* ed. Jill G. Morawski, 188–212. New Haven: Yale University Press, 1988.
Harris, Neil. *Humbug: The Art of P. T. Barnum.* Boston: Little, Brown, 1973.
Harvey, Charles E. "John D. Rockefeller, Jr., and the Interchurch World Movement of 1919–1920: A Different Angle on the Ecumenical Movement." *Church History* 51, no. 2 (1982): 198–209.
Hawley, Ellis W. "Herbert Hoover, the Commerce Secretariat, and the Vision of an 'Associative State,' 1921–1928." *Journal of American History* 61, no. 1 (1974): 116–40.
Hegarty, Peter. "From Genius Inverts to Gendered Intelligence: Lewis Terman and the Power of the Norm." *History of Psychology* 10, no. 2 (2007): 132–55.
———. "'More Feminine than 999 Men out of 1,000': Measuring Sex Roles and Gender Nonconformity in Psychology." In *Gender Non-conformity, Race, and Sexuality: Charting the Connections,* ed. Toni Lester, 62–83. Madison: University of Wisconsin Press, 2003.
Herman, Ellen. "Families Made by Science: Arnold Gesell and the Technologies of Modern Child Adoption." *Isis* 92, no. 4 (2001): 684–715.
———. *Kinship by Design: A History of Adoption in the Modern United States.* Chicago: University of Chicago Press, 2008.
———. *The Romance of American Psychology: Political Culture in the Age of Experts.* Berkeley: University of California Press, 1995.
Herzig, Rebecca. *Suffering for Science: Reason and Sacrifice in Modern America.* New Brunswick, NJ: Rutgers University Press, 2005.

Higbie, Tobias. *Indispensable Outcasts: Hobo Workers and Community in the American Midwest, 1880–1930*. Urbana: University of Illinois Press, 2003.

Hilkey, Judy. *Character Is Capital: Success Manuals and Manhood in Gilded Age America*. Chapel Hill: University of North Carolina Press, 1997.

Hirschman, Albert O. *The Passions and the Interests: Political Arguments for Capitalism before Its Triumph*. Princeton, NJ: Princeton University Press, 1977.

Hochfelder, David. "'Where the Common People Could Speculate': The Ticker, Bucket Shops, and the Origins of Popular Participation in Financial Markets, 1880–1920." *Journal of American History* 93, no. 2 (2006): 335–58.

Hopkins, Albert A., ed. *Magic, Stage Illusions and Scientific Diversions*. New York: Benjamin Blom, 1967.

Horn, David G. *The Criminal Body: Lombroso and the Anatomy of Deviance*. New York: Routledge, 2003.

Hornstein, Gail A. "The Return of the Repressed: Psychology's Problematic Relations with Psychoanalysis, 1909–1960." *American Psychologist* 47 (1992): 254–63.

Horwitz, Morton J. *The Transformation of American Law, 1780–1860*. Cambridge, MA: Harvard University Press, 1977.

Howell, Joel D. *Technology in the Hospital: Transforming Patient Care in the Early Twentieth Century*. Baltimore: Johns Hopkins University Press, 1995.

Hyde, Alan. *Bodies of Law*. Princeton, NJ: Princeton University Press, 1997.

Igo, Sarah E. *The Averaged American: Surveys, Citizens, and the Making of a Mass Public*. Cambridge, MA: Harvard University Press, 2007.

Jacobs, Meg. *Pocketbook Politics: Economic Citizenship in Twentieth-Century America*. Princeton, NJ: Princeton University Press, 2005.

James, William. "The Confidences of a 'Psychical Researcher.'" In *Essays in Psychical Research*, 361–75. Cambridge: Cambridge University Press, 1986.

———. *The Will to Believe and Other Essays in Popular Philosophy*. Cambridge, MA: Harvard University Press, 1979.

Jasanoff, Sheila. *Science at the Bar: Law, Science, and Technology in America*. Cambridge, MA: Harvard University Press, 1995.

Jones, Kathleen W. *Taming the Troublesome Child: American Families, Child Guidance, and the Limits of Psychiatric Authority*. Cambridge, MA: Harvard University Press, 1999.

Kaptchuk, Ted J. "Intentional Ignorance: A History of Blind Assessment and Placebo Controls in Medicine." *Bulletin of the History of Medicine* 72, no. 3 (1998): 389–433.

Kasson, John F. *Amusing the Million: Coney Island at the Turn of the Century*. New York: Hill and Wang, 1978.

———. *Houdini, Tarzan, and the Perfect Man: The While Male Body and the Challenge of Modernity in America*. New York: Hill and Wang, 2001.

———. *Rudeness and Civility: Manners in Nineteenth-Century Urban America*. New York: Hill and Wang, 1990.

Kay, Gwen. "Healthy Public Relations: The FDA's 1930s Legislative Campaign." *Bulletin of the History of Medicine* 75, no. 3 (2001): 446–87.

Kennedy, Samuel V., III. *Samuel Hopkins Adams and the Business of Writing*. Syracuse, NY: Syracuse University Press, 1999.

Kintner, Earl. *A Primer on the Law of Deceptive Practice: A Guide for Businessmen*. New York: Macmillan, 1971.

Knupfer, Anne Meis. *Reform and Resistance: Gender, Delinquency, and America's First Juvenile Court*. New York: Routledge, 2001.

Korn, James H. *Illusions of Reality: A History of Deception in Social Psychology*. Albany: SUNY University Press, 1997.

Kuna, David P. "The Concept of Suggestion in the Early History of Advertising Psychology." *Journal of the History of the Behavioral Sciences* 12 (1976): 347–53.

Kusch, Martin. "Recluse, Interlocutor, Interrogator: Natural and Social Order in Turn-of-the-Century Psychological Research Schools." *Isis* 86 (1995): 419–39.

Kwolek-Folland, Angel. *Engendering Business: Men and Women in the Corporate Office, 1870–1930*. Baltimore: Johns Hopkins University Press, 1994.

Lachapelle, Sofie. "From the Stage to the Laboratory: Magicians, Psychologists, and the Science of Illusion." *Journal of the History of the Behavioral Sciences* 44, no. 4 (2008): 319–34.

Lamont, Peter. "Magician as Conjuror: A Frame Analysis of Victorian Mediums." *Early Popular Visual Culture* 4, no. 1 (2006): 21–33.

Lawrence, Christopher. "Incommunicable Knowledge: Science, Technology and the Clinical Art in Britain, 1850–1914." *Journal of Contemporary History* 20, no. 4 (1985): 503–20.

Lears, T. J. Jackson. *Fables of Abundance: A Cultural History of Advertising in America*. New York: Basic Books, 1994.

———. "From Salvation to Self-Realization: Advertising and the Therapeutic Roots of Consumer Culture, 1880–1920." In *The Culture of Consumption: Critical Essays in American History, 1880–1980*, ed. Richard Wightman Fox and T. J. Jackson Lears, 1–38. New York: Pantheon Books, 1983.

———. *Something for Nothing: Luck in America*. New York: Viking, 2003.

Leja, Michael. *Looking Askance: Skepticism and American Art from Eakins to Duchamp*. Berkeley: University of California Press, 2004.

Levine, Lawrence. *Black Culture and Black Consciousness: Afro-American Folk Thought from Slavery to Freedom*. New York: Oxford University Press, 1977.

———. *Highbrow/Lowbrow: The Emergence of Cultural Hierarchy in America*. Cambridge, MA: Harvard University Press, 1988.

Levy, Jonathan Ira. "Contemplating Delivery: Futures Trading and the Problem of Commodity Exchange in the United States, 1875–1905." *American Historical Review* 111, no. 2 (2006): 307–35.

Lindberg, Gary. *The Confidence Man in American Literature*. New York: Oxford University Press, 1982.

Lindberg, Richard. *To Serve and Collect: Chicago Politics and Police Corruption from the Lager Beer Riot to the Summerdale Scandal*. New York: Praeger, 1991.
Livingston, James. *Pragmatism and the Political Economy of Cultural Revolution, 1850–1940*. Chapel Hill: University of North Carolina Press, 1994.
Lucier, Paul. "The Professional and the Scientist in Nineteenth-Century America." *Isis* 100, no. 4 (2009): 699–732.
Lukas, J. Anthony. *Big Trouble: A Murder in a Small Western Town Sets Off a Struggle for the Soul of America*. New York: Simon and Schuster, 1997.
Lunbeck, Elizabeth. *The Psychiatric Persuasion: Knowledge, Gender, and Power in Modern America*. Princeton, NJ: Princeton University Press, 1994.
Lutes, Jean Marie. "Into the Madhouse with Nellie Bly: Girl Stunt Reporting in Late-Nineteenth-Century America." *American Quarterly* 54, no. 2 (2002): 217–53.
Lutz, Catherine. "Epistemology of the Bunker: The Brainwashed and Other New Subjects of Permanent War." In *Inventing the Psychological: Toward a Cultural History of Emotional Life in America*, ed. Joel Pfister and Nancy Schnog, 245–67. New Haven, CT: Yale University Press, 1997.
Macleod, David I. *Building Character in the American Boy: The Boy Scouts, YMCA, and Their Forerunners, 1870–1920*. Madison: University of Wisconsin Press, 1983.
Marchand, Roland. *Advertising the American Dream: Making Way for Modernity, 1920–1940*. Berkeley: University of California Press, 1985.
Marks, Harry M. *Progress of Experiment: Science and Therapeutic Reform in the United States, 1900–1990*. Cambridge: Cambridge University Press, 1997.
McCormick, Richard L. "The Discovery that Business Corrupts Politics: A Reappraisal of the Origins of Progressivism." *American Historical Review* 86, no. 2 (1981): 247–74.
McGovern, Charles. *Sold American: Consumption and Citizenship, 1890–1945*. Chapel Hill: University of North Carolina Press, 2006.
Menand, Louis. *The Metaphysical Club*. New York: Farrar, Straus and Giroux, 2001.
Mihm, Stephen. *A Nation of Counterfeiters: Capitalists, Con Men, and the Making of the United States*. Cambridge, MA: Harvard University Press, 2007.
Miller, Geoffrey P. "Public Choice at the Dawn of the Special Interest State: The Story of Butter and Margarine." *California Law Review* 77, no. 1 (1989): 83–131.
Miller, Peter, and Nikolas Rose. "Mobilizing the Consumer: Assembling the Subject of Consumption." *Theory, Culture and Society* 14, no. 1 (1997): 1–36.
Mnookin, Jennifer L., and Michelle West. "Theaters of Proof: Visual Evidence and the Law in *Call Northside 777*." *Yale Journal of Law and Humanities* 13 (2001): 329–402.

Modell, John. *Into One's Own: From Youth to Adulthood in the United States, 1920–1975.* Berkeley: University of California Press, 1989.
Mol, Annemarie. *The Body Multiple: Ontology in Medical Practice.* Durham, NC: Duke University Press, 2002.
Monroe, John Warne. *Laboratories of Faith: Mesmerism, Spiritism, and Occultism in Modern France.* Ithaca, NY: Cornell University Press, 2008.
Moore, R. Laurence. *In Search of White Crows: Spiritualism, Parapsychology, and American Culture.* New York: Oxford University Press, 1977.
Morawski, Jill G. "The Measurement of Masculinity and Femininity: Engendering Categorical Realities." *Journal of Personality* 53 (1985): 193–223.
———. "There Is More to Our History of Giving: The Place of Introductory Textbooks in American Psychology." *American Psychologist* 47, no. 2 (1992): 161–69.
Morawski, Jill G., and Gail A. Hornstein. "Quandary of the Quacks: The Struggle for Expert Knowledge in American Psychology, 1890–1940." In *The Estate of Social Knowledge,* ed. JoAnne Brown and David K. van Kuren, 106–33. Baltimore: Johns Hopkins University Press, 1991.
Mülberger, Annette. "Teaching Psychology to Jurists: Initiatives and Reactions Prior to World War I." *History of Psychology* 12, no. 2 (2009): 60–86.
Mumford, Lewis. *The Golden Day: A Study of American Literature and Culture.* New York: Dover, 1968.
Muncy, Robyn. *Creating a Female Dominion in American Reform, 1890–1935.* New York: Oxford University Press, 1990.
Münsterberg, Hugo. *Psychology and Industrial Efficiency.* New York: Arno Press, 1973.
Murphy, Michelle. *Sick Building Syndrome and the Problem of Uncertainty: Environmental Politics, Technoscience, and Women Workers.* Durham, NC: Duke University Press, 2006.
Nasaw, David. *Going Out: The Rise and Fall of Public Amusements.* New York: Basic Books, 1993.
Nicholson, Ian A. M. "Academic Professionalization and Protestant Reconstruction, 1890–1902: George Albert Coe's Psychology of Religion." *Journal of the History of the Behavioral Sciences* 30, no. 4 (1994): 348–68.
———. *Inventing Personality: Gordon Allport and the Science of Selfhood.* Washington, DC: APA Press, 2003.
Noakes, Richard. "'Instruments to Lay Hold of Spirits': Technologising the Bodies of Victorian Spiritualism." In *Bodies/Machines,* ed. Iwan Rhys Morus, 125–63. Oxford: Berg Publishers, 2002.
Novak, William J. *The People's Welfare: Law and Regulation in Nineteenth-Century America.* Chapel Hill: University of North Carolina Press, 1996.
O'Donnell, Sean Tath. "Courting Science, Binding Truth: A Social History of Frye v. U.S." PhD diss., Harvard University, 2007.

O'Malley, Michael. "Specie and Species: Race and the Money Question in Nineteenth-Century America." *American Historical Review* 99, no. 2 (1994): 369–95.

O'Malley, Pat. "Risk and Responsibility." In *Foucault and Political Reason*, ed. Andrew Barry, Thomas Osborne, and Nikolas Rose, 189–207. London: UCL Press, 1996.

Ohmer, Susan. *George Gallup in Hollywood*. New York: Columbia University Press, 2006.

Oppenheim, Janet. *The Other World: Spiritualism and Psychical Research in England, 1850–1914*. Cambridge: Cambridge University Press, 1985.

Ordover, Abraham P. "Admissibility of Patterns of Similar Sexual Conduct: The Unlamented Death of Character for Chastity." *Cornell Law Review* 63 (1978): 90–126.

Orvell, Miles. *The Real Thing: Imitation and Authenticity in American Culture, 1880–1940*. Chapel Hill: University of North Carolina Press, 1989.

Owen, Alex. *The Darkened Room: Women, Power, and Spiritualism in Late Victorian England*. Philadelphia: University of Pennsylvania Press, 1990.

———. *The Place of Enchantment: British Occultism and the Culture of the Modern*. Chicago: University of Chicago Press, 2004.

Palfreman, Jon. "Between Scepticism and Credulity: A Study of Victorian Scientific Attitudes to Modern Spiritualism." In *On the Margins of Science: The Social Construction of Rejected Knowledge*, ed. Roy Wallis, 201–36. Keele: University of Keele, 1979.

Pandora, Katherine. *Rebels within the Ranks: Psychologists' Critique of Scientific Authority and Democratic Realities in New Deal America*. Cambridge: Cambridge University Press, 1997.

Pastore, Nicholas. "The Army Intelligence Tests and Walter Lippmann." *Journal of the History of the Behavioral Sciences* 14, no. 4 (1978): 316–27.

Patterson, Martha H. *Beyond the Gibson Girl: Reimagining the American New Woman, 1895–1915*. Urbana: University of Illinois Press, 2005.

Peck, Gunther. *Reinventing Free Labor: Padrones and Immigrant Workers in the North American West, 1880–1930*. New York: Cambridge University Press, 2000.

Peiss, Kathy Lee. *Cheap Amusements: Working Women and Leisure in Turn-of-the-Century New York*. Philadelphia: Temple University Press, 1986.

Pendergrast, Mark. *For God, Country, and Coca-Cola: The Unauthorized History of the Great American Soft Drink and the Company that Makes it*. New York: Scribner's, 1993.

Pickren, Wade E. "Indigenization and the History of Psychology." *Psychological Studies* 54, no. 2 (2009): 87–95.

Pittenger, Mark. "A World of Difference: Constructing the 'Underclass' in Progressive America." *American Quarterly* 49, no. 1 (1997): 26–65.

Platt, Anthony M. *The Child Savers: The Invention of Delinquency*. Chicago: University of Chicago Press, 1969.

Platt, Jennifer. "The Development of the 'Participant Observation' Method in Sociology: Origin Myth and History." *Journal of the History of the Behavioral Sciences* 19, no. 4 (1983): 379–93.

Poovey, Mary. *A History of the Modern Fact: Problems of Knowledge in the Sciences of Wealth and Society.* Chicago: University of Chicago Press, 1998.

Porter, Theodore M. "The Objective Self." *Victorian Studies* 50, no. 4 (2008): 641–47.

———. *Trust in Numbers: The Pursuit of Objectivity in Science and Public Life.* Princeton, NJ: Princeton University Press, 1995.

Potter, Claire Bond. *War on Crime: Bandits, G-Men, and the Politics of Mass Culture.* New Brunswick, NJ: Rutgers University Press, 1998.

Pressman, Jack D. "Psychiatry and Its Origins." *Bulletin of the History of Medicine* 71 (1997): 129–39.

Preston, Ivan L. *The Great American Blowup: Puffery in Advertising.* Madison: University of Wisconsin Press, 1996.

Proctor, Robert N., and Londa Schiebinger, eds. *Agnotology: The Making and Unmaking of Ignorance.* Stanford, CA: Stanford University Press, 2008.

Rabinow, Paul. *Reflections on Fieldwork in Morocco.* Berkeley: University of California Press, 1977.

Rabinowitz, Lauren. *For the Love of Pleasure: Women, Movies, and Culture in Turn-of-the-Century Chicago.* New Brunswick, NJ: Rutgers University Press, 1998.

Raia, Courtenay Grean. "From Ether Theory to Ether Theology: Oliver Lodge and the Physics of Immortality." *Journal of the History of the Behavioral Sciences* 43, no. 1 (2007): 19–43.

Rauchway, Eric. *Blessed among Nations: How the World Made America.* New York: Hill and Wang, 2006.

Reiss, Benjamin. *The Showman and the Slave: Race, Death, and Memory in Barnum's America.* Cambridge, MA: Harvard University Press, 2001.

Robinson, David R. *Illegal Tender: Counterfeiting and the Secret Service in Nineteenth-Century America.* Washington, DC: Smithsonian Institution Press, 1995.

Rodgers, Daniel T. *Atlantic Crossings: Social Politics in a Progressive Age.* Cambridge, MA: Harvard University Press, 1998.

Rodriguez, Julia. "South Atlantic Crossings: Fingerprints, Science, and the State in Turn-of-the-Century Argentina." *American Historical Review* 109, no. 2 (2004): 387–416.

Roosevelt, Theodore. "The Man with the Muck Rake." In *The Muckrakers,* ed. Arthur Weinberg and Lila Weinberg, 58–65. Urbana: University of Illinois Press, 2001.

Rose, Anne C. *Psychology and Selfhood in the Segregated South.* Chapel Hill: University of North Carolina Press, 2009.

Rose, Nikolas. *Governing the Soul: The Shaping of the Private Self.* London: Routledge, 1990.

———. *Inventing Ourselves: Psychology, Power, and Personhood.* Cambridge: Cambridge University Press, 1996.

———. *Powers of Freedom: Reframing Political Thought.* Cambridge: Cambridge University Press, 1999.

Rosenberg, Charles. *The Trial of the Assassin Guiteau: Psychiatry and the Law in the Gilded Age.* Chicago: University of Chicago Press, 1968.

Rosenzweig, Roy. *Eight Hours for What We Will: Workers and Leisure in an Industrial City, 1870–1920.* Cambridge: Cambridge University Press, 1983.

Rubin, Joan Shelley. *The Making of Middle-Brow Culture.* Chapel Hill: University of North Carolina Press, 1992.

Rydell, Robert. *All the World's a Fair: Visions of Empire at American International Expositions, 1876–1916.* Chicago: University of Chicago Press, 1984.

Sandage, Scott A. *Born Losers: A History of Failure in America.* Cambridge, MA: Harvard University Press, 2005.

Satter, Beryl. *Each Mind a Kingdom: American Women, Sexual Purity, and the New Thought Movement, 1875–1920.* Berkeley: University of California Press, 1999.

Saxon, A. H. *P. T. Barnum: The Legend and the Man.* New York: Columbia University Press, 1989.

Schickore, Jutta. "Misperception, Illusion, and Epistemological Optimism: Vision Studies in Early Nineteenth-Century Britain and Germany." *British Journal for the History of Science* 39, no. 3 (2006): 383–405.

Schmidt, Leigh Eric. *Hearing Things: Religion, Illusion, and the American Enlightenment.* Cambridge, MA: Harvard University Press, 2000.

Schweitzer, Marlis. "'The Mad Search for Beauty': Actresses' Testimonials, the Cosmetics Industry, and the 'Democratization of Beauty.'" *Journal of the Gilded Age and Progressive Era* 4, no. 3 (2005): 255–92.

Scott, Joan W. "The Evidence of Experience." *Critical Inquiry* 17, no. 4 (1991): 773–97.

Sekula, Allan. "The Body and the Archive." *October* 39 (1986): 3–64.

Sellers, Christopher G. *The Market Revolution: Jacksonian America, 1815–1846.* New York: Oxford University Press, 1991.

Serlin, David. *Replaceable You: Engineering the Body in Postwar America.* Chicago: University of Chicago Press, 2004.

Shamdasani, Sonu. "'Psychotherapy': The Invention of a Word." *History of the Human Sciences* 18 (2005): 1–22.

Shapin, Steven. *The Scientific Life: A Moral History of a Late Modern Vocation.* Chicago: University of Chicago Press, 2008.

———. *A Social History of Truth: Civility and Science in Seventeenth-Century England.* Chicago: University of Chicago Press, 1994.

Sklansky, Jeffrey. *The Soul's Economy: Market Society and Selfhood in American Thought, 1820–1920.* Chapel Hill: University of North Carolina Press, 2002.

Sklar, Martin J. *The Corporate Reconstruction of American Capitalism, 1890–1916: The Market, the Law, and Politics.* Cambridge: Cambridge University Press, 1988.

Skrupskelis, Ignas K., and Elizabeth M. Berkeley, eds. *The Correspondence of William James.* Vol. 12. Charlottesville: University of Virginia Press, 2004.

Smith, Roger. "Does the History of Psychology Have a Subject?" *History of the Human Sciences* 1, no. 2 (1988): 147–77.

———. *The Norton History of the Human Sciences.* New York: Norton, 1997.

Smith, Shawn Michelle. *American Archives: Gender, Race, and Class in Visual Culture.* Princeton, NJ: Princeton University Press, 1999.

Sokal, Michael M. "Biographical Approach: The Psychological Career of Edward Wheeler Scripture." In *Historiography of Modern Psychology: Aims, Resources, Approaches,* ed. Josef Broszek and Ludwig J. Pongratz, 255–78. Toronto: C. J. Hogrefe, 1980.

———, ed. *An Education in Psychology: James McKeen Cattell's Journal and Letters from Germany and England, 1880–1888.* Cambridge, MA: MIT Press, 1981.

———. *Psychological Testing and American Society, 1890–1930.* New Brunswick, NJ: Rutgers University Press, 1987.

Stanley, Amy Dru. *From Bondage to Contract: Wage Labor, Marriage, and the Market in the Age of Slave Emancipation.* New York: Cambridge University Press, 1998.

Stanziani, Alessandro. "Negotiating Innovation in a Market Economy: Foodstuffs and Beverage Adulteration in Nineteenth-Century France." *Enterprise and Society* 8, no. 2 (2007): 375–412.

Stark, Laura. "The Science of Ethics: Deception, the Resilient Self, and the APA Code of Ethics, 1966–1973." *Journal of the History of the Behavioral Sciences* 46, no. 4 (2010): 337–70.

Stigler, Stephen M. *The History of Statistics: The Measurement of Uncertainty before 1900.* Cambridge, MA: Belknap Press of Harvard University Press, 1986.

Strasser, Susan. *Satisfaction Guaranteed: The Making of the American Mass Market.* New York: Pantheon Books, 1989.

Tanenhaus, David S. *Juvenile Justice in the Making.* Oxford: Oxford University Press, 2005.

Taves, Ann. *Fits, Trances, and Visions: Experiencing Religion and Explaining Experience from Wesley to James.* Princeton, NJ: Princeton University Press, 1999.

Taylor, Charles. *Sources of the Self.* Cambridge, MA: Harvard University Press, 1989.

Tedlow, Richard S. *New and Improved: The Story of Mass Marketing in America.* New York: Basic Books, 1990.

Thomas, Ronald R. *Detective Fiction and the Rise of Forensic Science.* New York: Cambridge University Press, 1999.

Thomson, Rosemarie Garland, ed. *Freakery: Cultural Spectacles of the Extraordinary Body.* New York: New York University Press, 1996.

Tomes, Nancy. "The Great American Medicine Show Revisited." *Bulletin of the History of Medicine* 79 (2005): 627–63.

———. "Merchants of Health: Medicine and Consumer Culture in the United States, 1900–1940." *Journal of American History* 88 (2001): 519–47.

Tomlins, Christopher. *Law, Labor, and Ideology in the Early Republic.* New York: Cambridge University Press, 1993.

Trachtenberg, Alan. *The Incorporation of America: Culture and Society in the Gilded Age.* New York: Hill and Wang, 1982.

Treitel, Corinna. *A Science for the Soul: Occultism and the Genesis of the German Modern.* Baltimore: Johns Hopkins University Press, 2004.

Valverde, Mariana. "'Despotism' and Ethical Liberal Government." *Economy and Society* 25, no. 3 (1996): 357–72.

———. *The Law's Dream of a Common Knowledge.* Princeton, NJ: Princeton University Press, 2003.

Van Dover, J. K. *You Know My Method: The Science of the Detective.* Bowling Green, OH: Bowling Green State University Popular Press, 1994.

Walkowitz, Judith. *City of Dreadful Delight: Narratives of Sexual Danger in Late Victorian London.* Chicago: University of Chicago Press, 1992.

Wallis, Roy, ed. *On the Margins of Science: The Social Construction of Rejected Knowledge.* Keele: University of Keele, 1979.

Warren, Heather A. "Character, Public Schooling, and Religious Education, 1920–1934." *Religion and American Culture* 7, no. 1 (1997): 61–80.

Welke, Barbara Young. *Recasting American Liberty: Gender, Race, Law, and the Railroad Revolution, 1865–1920.* New York: Cambridge University Press, 2001.

White, Christopher G. *Unsettled Minds: Psychology and the American Search for Spiritual Assurance, 1830–1940.* Berkeley: University of California Press, 2009.

White, Paul. *Thomas Huxley: Making the "Man of Science."* Cambridge: Cambridge University Press, 2003.

White, Richard. "Information, Markets, and Corruption: Transcontinental Railroads in the Gilded Age." *Journal of American History* 90, no. 1 (2003): 19–43.

Wiebe, Robert H. *The Search for Order, 1877–1920.* New York: Hill and Wang, 1967.

Williams, Bernard. *Truth and Truthfulness: An Essay in Genealogy.* Princeton, NJ: Princeton University Press, 2002.

Willrich, Michael. *City of Courts: Socializing Justice in Progressive Era Chicago.* Cambridge: Cambridge University Press, 2003.

Winston, Andrew S. "'As His Name Indicates': R. S. Woodworth's Letters of Reference and Employment for Jewish Psychologists in the 1930s." *Journal of the History of the Behavioral Sciences* 32 (1996): 30–43.

———. "Cause into Function: Ernest Mach and the Reconstruction of Explanation in Psychology." In *The Transformation of Psychology: Influences of*

Nineteenth-Century Philosophy, Technology, and Natural Science, ed. Christopher D. Green, Thomas Teo, and Marlene Shore, 107–31. Washington, DC: APA Press, 2001.

Winter, Alison. "The Making of 'Truth Serum.'" *Bulletin of the History of Medicine* 79, no. 3 (2005): 500–533.

———. *Mesmerized: Powers of Mind in Victorian Britain.* Chicago: University of Chicago Press, 1998.

Young, James Harvey. *The Medical Messiahs: A Social History of Health Quackery in Twentieth-Century America.* Princeton, NJ: Princeton University Press, 1967.

———. *Pure Food: Securing the Federal Food and Drugs Act of 1906.* Princeton, NJ: Princeton University Press, 1989.

———. "This Greasy Counterfeit: Butter versus Oleomargarine in the U.S. Congress, 1886." *Bulletin of the History of Medicine* 53, no. 3 (1979): 392–414.

Zimmerman, David A. *Panic!: Markets, Crises, and Crowds in American Fiction.* Chapel Hill: University of North Carolina Press, 2006.

Zunz, Olivier. *Making America Corporate, 1870–1920.* Chicago: University of Chicago Press, 1990.

Index

accuracy: defined, 13–15; as feature of scientific reports, 56–57, 86, 96, 100, 104–6, 180, 182; magicians as models of, 68–71; in labeling, 137; in medical diagnosis, 163; in scale for consumer deception, 149–51
Adams, Samuel Hopkins, 136, 138, 163
Addams, Jane, 166
Adler, Herman, 174, 189
adulteration, 5, 13, 132–40
advertising, 1, 4, 6, 38, 79–80, 122–25, 131, 148–54, 161
advice literature, 23, 28–29, 82, 192–93
aesthesiometer, 65, 69, 109
Allport, Gordon W., 215–18, 222
Amalgamated Copper Mining Company, 46
American Society for Psychical Research, 90–91, 106
amusements and leisure, 9–11, 51, 63–71, 75, 197–98, 211
Angell, Frank, 75
attention, 3, 50, 59, 64, 78, 96, 108; versus recognition, 147–48; and stage magic, 67–72; in the unwary purchaser, 128–29, 139
Attitude-Interest Analysis. *See* Masculinity-Femininity scale
Aussage test (testimony test), 172–73

Barnum, P. T., 9–12, 19, 51,119, 161, 228–30, 234
Beatty, Bessie, 84–85, 112–13
Berkeley, CA, 180–82
Bertillon, Alphonse, 157
Bigelow, Melville, 2

Binet, Alfred, 68
Black Art, the (magic trick), 98–99
Black, Hugo, 121–122
Blackford, Katherine, 230
Blake, Doris, 192–193
blood pressure, 177–79, 182, 189, 191, 193
Bly, Nellie, 38
Boas, Franz, 65
Boy Scouts of America, 197, 214–15
Bradley, Milton, 75
brainwashing, 194, 224–25
Brewster, David, 52, 54
Brown, Robert, 183
Burtt, Harold, 142
butter, 5–6, 60, 132–36
Byrnes, Thomas, 32, 157

Cabot, Richard, 163, 174
Cady, Vernon M., 204–5, 226
Carnegie, Dale, 214
Carpenter, William B., 96
Carrington, Hereward, 93–95, 101–2
Cattell, James McKeen, 55, 58, 111, 233
caveat emptor, 11, 17, 25–28, 122–23
Century of Progress Fair, 187
Chapman, John Jay, 102
character: moral, 25, 30, 89, 100, 106, 157, 159–60, 195–96; compared to intelligence, 201–5; versus personality, 214–15
character education, 196–199, 214–15
Character Education Inquiry (CEI), 206–20
character tests, 202–6, 209–12
Charcot, Jean-Martin, 53
Charpentier, Augustin, 71
Chase, Stuart, 151

Chero-Cola, 143–146, 155
Chicago, 1, 21, 43–44, 64–65, 79, 91, 116–17, 165–66, 169, 185–87
Chicago Board of Trade, 44, 46
child psychology, 62–63, 73–74, 112, 166–78, 196, 199, 201–6, 209–24
civilization: consumerism as a new basis for, 4, 62; as racial ideology, 12–13, 62–63, 78
class passing, 38–41, 112–13
Coca-Cola, 137–138, 143–147, 155, 183
Coca-Cola Company v. The Koke Company of America (1920), 144
Coca-Cola v. Chero-Cola (1915), 143–46, 155, 183
Coe, George Albert, 208
Collier's Magazine, 95, 98, 136, 138, 149
color, 70, 116–117, 133–135
Columbia University, 17, 40, 58, 90, 93–94, 113, 138, 143, 152, 202, 208. *See also* Teachers College
common sense, 8, 58; consumers lacking in, 130–31; legal, 122–23, 127, 139, 153–55; of the police officer, 180; psychologists' opposition to legal, 142–44
Coney Island, 66
confederate (role in confidence game), 28–33, 35
confederate (role in psychological experiment), 97–100, 105, 203–4
confidence games: bunco, 25–26; gold brick, 26; green goods, 26; the wire, 31
confidence man: the original, 22–23, 25; compared to salesman, 36–38; self-reliance of, 35; as threat to freedom of others, 33–34
consumer citizenship, 122
contract law. *See* contracts
contracts, 26–28, 33
counterfeiting, 5–7, 11, 26, 31, 46, 134
Crandon, Margery, 111
Crane, Stephen, 40
credulity, 11–12, 14, 31, 51; as aspect of psychology of deception, 59–64, 73–74; of test takers, 123, 194–95; of witnesses of psychical phenomena, 84–86, 89–92, 100–1, 106, 113–20
crime control, 185–88
Cushing, Frank Hamilton, 40
Cushing, Harvey, 178

Dana, Charles, 93–94
Darrow, Clarence, 161
Davis, Katherine Bement, 221
Davis, W.S., 97, 101, 115–16
deceitful self: defined, 7–8
deceitful situations, 214, 217
deceivable self: defined, 7–8
deception test. *See* lie detector; polygraph
deceptive things, 2, 5, 52, 143, 228
Department of Justice, 152
department store, 130, 143, 185
detective, 1–2, 15, 23–24, 29, 32, 34, 36, 41, 43, 84–7, 95–6, 100, 103, 106–7, 119, 182–83, 193, 203, 212, 223, 233
developmental psychology. *See* child psychology
Dewey, John, 55, 58
Dorr, George, 107, 109
doubt, 11–12, 59–60, 63–64, 113. *See also* credulity
duck-rabbit, 76–80

Eastman, Max, 163–64
economy of uncertainty, 3–4, 9–11
economy, law of, 16, 50–51, 61–63, 79, 83, 232
Edison, Thomas, 200
Emmanuel Movement, 162–163
emotion: as an aspect of consumption, 37–38, 131; as index for deceitfulness, 164, 178–80, 190–91; as sustaining belief, 118, 209; as undercutting autonomy, 22–23
epistemological optimism, 52–54
evolution, 60–63
expectation, 25, 50–1, 62, 64, 67, 71–75, 80, 93, 102, 108, 137, 218, 232
expert witness, 147; the psychologist as, 137, 143–46, 183–84

Fairbank v. Bell (1896), 127
Fairchild, Henry Pratt, 214–215
fakir, 28, 36–37, 45
Fechner, Gustav, 56
Federal Trade Commission (FTC), 6, 17, 121–22, 138–40, 150
Feingold, Gustave, 142–43, 148, 154
femininity: as credulous, 12, 100–1, 161; consumption gendered as, 38, 129; the measurement of 220–21
feminism, 110–13

INDEX   307

fibs. *See* lies, white
finance, 25, 34, 38
Fleming, Cochrane, 125
Fleming, John, 125
flour, 123, 128
Flynt, Josiah. *See* Willard, Josiah
Folkers v. Chadd (1782), 147
Forbush, William Byron, 198
Forer, Bertram, 229–230
Fourteenth Amendment, 134
Frankfurter, Felix, 154
Franklin, Benjamin, 231
fraud (legal), 1–2, 4–5, 9, 11, 13–15, 26–27, 29–32
Freud, Sigmund, 159, 165, 169, 177, 219
Frye v. United States (1923), 183–84
Frye, James Alphonso, 184
FTC v. Raladam Co. (1931), 6

Gallup, George, 148–49
Galton, Francis, 53, 58, 157
gambling: compared to confidence games, 35–36; stock market compared to, 41, 44, 46; as metaphor for psychological subject, 56, 164
gangsters, 185, 188
Gault, Robert, 117
gender. *See* femininity; masculinity
genetic psychology. *See* child psychology
Gesell, Arnold, 223
Glueck, Bernard, 175–76
Goddard, Calvin, 185–86
Goddard, Henry, 164, 180, 200
graft, 16, 21–26, 34–38, 47, 49, 186, 188; coining of term 41–44
grafter. *See* confidence man
grocery store, 79, 141

habit, 50–51, 56, 80, 83, 122; in character education, 203, 205; and consumer psychology, 127–28; Hartshorne and May on, 209–10, 215–17; Healy on, 167, 171, 173; James on, 61–63; Marston on, 190; Peirce and Jastrow on, 59–60; Seashore on, 72
Hall, G. Stanley, 55, 57–58, 60, 159, 199; on children's lies 62–63; as amateur magician, 68; involvement in Palladino case, 90, 93–94; involvement in Piper case, 102, 107–112

hallucination, 72–73, 189–90
Harlan, John Marshall, 135
Harrison, Carter Jr., 42–43
Hartshorne, Hugh, 196, 206–17, 219–20, 222, 224, 226
Harvard University, 17, 18, 55, 75, 111–12, 142–43, 162–63, 178–79, 216
Haunted Swing, 65–67
Haywood, Big Bill, 160–61
Healy, Mary, 165–76
Healy, William, 165–76
Helmholtz, Hermann von, 52–53
Herrmann, Alexander, 68–71, 98
Heth, Joice, 9–10
Hewitt, Ephraim, 128
Hill, John Jr., 46
hobo. *See* tramp
Hodder, Alfred, 44
Hodgson, Richard, 90; memorial fund named after, 111–12; spirit of, 106–9
Holmes, Oliver Wendell Jr., 144, 153–54
homosexuality, 187, 220–21
Honesty Bureau, 197–98
Hoover, Herbert, 198
Houdini, Harry, 85, 111
Huggins, Willetta, 116–18
Hull House, 166
hypnosis, 22, 53, 108, 113, 231

ignorance: as part of experimental design, 55–56, 147, 221; as racial category, 12; willful, 136–137
illusions: decade of, 16, 50; European studies of 52–54, 71; intellectual property over, 75–76; and the law of economy 51–52; optical, 75–82; and psychology of testimony, 159–60; and racial difference, 78; and self-reliance, 81–82; spiritualist exposés compared to laboratory study of, 108, 119; and stage magic, 63–71; use on Torres Strait expedition, 77–78; uses in theories of advertising, 79–80. *See also* size-weight illusion; duck-rabbit
insanity, 72, 88, 189–90
Institute Social and Religious Research (ISRR), 206–8
intelligence: of the consumer, 127, 129, 140
intelligence tests, 9, 155, 164, 167–69; as model of character tests, 199–206
Interstate Commerce Commission, 5

James, William, 16, 54; 163; and the "cash value theory," 61; interest in psychical researcher, 87, 90–91; and the law of economy, 61–62; role in Palladino case, 95, 101–2, 105–6; role in Piper case, 106, 110; and the social self, 80–81; and the will to believe, 118–19
Jastrow, Joseph, 16, 50–51, 228–29; and his experiments on magicians, 67–73; and his psychology of deception, 55–65; his role during Huggins case, 115–19; his role during Palladino case, 90, 95–106
Jenkins, James, 127–28
Johns Hopkins University, 54–55, 181
Johnston, J. P., 21–22
journalism, 7–8, 15–16, 38–46, 49, 83–85, 92, 118, 151; as career for pathological liar, 173
Joyce, William B., 197
Jung, Carl, 159
Jungle, The (1906), 135–36
juvenile court, 158, 166
juvenile deliquency, 166–72, 204
Juvenile Psychopathic Institute, 166–73

Keeler v. Keeler (1889), 91
Keeler, Leonarde, 181, 184–88, 191
Kellar, Harry, 68–70, 98
Keller, Helen, 65, 69, 116
Kelley, Florence, 136–37, 151, 166
Kellogg, James L., 97, 101
Kelly, E. Lowell, 221
Kennedy, Craig, 177, 181
Kerns, Ben, 34–35, 37
Kinsey, Alfred, 224
kleptomania, 175

labels and labeling, 5, 123–27, 130–31, 135–38, 150–53
Lacombe, E. Henry, 127
Ladd, George Trumbull, 71
Laidlaw v. Organ (1817), 27
Lanham Act (1946), 149–53
Larson, John A., 177, 181–82, 187–91
law. *See* Supreme Court; *see also individual legal cases*
law of economy. *See* economy, law of
Lawson, Thomas W., 46
Leacock, Stephen, 201
Leeper, David H., 29

Legal Tender Act (1862), 26
Lewis, A. A., 156
lie detector, 2, 8, 18–19, 142, 147, 156–58, 176–93, 232, 235
lies, white, 14, 187, 192
Lippmann, Walter, 201
Lodge, Oliver, 92, 108, 110
Lombroso, Cesare, 92, 157
Los Angeles, 84
Los Angeles Herald, 84
lying, pathological, 164–76
Lyman, Harrison, 151–52
Lynd, Robert S., 207–8, 213

Mach, Ernst, 61
MacKinnon, Donald, 218–20
magic, stage, 51, 63–71, 74, 93–100, 223
magician, 15, 17, 60, 67–71, 74, 83, 85–6, 93–100, 103–4, 111, 115–16, 119, 156, 223, 233. *See also* Herrmann, Alexander; Kellar, Harry; Houdini, Harry; Rinn, J. L.; Robert-Houdin; *specific magicians*
magnetism, 23, 34, 90, 174; George Beard on 87–88; Franklin Commission on 231–32
margarine, 5–6, 132–36
mark, 28–33
Marston, William Moulton: development of lie detector, 176–80; on the lie detector as therapeutic 189–92; role during Frye case, 182–84
masculinity, 23, 94, 174; hardboiled, 86–87, 110–11, 113, 187; the measurement of 220–21; of salesmen, 37–38
Masculinity-Femininity scale, 221
masturbation, 171
May, Mark A., 196, 206–17, 219–20, 222, 224, 226
McCann v. Anthony (1886), 130
McClary v. Stull (1895), 91–92
McClure's Magazine: support of muckraking, 42, 44; on unwary purchaser, 141; on Orchard trial, 160–161
McCord, Henry, 29–30
McCoy, Irvin, 183–84
McLean v. Fleming (1878), 125–26, 131–32
McLean, James, 125–26
medium, spirit. *See* Palladino, Eusapia; Piper, Leonora; Reynolds, Elsie
Meehl, Paul, 229–30

Mège-Mouriès, Hippolyte, 133
memory, 109, 128–29, 132, 144, 147–49, 219
men of science: as gendered ideology 54, 85–86, 94, 101, 111, 104, 118
mesmerism, 59, 88, 231–32
Metropolitan Magazine, 105
middlebrow publishing, 51, 58, 62, 75, 251
Midwinter Fair, 65
Miles, Catharine Cox, 220–21
Milgram, Stanley, 225
Miller, Beulah, 113–16
Miller, Dickinson, 93–94, 97, 100, 102–3
Milton Bradley Company, 75
Montaigne, Michel de, 2
Moreau, William, 28, 45
muckraking: 38, 44–46, 160; the exposure of spiritualism as a genre of 83–84, 92, 118–20; concerning food and drugs 135. *See also* journalism
Mumford, Lewis, 61
Munsterberg, Hugo: and deception tests, 182, 184, 191; and illusions, 75, 77; and Marston, 176–78; role in Miller case 110–16; role in Orchard case, 159–62; role in Palladino case, 80, 93, 104–6; and the unwary purchaser, 141–43, 149, 154
Murray, Henry A., 214, 216, 220

neurathenia, 87
New Deal, 122, 150–53
New York City, 1, 10, 25, 29, 42, 170, 185, 190
New York Times, 51; on honesty tests, 204, 213; on intelligence tests, 200; interview with Goddard about lie detector, 164; role during Miller case, 114, 116; role during Palladino case, 94, 100, 102–3

objectivity, 11–15, 96, 100–3, 131–32, 149, 195–96, 207, 223–25, 233
observer (role in psychological experiments), 55–56
oleomargarine. *See* margarine
Oleomargarine Act (1886), 134
On the Witness Stand (1908), 159–60, 162
Orchard, Harry, 160–61
ordinary purchaser. *See* purchaser, ordinary
Organization Man, The (1956), 194–95
Orsdel, Josiah van, 183
O'Sullivan, Frank Dalton, 1–2

Palladino, Eusapia, 85, 92–106, 114
Parker, C. L., 146
participant-observation, 38–41
patent medicine, 18, 125–26, 136, 161–63
Paterson, Donald G., 230, 234
Patten, Simon, 4, 51, 62
Paynter, Richard, H. 143–47, 150, 154, 183
Peckham, Sr., Rufus Wheeler, 32–33
Peirce, Charles Sanders: on belief, 59–60; as professor at Johns Hopkins University, 54–56
People v. Livingston (1900), 31
People v. McCord (1871), 29–32
People v. Tompkins (1906), 31
People v. Williams (1842), 30–31
Pepsi Cola, 152
personality: the confidence man's magnetism as, 23, 34, 37, 174; medium's multiple, 106–8; tests, 194–6, 212–26, 229–30
Pillsbury v. Pillsbury (1894), 127–29
Pillsbury, Charles, 128
Pillsbury, L. F., 128
Pinkerton, Allan, 87
Piper, Leonora, 106–13
Plessey v. Ferguson (1896), 183
Plumley v. Massachusetts (1894), 135
police: and confidence games, 23, 25, 29; corruption, 42–43, 182; and spiritualists, 84; and the lie detector, 18, 180–82
police powers of the state, 5–6, 26, 132–35, 137–39, 171
polygraph, 2, 8, 18–19, 142, 147, 156–58, 176–93, 232, 235
Popular Science Monthly, 13, 58, 59, 68, 89
popular science writing, 58–59
populism, 44
Powell v. Pensylvannia (1888), 134
Prince, Morton, 105, 162, 163
probability, 59–61, 146
Protestantism, 206–10, 217
Pseudoptics (1894), 75
psychical research, 62, 72, 85–86, 89–93, 101–2, 107–12
psychoanalysis, 157, 163, 189–91, 219–20
psychological society, 9, 226, 232, 235
Psychology and Industrial Efficiency (1913), 141
Psychology Corporation, 149
psychophysics, 55–56, 71–74
psychotherapy, 159, 162–64, 177, 191–92

purchaser, ordinary, 123–32
Pure Food and Drugs Act (1906), 5, 135–17
Pyne, Warner, 98–100

race, 12–13, 26, 56–57, 62–63, 64, 88, 102, 183–84, 206
racial differences, 78
racial prejudice, 14
randomization, 56
rapport, 189, 224
recapitulation theory, 62–63
recognition, 143–149
Reeve, Arthur, 177
Reik, Theodore, 191
religion, 12, 73, 85, 91–92, 118, 197–99, 206–11, 216, 217
Reynolds, Elsie, 84
Rinn, J. L., 98–100, 103
Rivers, W. H. R., 77–79
Roback, A. A., 191, 205, 210, 215, 218
Robert-Houdin, Jean Eugene, 68, 70, 74
Robinson v. Adams (1870), 91
Rockafellow v. Baker (1861), 27–28
Rockefeller, John D., 45
Rockefeller, John D. Jr., 207, 210
Rogers, Edward S., 123, 140–47, 150–55
Roosevelt, Theodore, 38, 43
Rorschach, Hermann, 76
Rousseau, Jean-Jacques, 13–14

Schilder, Paul, 191
Schlink, Frederick J., 151
Scientific Crime Detection Laboratory, 1–2, 18, 185–88
scientific naturalism, 61
Scott, Walter Dill, 51, 131, 177; as president of Northwestern University, 186; on the psychology of advertising, 79–80
Scripture, E. W., 16, 68, 71–75
Sears, Roebuck Company, 139
Seashore, Carl, 16, 72–74, 82, 120
Secor, Alson, 34
Secret Six, 187–188
sexuality, 165, 170–71, 173–75, 187, 220–21, 224, 226
Sherman Anti-Trust Act (1890), 5
sincerity: as a demand on spirit mediums, 86–87; defined: 13–14; rejected by magicians, 69, 71; rejection by psychologists when studying mediums, 100–8, 114,

119; role in confidence games, 23–25, 28; utopia of universal, 189–94
Sinclair, Upton, 135–36
size-weight illusion, 71–74
Smith, Edward H., 34
Smyth, Constantine, 146–47
Social Darwinism, 22
social psychology, 214, 225–26. *See also* social self
social self, 51, 80–82, 214
speculation, 24–25, 34, 44, 45–46, 164
Spencer, Herbert, 61
spirit medium. *See* Palladino, Eusapia; Piper, Leonora; Reynolds, Elsie
spiritualism, 62, 68, 84–92
St. John, E. A., 197–98
Standard Education Society, 121
Standard Oil Company, 44–46, 207
State v. Crowley (1876), 31
statistics: in the census, 56–57; used in unwary purchaser experiments 142–44, 146; versus the case history 157, 174
Steffens, Lincoln, 43, 49
Stern, William, 158–59
stock market, 1, 24, 44–46, 163, 164, 198
Stratton, George, 81–82
Studies in Deceit (1928), 206–12, 215, 217
suggestion, 8, 53, 71–74, 82, 88, 108–9, 131, 173, 231; as therapeutic technique, 163
Sully, James, 53–54
Summers, Walter, 189–190
Supreme Court of the United States, 6, 27, 46, 121, 125, 126, 134–35, 138, 144, 153–54, 183
swindler. *See* confidence man

Talks to Teachers (1899), 61
Tanner, Amy, 106–9, 112–13
Tarbell, Ida, 44–45
Teachers College (Columbia University), 202, 206, 208, 212, 217, 222
Terman, Lewis M., 200–5, 208, 220–21, 224
testimony: the unreliability of, 88, 129–30, 158–60, 175; expert, 147; psychologist as offering, 137, 143–46, 183–84
testimony test. *See* Aussage test
Thematic Apperception Test, 220
Theory of Advertising (1903), 79–80
therapeutic criminology, 157–58, 171, 176, 189

therapeutic penology. *See* psychological criminology
Thomas, Dorothy Swain, 222
Thorndike, E. L., 201–2, 208
Thurstone, Louis L., 195
Titchener, E. B., 75
Torres Strait Expedition, 77–78
touch, 65, 69, 71, 81, 122
trademark, 122–31, 137–55
tramp, 39–42
trance, 88, 106–9, 114
Triplett, Norman, 16, 60–63, 68, 82
trusts, the, 5, 80–81, 136, 138–39
truth in advertising, 1, 6, 161
truthfulness, 11–13, 103
Tugwell, Rexwell, 151
Twain, Mark, 14
Tylor, E. B., 62

University of Wisconsin, 58, 68–70
unobserved but observing observer, 14, 98–103, 222, 233
unwary purchaser. *See* purchaser, ordinary

vaudeville, 11, 51, 114
Voelker, Paul, 202–5, 207, 226
Vollmer, August, 180–82, 185

war on crime. *See* crime control
Watson, Goodwin, 217–18, 221
Weldon, S. James, 36–37
Western Union Telegraph Company, 31
Wheeler-Lea Act (1938), 6, 150
whiteness, 12–13, 23, 57–58, 62, 206. *See also* race
Whyte, William H. Jr., 194–96, 218
Wigmore, John H., 161–62, 174–75, 184–86, 266
Wiley, Harvey Washington, 137–38
Willard, Josiah, 38–41, 113; coining of word graft, 41–43
Williams v. Brooks (1882), 131
Williams, Bernard, 13
Williams, T. J., 116–17
Wirtz v. Eagle Bottling Co. (1892), 127
Witmer, Lighter, 162
wonder, 11–12
Wood, Robert W., 66–67, 103–4
Wooldridge, Clifton, 24, 36, 43
World of the Graft, The (1901), 42–43
World's Columbian Exposition, 64–65
Wundt, Wilhelm, 49, 52–56, 71, 75, 81
Wyckoff, Walter, 40–41, 113

X-ray, 103–4; as metaphor, 113, 220

Yale University, 71–74, 223
Youmans, Edward L., 13
Young Men's Christian Association (YMCA), 197, 198, 217
Your Money's Worth (1927), 151